From
Invention to Patent

A Scientist
and Engineer's Guide

从发明到专利
科学家和工程师指南

[美] 史蒂文·H. 沃尔德曼（Steven H. Voldman）—— 著
李文宇 何琳琳 姜磊 张超 秦乐 —— 译

清华大学出版社
北京

北京市版权局著作权合同登记号　　图字：01-2018-5929

Steven H. Voldman
From Invention to Patent: A Scientist and Engineer's Guide
EISBN: 978-1119125259

Copyright © 2018 by John Wiley & Sons, Inc. All rights reserved.

Original language published by John Wiley & Sons, Inc. All rights reserved.
本书原版由 John Wiley & Sons, Inc. 出版。版权所有，盗印必究。

Tsinghua University Press is authorized by John Wiley & Sons, Inc. to publish and distribute exclusively this Simplified Chinese edition. This edition is authorized for sale in the People's Republic of China only (excluding Hong Kong, Macao SAR and Taiwan). Unauthorized export of this edition is a violation of the Copyright Act. No part of this publication may be reproduced or distributed by any means, or stored in a database or retrieval system, without the prior written permission of the publisher.

本中文简体字翻译版由 John Wiley & Sons, Inc. 授权清华大学出版社独家出版发行。此版本仅限在中华人民共和国境内（不包括中国香港、澳门特别行政区及中国台湾地区）销售。未经授权的本书出口将被视为违反版权法的行为。未经出版者预先书面许可，不得以任何方式复制或发行本书的任何部分。

本书封面贴有 Wiley 公司防伪标签，无标签者不得销售。
版权所有，侵权必究。举报：010-62782989，beiqinquan@tup.tsinghua.edu.cn。

图书在版编目(CIP)数据

从发明到专利：科学家和工程师指南 /（美）史蒂文•H. 沃尔德曼 (Steven H. Voldman) 著；李文宇等译. —北京：清华大学出版社，2020.9（2024.7重印）
（新时代•技术新未来）
书名原文：From Invention to Patent: A Scientist and Engineer's Guide
ISBN 978-7-302-53300-9

Ⅰ. ①从… Ⅱ. ①史… ②李… Ⅲ. ①科学技术－创造发明－指南 ②科学技术－专利－指南
Ⅳ. ① N19-62 ② G306-62

中国版本图书馆 CIP 数据核字 (2019) 第 227824 号

责任编辑：刘　洋
封面设计：徐　超
版式设计：方加青
责任校对：王荣静
责任印制：沈　露

出版发行：清华大学出版社
网　　址：https://www.tup.com.cn, https://www.wqxuetang.com
地　　址：北京清华大学学研大厦 A 座　　邮　编：100084
社 总 机：010-83470000　　邮　购：010-62786544
投稿与读者服务：010-62776969，c-service@tup.tsinghua.edu.cn
质 量 反 馈：010-62772015，zhiliang@tup.tsinghua.edu.cn

印 装 者：三河市龙大印装有限公司
经　　销：全国新华书店
开　　本：187mm×235mm　　印　张：20　　字　数：341 千字
版　　次：2020 年 9 月第 1 版　　印　次：2024 年 7 月第 5 次印刷
定　　价：99.00 元

产品编号：081000-01

内容简介

本书全面介绍了与专利相关的基本知识，系统论述了美国专利文件的各部分组成以及如何撰写发明的技术交底书。本书聚焦于怎样成为一名发明人和成为多产发明人的过程，教会读者如何有效地与专利律师和专利审查员沟通、如何理解和答复专利局发出的审查意见通知书。本书还讨论了企业产生专利的多种途径以及企业的专利战略，并进行了有关专利诉讼实践的讨论。

本书可作为知识产权专业师生或相关从业人员了解专利实用基本知识的参考用书，也可供技术研发工作者以及科研管理人员学习和培训使用。

致贝蒂·H. 布朗

作者简介

史蒂文·H.沃尔德曼博士是首位静电放电领域（ESD）的"CMOS 中 ESD 保护，绝缘体上硅与硅锗技术"的 IEEE 会员。他在布法罗大学获得工程学的理学学士（1979）；在麻省理工学院（MIT）获得第一学位电子工程硕士（1981）和第二学位工程师学位；在 IBM 留学生奖学金计划下，从佛蒙特大学获得工程物理硕士（1986）和电子工程博士（1991）。

1982 至 2007 年间，沃尔德曼任职于 IBM 半导体发展部。他也是 Qimonda 和 Intersil 公司开发团队的一员。他也是中国台湾地区新竹市的台湾半导体制造公司（TSMC）的顾问和韩国东滩三星电子公司的顾问。

史蒂文·H.沃尔德曼是首批被 IBM 委任为大师级发明家之一。他连续三年获得"IBM 顶级发明人"——该殊荣仅颁发给排名前十位的发明人。他是 259 件美国授权专利的受让人，并且撰写超过 150 篇技术论文。在 IBM，沃尔德曼博士获得过 67 次发明成就奖。

沃尔德曼博士在马来西亚、斯里兰卡、塞内加尔、斯威士兰和美国提供发明、创新和专利方面的导师辅导和短期课程。他为高校职员和国际学生开设了讲座课程并与之互动，开设讲座的大学超过 45 所（遍及美国、马来西亚、菲律宾、泰国、印度、中国、塞内加尔、贝宁和斯威士兰等国家）。

从 2007 年开始，他在专利诉讼中担任专家证人并成立了一家有限责任公司（LLC），主要就商业支撑专利、专利撰写和专利诉讼开展咨询。在他的公司里，他在 10 多个涉及 DRAM 开发、半导体开发、集成电路和电子放电等领域的案件中担任专家证人。他是 Hogan Lovells LLP、Dechert LLP、Quinn Emanuel Urquhart and Sullivan LLP、Sydney and Austin LLP 和 Latham and Watkins LLP 的专家证人。

沃尔德曼博士目前专注于专利检索、专利撰写、官方通知的回复和初步修改，现主要事务是为美国律师事务所撰写专利，同时也为专利、专利撰写和美国小企业专利组合开发提供咨询服务。

沃尔德曼博士也为"科学美国人"写作，已著有 10 本技术专著。

致　　谢

多年来，我对提供发明和专利、撰写开展讲座、受邀演讲和辅导的机会与支持表示感谢。

我要感谢位于伯灵顿和佛蒙特的 IBM 知识产权法律部门，他们教会我专利撰写、发明交底书和权利要求撰写。

我要感谢 Jerry Walters、Richard Kotulak、James Leas 和 Mark Chadurjian；感谢与我一起并肩作战完成 259 个授权专利的工作人员；感谢 IBM 知识产权法律部门在 2007 年 8 月授予我的"优秀捕鼠器奖"，上面写有"感谢 25 年来帮助 IBM 制造更好的捕鼠器"。

我要感谢 IBM 公司允许我开始讲述创新和专利结构课程。我的第一个课程的构建源于一名经理提出想要提高他 60 人团队的发明数量的要求。

我要感谢 Dreamcatcher Malaysia 和 Dreamcatcher Asia 在 2007 年开设了第一个外部课程，该课程从 2007 年教授到 2014 年。马来西亚政府希望提高专利数量和普及专利撰写知识。对我来说，能在 2008 年给马来西亚科学技术和创新部的员工（MOSTI）和在布城的马来西亚总理介绍课程是一种荣幸。另外，还为 Intel U（马来西亚）、马来西亚理科大学（USM）和其他马来西亚企业和院所开设了创新和发明课程。给化学家、生物技术专家和农业专家开设课程，也对我更加了解植物专利有所帮助。

我还要感谢 Shanta Yapa of EPIC Lanka（斯里兰卡）。在斯里兰卡信息技术工业联合会开设第一个发明课程——"发明、创新和专利"是我的荣幸。随后又开设了高级课程，聚焦于专利撰写。FITIS 的参与者主要是软件工程师，所以我们聚焦于软件和系统专利。本课程从 2008 年至 2015 年每年都在开设。在斯里兰卡，有人建议我出一本关于专利课程的书。每年去斯里兰卡旅游都是一件愉悦的事。

我要感谢 Saile Ackerman LLC 的 Stephen Ackerman，让我学到了专利检索、专利撰写和审查意见答复的细节。

我要感谢 Dechert LLP、Hogan Lovells LLP、Quinn Emanuel Urquhart and Sullivan LLP、Thompson and Knight LLP，以及 Lastly、Akin Gump、Strauss、Hauer and Feld LLP 律师事务所，为我提供作为专家证人和咨询顾问的机会。

作为技术团队的一员，我要感谢 IBM 公司、Qimonda、中国台湾半导体制造公司、Intersil 公司和三星电子公司。我很幸运可以在一个大技术团队中为大量客户工作，并有幸成为天才技术和设计团队的一员，该团队极有创新、智慧和创造力。

我要感谢在会议、专题讨论会、行业领域和大学允许我演讲和讲课的机构，它们启发了我撰写本书。我要感谢以下大学的工作人员：麻省理工学院、斯坦福大学、佛罗里达中部大学、伊利诺伊大学城分校、佛罗里达大学河滨分校、布法罗大学、中国台湾交通大学、清华大学、中国台湾科技大学、新加坡国立大学、南洋理工大学、北京大学、复旦大学、上海交通大学、浙江大学、华中科技大学、电子科技大学、首尔大学、马来西亚理科大学、博特拉大学、伯乐学院、朱拉隆功大学、马哈坎大学、卡塞塔尔大学、托马斯托大学、马布亚工程学院。

我要感谢让我教授发明课程的非洲大学。对达喀尔、塞内加尔、林国荣创意科技大学、姆巴巴纳、斯威士兰、塞尔科集团和科托努科托努阿波美卡拉维大学要特别致谢。

我要感谢 John Wiley 和让本书出版的员工。

致我的孩子们，Aaron Samuel Voldman 和 Rachel Pesha Voldman，希望未来你们都有好运气。

致谢 Betsy H.Brown，感谢她对本书撰写的帮助。

致谢我的父母——Carl 和 Blossom Voldman。

史蒂文·H. 沃尔德曼博士
IEEE 会员

前言

发明和专利逐渐成为技术领域中最为重要的问题，是全球经济的驱动因素。物联网和万物互联持续发展。未来，专利和发明将会驱动全球经济和贸易发展。电子、医药、能源、运输和其他领域将在未来繁荣发展。目前，整个世界正在就相同领域进行利益竞争或期望去竞争。

就目前的工作环境来看，对科学家和工程师有更高的期望：他们不仅要承担本职工作，还要进行创新。很久之前，企业或政府并不期望创新。

在20世纪80年代早期，IBM公司的期望是一名工程师或科学家在30年的职业生涯中能产生一件专利。因为这就是每个雇员产出专利的水平。在1990年之后，这一期望开始发生变化。个人和组织的目标建立在年度预期上。普通员工对于专利和创新的态度发生了变化。

《从发明到专利——科学家和工程师指南》为如何成为一个多产的发明人、理解专利和专利过程提供了一幅清晰的图画，从提交过程到最终草稿对于专利的发明交底书的撰写提供了具有启发性的深刻见解。本书将教你怎样有效地与专利律师和专利审查员沟通，教你怎样使用法律措辞。本书的特殊性在于它覆盖了早期的发明过程和最终的专利撰写，以提供技术领域高质量的专利。

本书有多重目的，下面就以下几个方面进行论述。

本书的首要目的是教会读者怎样成为发明人并理解到底什么是发明。第一步，鼓励读者开始发明创造。这需要一些关于塑造成为发明人的信心方面的建议。另外，发明的概念和定义也包含在内。第二步，想成为发明人需要做的准备工作。

本书的第二个目的是教会读者怎么做好专利检索。迄今为止，有很多提供专利学习和专利检索的引擎可用。本书将会列出不同专利检索引擎的优缺点。

本书的第三个目的是讲述专利的组成结构。充分理解专利局的要求十分重要。

第四个目的是解读怎样撰写专利。因为撰写专利时，版式、文字和图片都有特定要求。

第五个目的是阐述专利的语言和条款。为了与专利律师顺利沟通，理解专利法中使用的语言很有必要。

第六个目的是怎么理解和撰写给专利局的回复。怎样给专利局答复对发明过程也很重要。

最后一个目的是讲述怎样成为一个多产的发明人。一个人怎样才能从撰写第一件专利或发明交底书到拥有许多专利呢？

本书组织结构编排如下：

第1章奠定了探讨发明和专利的基础。介绍专利、版权和商标的基础知识，就专利、版权和商标对当今社会的影响进行探讨。主要列举IBM、苹果、英特尔、三星和Levi Strauss等公司的案例内容，也会选取一些现代案例，例如笔记本电脑和手机。上述案例均与现代生活息息相关。本章的主要焦点在电子行业和软件工程，次要焦点在生物技术人员、生物学家、化学家和药学团队。

第2章聚焦于怎样成为一名发明人和成为多产发明人的过程。选取的案例为著名发明人是怎样工作的和他们怎样在创造性和多产方面展示自己。本章讨论了从爱迪生到乔布斯等人的事例，并从Sir Lanka和Penang Malaysia课程中列举"发明、创新和专利"的案例。

第3章讨论专利检索引擎和专利使用的语言。为了理解这个过程，熟悉并掌握专利语言成为一种必要条件。本章将讨论专利局、专利检索引擎和专利审查、专利书籍和课程。

第4章提出怎么写专利。全球范围内的专利申请由专利局管理。美国专利商标局（USPTO）提供三种类型的专利：发明专利、外观专利和植物专利，本章将对这些专利展开讨论，重点强调三种类型专利之间的区别。

第5章聚焦于专利附图。本章以不同类型的附图展示作为案例，展示发明附图、设计和植物附图。讨论与说明书和权利要求相关的附图的要求和规则。

第6章聚焦于专利权利要求。本章讨论了不同类型的权利要求，怎样撰写独立和从属权利要求以及发明专利要求、设计权利要求和植物权利要求。

第 7 章聚焦于官方的意见和回复。本章首先讨论 USPTO 的审查意见。理解怎样从审查员的角度读懂审查意见，对于发明过程来说最关键的是撰写给专利局的答复。本章还讨论了欧盟专利局的审查意见以及其与 USPTO 的审查意见有什么不同。

第 8 章讨论公司产生专利的不同方法。一些公司产生专利的实践方法是创造性问题解决会议（CPS）。本章对 CPS 会议进行了讨论，目的是说明如何通过这些会议来产生专利、实现公司目标和设置培训新发明人的培训场景。本章展示了 CPS 会议的规则和过程。其中，马来西亚和斯里兰卡发明课程中的一部分集成了 CPS 会议方法的使用。第二个实践方法是使用系统创造性思维方法，叫作 TRIZ、SIT 和 USIT。

第 9 章提出了公司发明战略。本章提供示例企业专利战略、软件系统、发明交底书和奖励，讨论大企业的专利过程，重点强调软件系统、追踪系统、目标和奖励。企业战略式讨论以 IBM、TSMC 和三星为例。另外，还回顾了 AT&T 和英特尔的案例。发明提交的示例和模板在附录中提供。

第 10 章提出成为专家证人的条件和专家证人的职责。本章讨论了不同类型的专家证人，开展了专利诉讼的实践讨论。

上述介绍文字希望能激发你在发明和专利撰写领域的兴趣。

享受文字，享受发明、创新和专利旅程的开始。

史蒂文 · H. 沃尔德曼博士
IEEE 会员

目 录

第1章 介绍 ··· 1
1.1 引言 ·· 1
1.2 专利 ·· 2
1.2.1 什么是专利？ ································ 2
1.2.2 专利和美国宪法 ······························ 3
1.2.3 为什么要申请专利？ ························ 3
1.2.4 什么是可专利的？ ··························· 3
1.3 版权 ·· 4
1.4 商标 ·· 4
1.5 发明 ·· 4
1.6 界定发明的产生方式 ··························· 6
1.6.1 发明——组合 ································ 6
1.6.2 发明——省略 ································ 7
1.6.3 发明——重排 ································ 7
1.6.4 当 1+1=3 时才能称为发明 ················ 7
1.7 在工作中发现发明 ······························ 7
1.8 发明时间 ·· 8
1.9 从发明到产品化 ································· 9
1.10 专利的价值 ······································ 9
1.10.1 作为发明人，你将会获得什么？ ······· 9
1.10.2 发明为我做什么？ ······················· 10
1.10.3 公司可以从发明中获得什么？ ······· 10
1.10.4 公司可以从专利中获得什么？ ······· 10
1.11 专利示例 ·· 11
1.12 意见和总结 ····································· 11
问题 ·· 11
案例研究 ·· 12
参考文献 ·· 13
附录 1.A ·· 15

第2章 发明 ··· 37
2.1 引言 ··· 37
2.2 如何成为发明人 ································ 37
2.2.1 做什么才是有创造性的？ ················ 37
2.2.2 你如何思考？ ······························· 38
2.2.3 你在哪里思考？ ···························· 39
2.2.4 你什么时候思考？ ························· 39
2.2.5 捕捉你的想法和发明 ····················· 40
2.2.6 时间 ··· 40
2.3 研究发明人 ······································ 41

| 2.3.1 研究多产发明人 ································· 41
| 2.3.2 研究发明人的习惯 ····························· 41
| 2.3.3 研究发明人的目标 ····························· 42
| 2.4 创作周期 ·· 42
| 2.4.1 什么是创作周期？ ····························· 42
| 2.4.2 激发创作周期 ··································· 43
| 2.5 左脑和右脑 ·· 43
| 2.6 跳出盒子去思考 ······································· 44
| 2.7 横向思维与批判性思维 ···························· 44
| 2.8 在学科之间的界限发现发明 ····················· 45
| 2.9 结构化发明 TRIZ ···································· 45
| 2.10 专利示例 ·· 46
| 2.11 总结 ··· 46
| 问题 ·· 46
| 案例研究 ·· 47
| 参考文献 ·· 47
| 附录 2.A ·· 49

第 3 章 专利和专利语言 ···························· 72

| 3.1 引言 ··· 72
| 3.2 专利搜索引擎 ··· 72
| 3.2.1 美国专利商标局（USPTO） ············· 73
| 3.2.2 Pat2PDF ·· 73
| 3.2.3 Google 专利 ····································· 73
| 3.3 专利语言 ·· 73
| 3.3.1 说明书 ··· 73
| 3.3.2 权利要求 ·· 74
| 3.3.3 发明人 ··· 75
| 3.3.4 共同发明人 ······································ 75

| 3.3.5 临时申请 ··· 75
| 3.3.6 非临时申请 ······································ 76
| 3.3.7 分案申请 ··· 76
| 3.3.8 持续专利申请 ·································· 76
| 3.4 专利语言——状态和操作 ························ 76
| 3.4.1 审查意见通知书 ······························· 77
| 3.4.2 美国专利审查指南 ··························· 77
| 3.4.3 授权 ··· 78
| 3.4.4 驳回 ··· 78
| 3.4.5 撤回 ··· 78
| 3.4.6 授权通知书 ······································ 78
| 3.4.7 专利未决 ··· 79
| 3.5 专利草案 ·· 79
| 3.5.1 专利草案——结构 ·························· 79
| 3.5.2 专利草案——发明名称 ··················· 79
| 3.5.3 专利草案——背景技术部分 ············ 80
| 3.5.4 专利草案——技术领域部分 ············ 80
| 3.5.5 专利草案——发明内容 ··················· 80
| 3.5.6 专利草案——附图简要说明 ············ 80
| 3.5.7 专利草案——具体实施方式 ············ 80
| 3.5.8 专利草案——权利要求 ··················· 81
| 3.5.9 专利草案——摘要 ·························· 81
| 3.6 总结 ··· 81
| 问题 ·· 82
| 案例研究 ·· 82
| 参考文献 ·· 83

第 4 章 专利 ·· 85

| 4.1 引言 ··· 85

4.2 专利类型 ……………………… 85
  4.2.1 发明专利 ………………… 86
  4.2.2 外观专利 ………………… 86
  4.2.3 植物专利 ………………… 86
4.3 发明专利的专利结构 ………… 87
  4.3.1 发明专利——名称 ……… 87
  4.3.2 发明专利——背景技术部分 … 87
  4.3.3 发明专利——技术领域 … 87
  4.3.4 发明专利——发明内容 … 87
  4.3.5 发明专利——附图简要说明 … 88
  4.3.6 发明专利——具体实施方式 … 88
  4.3.7 发明专利——附图 ……… 88
  4.3.8 发明专利——权利要求 … 89
  4.3.9 发明专利——摘要 ……… 89
4.4 外观专利 ……………………… 89
  4.4.1 外观专利——专利名称设计 … 89
  4.4.2 外观专利——名称 ……… 90
  4.4.3 外观专利——发明人 …… 90
  4.4.4 外观专利——申请人和受让人 … 90
  4.4.5 外观专利——参考文献 … 91
  4.4.6 外观专利——外国专利文献 … 91
  4.4.7 外观专利——其他参考文献 … 91
  4.4.8 外观专利——简要说明 … 91
  4.4.9 外观专利——附图 ……… 92
  4.4.10 外观专利——权利要求 … 92
4.5 植物专利 ……………………… 92
  4.5.1 植物专利——专利名称设计 … 93
  4.5.2 植物专利——标题 ……… 94
  4.5.3 植物专利——相关申请的交叉引用 … 94
  4.5.4 植物专利——该属的拉丁名 … 94
  4.5.5 植物专利——品种名称 … 94
  4.5.6 植物专利——背景技术部分 … 94
  4.5.7 植物专利——发明内容部分 … 95
  4.5.8 植物专利——附图说明 … 95
  4.5.9 植物专利——植物学具体实施方式 … 96
  4.5.10 植物专利——权利要求 … 97
  4.5.11 植物专利——摘要 …… 97
4.6 专利示例 ……………………… 98
4.7 总结 …………………………… 98
问题 ………………………………… 99
案例研究 …………………………… 99
参考文献 …………………………… 100
附录 4.A …………………………… 102

第 5 章 专利说明书附图 ……… 139

5.1 引言 …………………………… 139
  5.1.1 绘图技巧——手绘 ……… 140
  5.1.2 绘图技术——通过计算机绘图 … 140
  5.1.3 绘图技术——通过照相机生成附图 … 140
5.2 专利附图——发明专利 ……… 141
5.3 专利附图——结构 …………… 141
5.4 专利附图——装置 …………… 143
5.5 专利说明书附图——电路图 … 145
5.6 专利附图——系统 …………… 147
5.7 专利附图——方法 …………… 148
  5.7.1 方法图规则 ……………… 148
  5.7.2 方法图与权利要求间的对应关系 … 149
5.8 专利附图——外观设计 ……… 149
5.9 专利附图——植物图 ………… 151

5.10　独特的专利附图——
　　　计算机可读介质权利要求 ······· 153
5.11　专利图和审查意见 ··············· 154
　5.11.1　初步审查专利图 ············ 155
　5.11.2　异议或驳回 ··················· 158
5.12　专利示例 ··························· 159
5.13　总结 ································· 159
问题 ··· 159
案例研究 ··································· 160
参考文献 ··································· 161
附录 5.A ··································· 163

第 6 章　权利要求 ··············· 181

6.1　引言 ································· 181
6.2　独立和从属权利要求 ··········· 181
　6.2.1　独立权利要求 ················ 182
　6.2.2　从属权利要求 ················ 182
6.3　结构权利要求 ····················· 183
6.4　装置权利要求 ····················· 183
6.5　方法权利要求 ····················· 184
6.6　混合权利要求 ····················· 185
6.7　方法加功能权利要求 ··········· 186
6.8　Beauregard 权利要求 ·········· 186
6.9　用尽的组合权利要求 ··········· 187
6.10　替代权利要求 ···················· 187
　6.10.1　马库什（Markush）权利要求 ··· 187
　6.10.2　两分法（Jepson）权利要求 ····· 187
　6.10.3　方法特征限定的产品权利要求 ··· 188

　6.10.4　计算机程序权利要求 ······ 188
　6.10.5　综合权利要求 ··············· 188
　6.10.6　信号权利要求 ··············· 188
　6.10.7　瑞士型权利要求 ············ 188
　6.10.8　延展性权利要求 ············ 188
6.11　总结 ································· 189
问题 ··· 189
案例研究 ··································· 190
参考文献 ··································· 191

第 7 章　审查意见通知书 ······· 193

7.1　引言 ································· 193
7.2　审查意见通知书——美国专利
　　 商标局 ······························ 193
7.3　权利要求的处理 ················· 196
　7.3.1　待决的权利要求 ············· 196
　7.3.2　允许的权利要求 ············· 196
　7.3.3　驳回的权利要求 ············· 196
　7.3.4　异议的权利要求 ············· 197
　7.3.5　受限制或要求选择的权利要求 ······· 197
7.4　申请文件 ··························· 197
　7.4.1　说明书的异议 ················ 197
　7.4.2　说明书附图状态——接受或反对 ··· 197
7.5　具体审查意见 ····················· 198
　7.5.1　权利要求的异议 ············· 198
　7.5.2　权利要求的驳回 ············· 198
　7.5.3　允许的主题 ··················· 201
　7.5.4　结论部分 ······················ 201

7.6 撰写审查意见通知书答复 …… 202
　7.6.1 审查意见通知书答复的概要 …… 202
　7.6.2 修改权利要求 …… 204
　7.6.3 修改驳回的权利要求 …… 205
　7.6.4 撤回权利要求 …… 206
　7.6.5 处理异议权利要求 …… 206
　7.6.6 审查意见通知书答复的结语 …… 207
7.7 预先修正 …… 207
7.8 最终审查意见通知书 …… 207
7.9 欧盟审查意见通知书 …… 208
　7.9.1 阅读欧盟审查意见通知书 …… 208
　7.9.2 欧盟审查意见通知书起始语段 …… 208
　7.9.3 欧盟专利审查员裁决 …… 208
　7.9.4 撰写欧盟审查意见通知书答复 …… 212
7.10 德国专利商标局（DPMA）审查意见通知书 …… 214
　7.10.1 审查文本 …… 214
　7.10.2 申请的主题 …… 215
　7.10.3 本领域技术人员 …… 215
　7.10.4 解释 …… 215
　7.10.5 形式缺陷 …… 216
　7.10.6 结论部分 …… 216
7.11 支持欧洲代理所和欧盟的答复 …… 217
7.12 总结 …… 217
问题 …… 217
案例研究 …… 218
参考文献 …… 219

第8章　发明产生的方法 …… 220

8.1 引言 …… 220
8.2 创造性问题解决会议 …… 220
　8.2.1 建设创造性解决问题（CPS）会议 …… 221
　8.2.2 组建CPS会议参与团队 …… 221
　8.2.3 CPS会议的指导准则 …… 221
　8.2.4 CPS会议主题 …… 222
　8.2.5 CPS会议发明产生的过程 …… 222
　8.2.6 CPS会议发明投票环节 …… 222
　8.2.7 激发团队与发明研发 …… 223
　8.2.8 CPS会议结束 …… 223
8.3 系统思考 …… 223
　8.3.1 TRIZ …… 223
　8.3.2 TRIZ—Altshuller 方法 …… 224
　8.3.3 TRIZ——消除矛盾 …… 224
8.4 系统化创造性思维（SIT） …… 225
8.5 整合式系统化创造性思维（USIT） …… 226
8.6 数据挖掘 …… 226
8.7 预测下一步的发明 …… 227
8.8 总结 …… 227
问题 …… 227
案例研究 …… 228
参考文献 …… 229

第9章　企业专利策略 …… 231

9.1 引言 …… 231
9.2 审核委员会体系 …… 231

9.3 数据库专利追溯系统——世界专利追溯系统（WPTS） ········ 232
9.4 记录发明想法与交底书 ········ 234
9.5 发明交底书的提交 ········ 234
9.6 发明审核与评估 ········ 236
 9.6.1 标题 ········ 236
 9.6.2 交底书编号 ········ 236
 9.6.3 发明人姓名 ········ 236
 9.6.4 审核人姓名 ········ 236
 9.6.5 发明——清晰度 ········ 236
 9.6.6 发明——范围 ········ 236
 9.6.7 审核人已知的现有技术 ········ 236
 9.6.8 发明的替代技术 ········ 237
 9.6.9 现有技术的改进 ········ 237
 9.6.10 可检索性 ········ 237
 9.6.11 发明的必要技术特征 ········ 237
 9.6.12 规避——替代的电路或方法 ········ 237
9.7 审核委员会 ········ 237
9.8 与专利律师合作 ········ 238
9.9 企业知识产权（IP）战略 ········ 238
 9.9.1 企业 IP 目标（Goals） ········ 238
 9.9.2 企业 IP 目的（Targets） ········ 239
 9.9.3 短期目标 ········ 240
 9.9.4 年度 IP 目标 ········ 240
 9.9.5 长期目标 ········ 240
9.10 激励 ········ 240
 9.10.1 发明交底书提交奖 ········ 241
 9.10.2 发明专利授权奖 ········ 241
 9.10.3 发明成就高原奖 ········ 241
 9.10.4 额外 20% 专利奖 ········ 241
 9.10.5 前 5% 专利奖 ········ 241
 9.10.6 部门奖 ········ 241
 9.10.7 企业奖 ········ 241
 9.10.8 年度发明人晚餐邀请 ········ 242
 9.10.9 顶级发明人企业技术认可奖邀请 ········ 242
 9.10.10 大师级发明人奖 ········ 242
9.11 总结 ········ 242
问题 ········ 242
案例研究 ········ 243
参考文献 ········ 244

第 10 章 专家证人 ········ 245

10.1 引言 ········ 245
10.2 专家证人 ········ 245
 10.2.1 专家证人的定义 ········ 245
 10.2.2 专家证人的角色 ········ 246
 10.2.3 专家证人的责任 ········ 246
10.3 专家证人的类型 ········ 246
 10.3.1 非作证证人 ········ 247
 10.3.2 咨询证人 ········ 247
 10.3.3 教育证人 ········ 247
 10.3.4 报告证人 ········ 247
 10.3.5 专家证人 ········ 247
10.4 与专利律师在诉讼中共事 ········ 248
10.5 专家证人报告 ········ 248
10.6 策略 ········ 248
 10.6.1 专利无效 ········ 248
 10.6.2 无效主张文件 ········ 249
10.7 获取公知领域内的资料 ········ 250

10.7.1　现有技术检索……251
10.7.2　新闻发布的文献……251
10.7.3　客户信息……251
10.7.4　产品和文档……251
10.8　获取非公知领域的资料……251
10.8.1　发明技术交底书记录本……252
10.8.2　设计手册……252
10.8.3　设计物理布局……252
10.8.4　工艺流程……252
10.9　科学证据……253
10.9.1　Frye规则……253
10.9.2　Daubert规则……253
10.10　总结……254
问题……254
案例研究……254
参考文献……255

附录……257

附录A　发明技术交底书……257
附录B　发明技术交底书评审员表格……259
附录C　新颖性检索报告……260
附录D　美国专利商标局审查意见通知书细节内容……261
附录E　美国专利商标局审查意见通知书部分……263
附录F　欧盟审查意见通知书……264
附录G　欧盟审查意见通知书答复……265
附录H　美国对欧盟代理人信件——审查意见通知书答复……268
附录I　申请提交彩色照片或附图……269
附录J　专利合作条约……276
附录K　补正书……285
附录L　补正通知书……287
附录M　授权通知书……289
附录N　初次修改……294
附录O　更正附图提交……297

术语表……299

第 1 章
介　　绍

1.1 引言

　　知识产权（Intellectual Property，IP）是全球企业实现成功目标的关键。发明和专利是当今世界创业和创新的途径之一。这个目标不仅仅在美国和欧洲受到关注，同样也受到整个南美、亚洲、中东和非洲的关注。随着对该目标关注度的提升，出现了越来越多的知识产权组织[1-7]、书籍[8-17,23-27]、搜索引擎[18,19]和专利短期课程[20-22]。

　　在 2006 年，我受 IBM 管理团队邀请就"如何发明"[20,21]进行了长达两个小时的讲座。一位经理指出：我在发明提交过程中的产出高于他的 60 人的供应链软件开发团队。于是，我为他的团队提供了两个小时的讲座，讲解如何发明以及如何成为发明家。我们为团队设定企业目标。那个经理询问他的团队应该设定什么目标，我告诉他："简单地说，你有 60 名员工，每年有 52 周，每周提交一个发明，大致相当于每名员工每年提交一个发明。"在确定这个目标之前，整个团队总共只提交了三个发明；在确定目标后，这个团队正在超额完成提交的任务。

　　2007 年，我被邀请到马来西亚讲授发明课程。据称，马来西亚政府希望改善其知识产权组合。我在马来西亚吉隆坡的 Putrajaya 综合中心为马来西亚科技部（MOSTI）的 80 名受训人讲授了一堂全面的发明课程[20,21]。该课程已扩展为马来西亚理科大学（USM）第一阶段为期两天的课程，受训人包括制药商、大学教师、生物技术专家和

从发明到专利——科学家和工程师指南．第一版．史蒂文・H. 沃尔德曼 ©2018 John Wiley & Sons 公司，2018 年由 John Wiley & Sons 公司出版。

化学家。该课程也引起了斯里兰卡科伦坡各界人士的兴趣,并于次年开始面向软件开发者。目前该课程已经扩展到第二阶段的高级课程[22]。

在那段时间里,有很多受训人要求我根据课程内容编写文稿。在此,我将课程的有关内容编撰成书,希望帮助更多的朋友!

什么是知识产权?

在形式上,知识产权是思维创造。思维创造是创造性的作品或思想,体现在可以共享的形式中,或者可以使其他人重新创造、模仿或制造它们。知识产权的保护方式有以下四种(如图 1-1 所示)[1]。

- 专利
- 商标
- 版权
- 商业秘密

图 1-1　知识产权的保护方式

1.2　专利

在本节中,我们将对专利进行讨论:专利是什么?以及为什么应该对你的创新、想法和发明申请专利?

1.2.1　什么是专利?

什么是专利?

专利是政府授予发明人的所有权,"排除他人制造、使用、许诺销售、销售或进口其发明"。

美国对专利的定义是"专利是美国政府授予发明人的权利,以公开披露换取在专利授权的一定时期内排除他人在美国制造、使用、许诺销售或者销售专利发明,或将专利发明进口到美国"[1]。

1.2.2 专利和美国宪法

美国宪法包含了为发明人提供保护的概念。美国宪法规定如下[1]：

国会应该有权力通过确保作者和发明人在一定时期内对自己的作品和发现拥有排他权，来促进科学和有用技术的进步（1790年，美国宪法第1条，第8部分）。

从此，保护专利、版权、商标和商业秘密的知识产权的思想被纳入美国宪法。

1.2.3 为什么要申请专利？

人们通常会问：为什么应该为发明申请专利？

专利是政府保护公民对发明拥有权利的一种手段。

专利也是政府保护公司对发明拥有权利的一种手段[1]。

专利局是政府保护您的创意作品的一种方式。在多数情况下，会有投资公司用于开发或制造您的发明，并且投资者期望能产生回报。

为发明申请专利将保护您的想法免受其他个人或企业的抄袭。作为企业，则希望保护其投资及其员工。

如果不为发明申请专利，您的想法则存在在国内被采用、抄袭、生产销售的风险。可口可乐公司虽然没有为可口可乐糖浆配方申请专利，但是选择将其作为商业秘密进行保护。

1.2.4 什么是可专利的？

什么是可专利的？

人们经常问到的一个问题是他们的发明是否可以获得专利。具有专利性的想法或发明可以申请不同类型的专利（如图1-2所示）[1]：

- 发明专利
- 外观专利
- 植物专利

发明专利：是"保护有用的过程、机器、制造物品和物质构成的专利"[1]。

外观专利：是"保护制造产品的新的、原始的或装饰性的设计的专利"[1]。

植物专利：是"保护发明或发现的无性繁殖植物品种的专利"[1]。

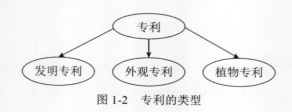

图1-2 专利的类型

1.3 版权

版权是知识产权的一种类型。版权保护文学、艺术和音乐作品的思想表达[1]。需要版权保护的有书籍、剧本、记录和专辑。书籍、剧本或音乐的版权使用需要所有者的批准，并且有时需要支付版税。在当今社会，收取剧本和音乐的版税是一种常见做法。当版权被侵犯时，其使用行为则被视为盗版。

1.4 商标

商标是知识产权的一种类型。我们并不是十分了解周围的商标。

商标是与商品或服务相关的、用来识别商品或服务来源的单词、名称、符号、设备或图像[1]。例如，耐克公司的"旋风"、Levi-Strauss牛仔裤后袋上的红色标签都是商标。在计算机领域，英特尔拥有"Intel Inside"商标。这些商标用于保护公司免受仿冒和盗版侵害。然而，即使在口袋上有红色标签，许多牛仔裤公司还是抄袭了Levi-Strauss牛仔裤的设计。

"服务标志"与商标相同，用于区分商品的来源，但它不能区分产品。

1.5 发明

人们经常使用发明这个词，但什么是发明？谁是发明家？

在本节中，我们将正式讨论这些问题。

1. 你是发明家吗？

你是发明家吗[20-22]？谁是发明家？是什么让你成为发明家？

是不是拥有专利才能称自己为发明家？你需要拥有专利才能成为发明家吗？你需要多少专利才能称自己为发明家？

我发现因为文化和社会的因素，我们不敢称自己为发明家。我们害怕承认自己是发明家，也害怕被评判。因为别人的意见，我们不愿意承认自己发明家的身份。

我发现拥有一项专利并不能缓解这种顾虑，因为人们会质疑专利的价值，或者质疑你是不是"发明家"。那我们如何判断专利的价值呢？在很多情况下，如果一家公司在产品中使用了专利，该专利就被认为是有价值的[20-22]。

就我个人而言，我认为获得一百项专利才可以称为"发明家"。当我在马来西亚教授发明课时，我告诉全班的人，他们必须走进洗手间，看着镜子，然后告诉自己他们可以成为发明家。

2. 你知道他们是发明家吗？

许多来自各行各业的人都拥有专利。以下是一些拥有美国专利[20-22]的人：

- 亚伯拉罕·林肯
- 马克·吐温
- 托马斯·阿尔瓦·爱迪生
- 本杰明·富兰克林
- 托马斯·杰斐逊
- 奥维尔和威尔伯·赖特
- 阿尔伯特·爱因斯坦
- 乔治·伊士曼
- 乔治·威斯汀豪斯
- 理查德·加特林
- 尼古拉·特斯拉

难以置信，即使是美国总统，如亚伯拉罕·林肯和托马斯·杰斐逊也有时间和兴趣去发明并拥有美国专利。同样，马克·吐温也对美国专利商标局感兴趣，并提交了发明。

我们都知道莱特兄弟发明了飞机，但他们也拥有标题为"飞行机器"的专利。

阿尔伯特·爱因斯坦曾在美国专利商标局工作，并对发明和专利产生了兴趣。

一些发明家非常多产，例如托马斯·阿尔瓦·爱迪生拥有1 069项美国专利。当讨论如何成为一位多产的发明家时,我们将讨论爱迪生的工作习惯,此外还有乔治·伊士曼、乔治·威斯汀豪斯、理查德·加特林以及物理学家尼古拉·特斯拉[20-22]。

3. 谁是年轻的发明家?

有一个关键问题:拥有美国专利是否有年龄限制?

拥有美国专利并没有年龄限制。1989年6月27日,一名儿童拥有了USPTO 4 842 157[20-22]号美国专利。在美国,只有提交专利的发明人,才能将其名字著录在专利上。乔纳森·斯通·帕克小时候,在喝一杯加冰的水时,发现倾斜玻璃杯冰块会冒出来。这让他很沮丧。所以,他需要发明一个固定器放在杯子里,避免冰块掉出来。

饮用容器的固定器
饮用容器有一个固定装置,能将像冰筒那样的较大物体固定在底部。固定装置具有中央部分,该中央部分印有标记,并有径向向外延伸的多个弹性臂与容器连接。中央部分的底部安装有多孔袋,来放置释放到液体中的添加剂。

1.6 界定发明的产生方式

什么样的产生方式能被视为一项发明?我们多数人都没有明确的定义。本节内容有助于我们理解发明过程是什么样的,怎样的产生方式可以构成一项有效发明(如图1-3所示)。

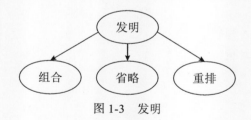

图1-3 发明

1.6.1 发明——组合

"组合"是发明的第一种产生方式。将两个相似单元结合在一起构成发明,这就是组合的形式。换言之,把第一部分和第二部分结合起来,就是一个发明的产生过程[1,20-21]。

组合是合并两部分内容的处理方式。

发明可能是第一部分和第二部分的结合。如果将第一部分和第二部分合并,能够产生比没有合并时更优越的效果,如果不合并该效果就不会产生,这种组合方式就构成

了发明 [1,20-22]。

如果将第一部分和第二部分结合后没有产生额外的效果，那么这种组合方式就不能构成发明。

1.6.2 发明——省略

"省略"是发明的第二种产生方式，是通过删除某个特征而构成发明。通过省略或删除某特征可以简化结构，这种简化方式也可能构成一项发明 [1,20-22]。

1.6.3 发明——重排

"重排"是发明的第三种产生方式 [1,20-22]。调整位置是重排的一种方式。因此，第一部分和第二部分位置的改变可能构成一项发明。

1.6.4 当1+1=3时才能称为发明

关于发明的三种产生方式，还需要考虑其他因素。针对具有发明价值的事物，它必须比现有技术具有更好的效果。如果将第一部分和第二部分合并没有带来更优越的效果，则这种合并就是没有价值的。还有，重排或者删除方式，如果没有带来更优越的效果，则这些方式也是没有新颖性或创造性的。因此，如果将第一部分和第二部分合并，并且产生了更优越的效果，即当1+1=3时才能称为发明 [20,21]。

1.7 在工作中发现发明

对于科学家和工程师，创造性行为是我们日常工作的一部分。我们工作中的创造性行为可能成为发明。识别工作中的发明，培养这项技能是很重要的。有时候，你不能在工作中发现发明，就不能做出最好的判断 [20,21]。

我在IBM有一位同事，他在做一项创造性工作，但他在自己的工作中找不到发明。他不能确定自己所做工作的新颖性程度，因而不敢向发明委员会提交自己的工作成果。

这样的一些能力可以来自培训。

麻省理工学院就有独特的教学方法。麻省理工学院的老师会提出问题并给出一个解决方案。但是，在麻省理工学院，这不是结束，而是开始。他们会说："好吧，现在我们有了解决方案，我们可以用它做什么？"许多教学内容以获得答案作为结束，而在麻省理工学院，这是创造性过程的开始。

有了这种思路，在 IBM 的许多实验室的发现成了发明。我相信最好的发明来自自己的工作和发现。一些最好的发现也可能来自实验室或工厂的事故。有一天，一位工艺技术员安装了三块晶圆进行高能量 MeV 植入实验。无意中，他拿错了晶圆，从其中一个晶圆上取下了 MeV 植入体，从而获得了三种实验结果：没有 MeV 植入、正常的 MeV 植入剂量和两倍的 MeV 植入剂量。从这项工作中，我们能够研究植入剂量对 4 Mb 动态可读存储器（DRAM）芯片器件特性的影响。所以，一些好的事故过程能够成为引导发明的最佳实验。

1.8　发明时间

在工作中，人们普遍认为不可能发明或找到发明的时间，因为他们太忙了。工程师是开发团队的一部分，总是显得过度劳累和超大压力。我的同事们常常认为他们没有时间写下自己的想法、发明或者思路[20, 21]。

工程师常常说"我没有时间去发明"。他们还认为没有时间撰写发明披露信息、参加委员会、并与专利律师合作进行专利申请。

导致这个问题的部分原因是他们没有设定专门的时间用于发明，解决的方式之一是设定一个"发明时间"。

设定"发明时间"[20,21]有很多方法：
- 用一个记事本来记录工作日里的发明。
- 每天晨练的时候思考发明。
- 在日历中设定一个时间，思考发明。
- 在日历中设定一天或者一个下午用于发明。
- 在午餐时思考发明。

就我个人而言，我做到了以上所有事情。

在 IBM，我们有小巧的皮质便签本，装有一些便签纸，适合放在白衬衫的口袋里。它被称为思考便签，封面有一个词——思考。

许多工程师和科学家问什么时候是提交专利的最佳时机。我认为最好能积极地提交发明，因为全世界有很多人都在关注着这一主题[20,21]。

我认识一位在一家小型制药公司工作的化学家，这家公司正在开发一种非常重要的药物。她的小公司每天工作一班，而在大公司工作的竞争对手每天工作三班。这家小公司没有先完成专利申请，输给了大公司。这家大公司因为这种药物盈利数百万美元。

就我个人而言，我起初对于提交自己的想法并不积极，然而在三周之后申请专利时受到了来自同一公司的另一个研究小组的严重打击。

当我和同事产生一个新的想法时，我建议我的同事不要拖延，就在想到的当天提交发明。

人们问我什么时候是提交发明的最佳时机，我的答案是"昨天"[20,21]。

1.9 从发明到产品化

将发明转变为产品的过程是公司成功的重要一步。一些非常成功的公司已经学会了如何做到这一点。这将在后面的章节中详细讨论。

1.10 专利的价值

专利对发明人和公司都有价值。在下一节中，我们将提供个人和公司的案例。

1.10.1 作为发明人，你将会获得什么？

作为发明人你会获得什么？相关的收益如图 1-4 [20,21] 所示。

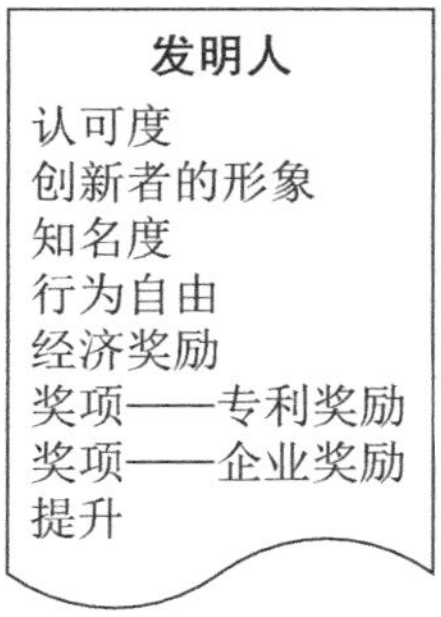

发明人
认可度
创新者的形象
知名度
行为自由
经济奖励
奖项——专利奖励
奖项——企业奖励
提升

图 1-4 发明人的收益

1.10.2 发明为我做什么？

发明和专利可以对个人的态度、职业和认可度产生重大影响。

从个人角度来看，发明是一个创造性的产出。它涉及思考、写作和绘图。在工作环境中，它会影响员工对工作和公司的态度。它可以使工作变得更有趣[20,21]。

从职业角度来看，发明可以影响同行的认可度、企业的认可度以及全行业的认可度。如果公司设有专利奖励制度，还可以获得经济奖励。

1.10.3 公司可以从发明中获得什么？

创新性企业可以树立创新的形象。一家被认为具有创新性的公司将提高它成功推出新产品的能力，从而吸引新客户。新员工也将被吸引到其认为具有创新性的企业。发布发明产品和专利是证明公司实力的最佳方式之一，在投资、品牌和股票市场上都对公司有很大帮助[20,21]。

从员工角度来看，发明是员工的创造性产出。在工作中，它会影响员工对工作和公司的态度。当员工合作发明或者参与企业项目和智囊决策时，可以促进团队工作和团队建设。它可以使工作变得更有趣，缓解员工在企业中的压力[20,21]。

1.10.4 公司可以从专利中获得什么？

凭借专利组合，公司将获得许多业务优势。这些优势如图 1-5 [20,21] 所示。

图 1-5 公司的收益

认可度：拥有强大专利组合的公司将获得公司、大学和员工的认可。

创新者的形象：产生专利的公司会被视为具有创新性的公司。

知名度：创新的公司或个人可以具有更高的知名度，从而产生更高的认可度和商机。

品牌：具有创新性的公司将获得愿意投入更多费用去购买该公司产品的客户。

行为自由：拥有专利的公司可以自由地生产产品，不必担心会对其他公司侵权。

收入：拥有专利组合的公司可以通过出售使用权、许可费用以及出售专利，从专利上盈利。

联盟：拥有专利组合的公司可以在联盟组织内进行内部交易、出售、共享专利组合。

1.11 专利示例

在本节中，给出了一些专利的示例，图 1.A.1 和图 1.A.2 中的这些专利详细描述了 DRAM 单元的结构和方法。在本节中，您将初步接触到专利，未来的章节会有更多的内容。

1.12 意见和总结

本章介绍了发明和专利的主题，简要论述了为什么要申请专利或不申请专利。此外，还引入了发明的概念。许多发明人起初都不确定是否存在发明以及其工作内容是否可以申请专利。

在第 2 章，本书将深入探讨如何成为发明家、如何激发创造力以及专利的产生过程。讨论多产的发明人，研究他们的创造周期、他们的习惯以及产生发明的方法，此外，还将讨论创意的产生方法。

问题

1. 知识产权是什么？为什么在当今世界知识产权如此重要？
2. 知识产权的三种类型是什么？
3. 商业秘密是什么？它与专利、版权和商标有什么不同？
4. 列举一项对过去世界有重大影响的专利。
5. 列举一项对当今世界具有重要意义的专利。
6. 美国的专利在哪里保护其内容？

7. 美国专利在哪些国家保护发明人？
8. 第一个专利局是什么？
9. 美国专利商标局有多久的历史？
10. 列举一个有版权的项目的例子。
11. 列出当今世界上使用的一些商标。
12. 发明是什么？什么使某些事物成为发明？
13. 描述合并、省略和重排，并提供一些示例。
14. 列举一些拥有专利的美国总统。他们发明了什么？
15. 列举一些成立公司的发明人。
16. 列举一些由于其专利而闻名的发明人。
17. 托马斯·阿尔瓦·爱迪生拥有多少专利？列举一些最重要的专利。
18. 发明人能从专利中获得什么？
19. 一家公司能从拥有的专利组合中获益多少？
20. IBM 的专利组合有多大？
21. 苹果公司的专利组合有多大？

案例研究

案例研究 A

提出一项发明，并展示它如何实现合并、省略或重排。

案例研究 B

提出三项发明，每一项发明分别对应合并、省略和重排的情况。

案例研究 C

贵公司正在制定知识产权策略。确定一个专利，涉及以下列表中的所有项目。说明你的发明将如何实现所有目标：

- 认可度
- 创新者的形象
- 知名度

- 品牌
- 行为自由
- 收入
- 联盟

参考文献

1. U.S. Patent Office (USPTO). https://www.uspto.gov (accessed 19 December 2017).
2. European Patent Office (EPO). https://www.epo.org (accessed 19 December 2017).
3. Japan Patent Office. https://www.jpo.go.jp (accessed 19 December 2017).
4. Malaysian Patent Office (MyIPO). https://www.myipo.gov/my (accessed 19 December 2017).
5. World Intellectual Property Organization (WIPO). https://www.wipo.int (accessed 19 December 2017).
6. Organisation Africaine de la Propriete Intellectuelle (OAPI). https://www.oapi.int (accessed 19 December 2017).
7. African Regional Intellectual Property Organization (ARIPO). https://www.airpo.org (accessed 19 December 2017).
8. U.S.Department of Commerce (2000). *Patents and How to Get One: A Practical Handbook*. Courier Corporation.
9. Mueller, J.M. (2016). *Patent Law,* 5e. Wolter Kluwer Publications.
10. Stim, R. (2016). *Patent, Copyright, and Trademark: An Intellectual Property Desk Reference*. Nolo. ISBN: 978-1-4133-2221-7.
11. Slusky, R. (2013). *Invention Analysis and Claiming: A Patent Lawyer's Guide,* 2e. ISBN:13 978-1614385615.
12. Rosenberg, M. (2016). *Essentials of Patent Claim Drafting,* LexisNexis IP Law and Strategy Series. Matthew Bender.
13. Adams, D.O. (2015). *Patents Demystified: An Insider's Guide to Protecting Ideas and Invention*. ISBN: 13 978-163425679.
14. Pressman, D. and Tuytschaevers, T. (2016). *Patent it yourself: your step-by-step guide to filing at the U.S. Patent Office*.
15. Charmsson, H.J.A. and Buchaca, J. (2008). *Patents, Copyrights, and Trademarks for Dummies*. Wiley.
16. Stim, R. and Pressman, D. (2015). *Patent Pending in 24 Hours,* 7e. Nolo. ISBN: 978-1-4133-2201-9.

17. Lo, J. and Pressman, D. (2015). *How to Make Patent Drawings*, 7e. Nolo. ISBN:978-1-4133-2156-2.

18. Google Search Engine. https://www.google.com/patents (accessed 19 December 2017).

19. Pat2PDF Search Engine. https://www.pat2pdf.org (accessed 19 December 2017).

20. Voldman, S. (2014).Short Course, *Innovating, Inventing, and Patenting*, Dr. Steven H. Voldman LLC, Ministry of Science Technology and Innovation（MOSTI）, Putrajaya, Malaysia, May 2014.

21. Voldman, S.（2015).Short Course, *Innovating, Inventing, and Patenting*, Dr. Steven H. Voldman LLC, FITIS. Sri Lanka, February 2015.

22. Voldman, S.（2016）.Short Course, *Writing and Generating Patents*, Dr. Steven H. Voldman LLC, FITIS. Sri Lanka, February 2016.

23. Amernick, B.A. (1991). *Patent Law for the Nonlawyer: A Guide for the Engineer, Technologist, and Manager*, 2e. Von Nostrand Reinhold. ISBN: 13 978-0442001773.

24. Durham, A.L. (2013). *Patent Law Essentials: A Concise Guide*, 4e. Oxford: Praeger, ABC-CLIO, LLC. ISBN: 13 978-1440828782.

25. DeMatteis, B., Gibb, A., and Neustal, M. (2006). *The Patent Writer: How to Write Successful Patent Applications*. Garden City Park, NY: Patents for Commerce, Square One Publishers.

26. Sutton, E. (2016). *Software Patents: A Practical Perspective*. CreateSpace Independent Publishing Platform.

27. Jackson Knight, H. (2013). *Patent Strategy for Researchers and Research Managers Wiley*, 3e. England: Chichester.

附录 1.A

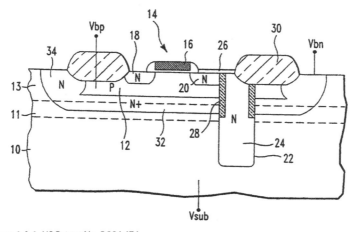

Figure 1.A.1 US Patent No. 5,384,474.

16 | 从发明到专利——科学家和工程师指南

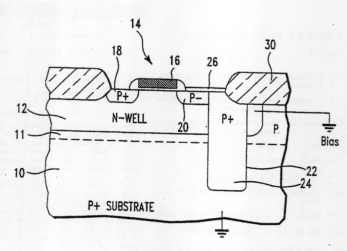

FIG. 1 PRIOR ART

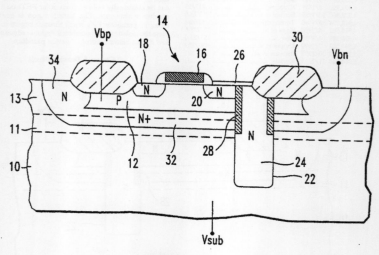

FIG. 2

Figure 1.A.1 (Continued)

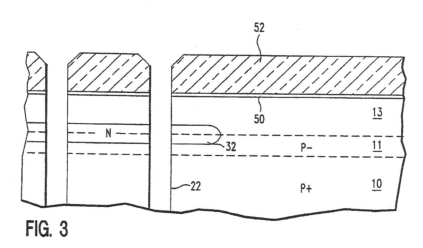

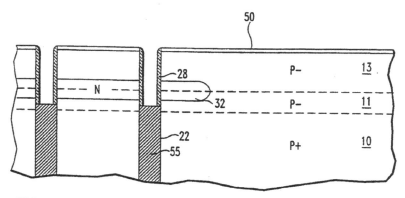

Figure 1.A.1 (Continued)

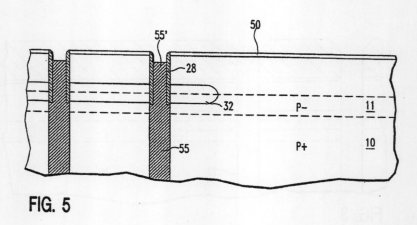

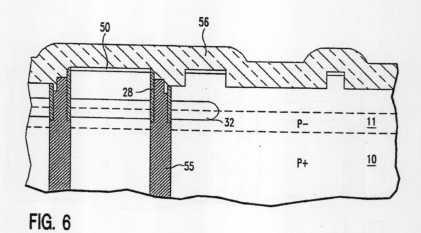

Figure 1.A.1 (*Continued*)

U.S. Patent Jan. 24, 1995 Sheet 4 of 5 **5,384,474**

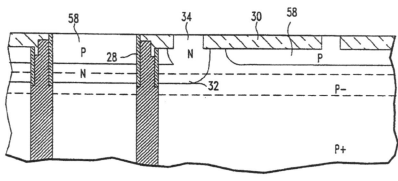

FIG. 7

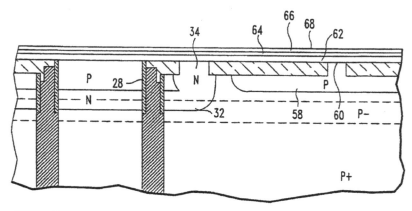

FIG. 8

Figure 1.A.1 *(Continued)*

U.S. Patent Jan. 24, 1995 Sheet 5 of 5 **5,384,474**

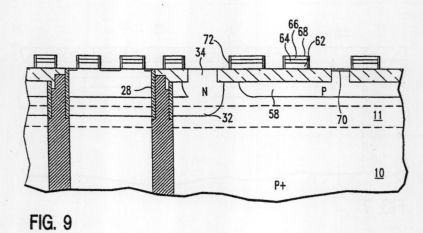

FIG. 9

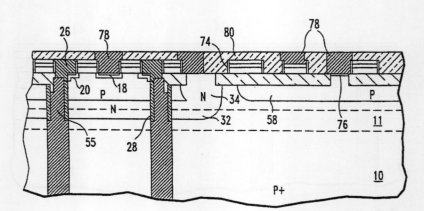

FIG. 10

Figure 1.A.1 (*Continued*)

5,384,474

DOUBLE GRID SUBSTRATE PLATE TRENCH

This is a continuation of copending application Ser. No. 07/819,159 filed on Jan. 9, 1992, now abandoned.

RELATED APPLICATIONS

This application is related to co-pending application Ser. Nos. 07/819,148 and 07/818,668 filed concurrently herewith and entitled "Diffused Buried Plate Trench DRAM Cell Array" by D. M. Kenney and "Double Well Substrate Plate Trench DRAM Cell Array" by G. B. Bronner et al, respectively, which issued as U.S. Pat. Nos. 5,254,716 and 5,250,829, respectively.

BACKGROUND OF THE INVENTION

1. Field of the Invention

This invention relates to semiconductor memory devices and particularly to high density dynamic random access memory cells and methods for their manufacture in sub-micron technologies.

2. Description of the Prior Art

Designers of technologies for producing semiconductor devices have been continually pressured to increase effective device densities in order to remain cost and performance competitive. As a result, VLSI and ULSI technologies have entered the sub-micron realm of structural dimensions and now are designing technologies in the deep submicron feature size range. In the foreseeable future absolute atomic physical limits will be reached in the conventional two-dimensional design approach to semiconductor device design. Traditionally, Dynamic Random Access Memory (DRAM) designers have met the severest of challenges in advancing technologies by pushing the limits of feature size resolution with each generation of DRAM. For example, designers of 64K bit DRAMs were perplexed to learn that a practical physical limit to charge capacity of storage capacitors in planar cell layouts had already been reached due to the minimum charge capacity required to allow reliable data signal sensing in the presence of naturally occurring atomic particle radiation inherently present in fabrication materials and the operating environment. Storage capacitors in the range of about 50 femtofarads were considered to be a physical limit. From a practical view, this limitation prevented a continuation of the scaling of DRAM dimensions and voltages initiated in the early 1980s. Reduction in the surface area of semiconductor substrate utilized by the DRAM storage capacitor has been severely restricted. Due to deceases in the thickness of reliable capacitor dielectric materials, existing 1 Megabit (1 Mb) DRAM technologies continue to enjoy the freedom of planar, two-dimensional device and circuit design. Beginning with 4 Mb DRAMs, the world of three-dimensional design has been utilized to the extent that the simple single device/capacitor memory cell has been altered to provide the capacitor in a vertical dimension. In such designs, the capacitor has been formed in a trench formed in the surface of the semiconductor substrate. In yet denser designs, other forms of three-dimensional capacitors have been proposed, such as stacking the plates of the capacitor above the transfer device. Such designs, however, present difficulties in forming the interconnections to the required word access and data bit lines to the DRAM memory cell. Additional designs have been proposed in which the transfer device and its associated capacitor are both formed within a trench of preferably minimum feature size. Currently, insurmountable processing difficulties make such designs impractical for product manufacturing processes.

A large number of proposals for 16 Mb and greater density DRAM cell designs have avoided continuing development of trench cell technology because of the existence of charge leakage mechanisms known to be present in trench capacitor structures. As these leakage mechanisms have become known, extensions of trench DRAM cells designs have been used successfully in 16 Mb designs.

The following references describe various aspects of prior art techniques used in DRAM and other semiconductor technologies.

The article "Trench and Compact Structures for DRAMs" by P. Chatterjee et al., International Electron Devices Meeting 1986, Technical Digest paper 6.1, pp. 128–131, describes variations in trench cell designs through 16 Mb DRAM designs, including the Substrate Plate Trench (SPT) cell described in more detail in U.S. Pat. No. 4,688,063 issued Aug. 18, 1987 to Lu et al. and assigned to the assignee of the instant invention. The SPT cell uses a highly conductive substrate as the DRAM cell plate. The storage node of each cell is formed in a deep trench in the substrate. U.S. Pat. No. 4,801,988 issued Jan. 31, 1989 to Kenney and assigned to the assignee of the instant invention, describes an improved SPT cell which includes a thick isolation region formed within the trench to enable higher density packing of DRAM cells. The article "CMOS Semiconductor Memory Structural Modification to Allow Increased Memory Charge" anonymous, IBM Technical Disclosure Bulletin, Vol. 31, No. 11, April 1989, pp. 162–5, teaches a method of isolating the substrate plate of an SPT cell from support devices by providing a buried region under support devices in order to allow the plate reference voltage to be separately biased at an optimum Vdd/2 volts.

U.S. Pat. 4,912,054 issued Mar. 27, 1990 to Tomassetti describes methods of isolating bipolar-CMOS circuit devices through the use of various epitaxial layers as commonly found in bipolar device technologies. The article "A 45-ns 16-Mbit DRAM with Triple-Well Structure" by S. Fujii et al., IEEE Journal of Solid-State Circuits, Vol. 24, No. 5, October 1989, pp. 1170–1175, describes techniques for isolating various different functional device types in which the entire array of trench DRAM cells is formed within a surface implanted P-well.

U.S. Pat. No. 4,829,017 issued May 9, 1989 to Malhi describes a method of forming a buried doped layer in a substrate by forming a shallow trench, protecting its sidewalls, further extending the trench and finally doping the walls of the extended trench to form a continuous doped region useful as the storage node of a trench DRAM.

The article "New Well Structure for Deep Submicron CMOS/BiCMOS Using Thin Epitaxy over Buried Layer and Trench Isolation" by Y. Okazaki et al., 1990 Symposium on VLSI Technology, Digest of Technical Papers, paper 6C-4, pp. 83–4, describes the use of buried epitaxial layers to isolate surface devices from the substrate.

The following references relate specifically to variations in SPT DRAM cells in which a buried region of opposite conductivity type from the substrate is used a one plate of the DRAM storage capacitor. U.S. Pat. No. 4,918,502 issued Apr. 17, 1990 to Kaga et al. describes a

Figure 1.A.1 (*Continued*)

buried plate trench DRAM cell in which the storage node of the cell and a sheath plate are formed in a single trench. At the bottom of the trench a diffusion of opposite type from the substrate is formed such that the diffusions of adjacent cells interconnect forming a grid-like structure. One or more trenches not associated with a DRAM cell is formed to act as a reach through to enable the doped region to be biased at a suitable reference voltage. FIG. 12, thereof, clearly illustrates the grid-like aspect of the buried region. European published application 0 283 964, published Sep. 28, 1988 describes a buried plate SPT DRAM cell in which an out-diffused region from the DRAM trenches, similar to that in Kaga et al., in which the diffused region forms the plate of the SPT cell. As in Kaga et al. a grid-like region is formed and is contacted by a non-cell trench. U.S. Pat. No. 4,873,560 issued Oct. 10, 1989 to Sunami et al, describes yet another buried plate SPT cell in which the access transistor is formed in the cell trench. FIG. 30, thereof, and its related text, describes the importance of maintaining the grid-like structure of the buried region in order to enable proper operation of the cell transfer device. Sunami et al, further cautions that in the event that opening in the grid-like buried region should be "filled by the depletion layer" isolating the surface devices from the substrate a separate connection can be made to the "isolated" surface region in order to bias it to the same potential as the substrate. UK Patent Application GB 2 215 913 A, published Sep. 27, 1989 describes yet another variation in the buried SPT DRAM cell design in which the dopant for the buried region is provided by ion implantation into the sidewalls of the deep trench of the DRAM cell. Finally, U.S. Pat. No. 4,794,434 issued Dec. 27, 1988 to Pelley, describes a buried plate SPT DRAM cell formed using bipolar device processing methods in which the buried plate region is formed from a buried sub-collector structure normally part of a bipolar transistor.

While the above cited references illustrate the diverse and concentrated efforts made by DRAM designers in attempting to overcome the inherent barriers in continuing to reduce the size, and increase the density, of DRAM cells, none provide the capability to carry DRAM technology into the sub-0.5 micron feature size range, a feat which must be achieved in order to continue the two decade "tradition" of providing ever increasing density of DRAM technology. DRAM designers have turned to the process-complicating use of "stacked capacitor" DRAM cells knowing that the addition of processing steps decreases the manufacturability of a specific design.

Referring to FIG. 1, there is shown a schematic cross-sectional view of the basic Substrate Plate Trench (SPT) DRAM cell described in U.S. Pat. No. 4,688,063 to Lu et al. entitled "Dynamic RAM Cell with MOS Trench Capacitor in CMOS". A P+ type semiconductor substrate 10 is provided with a lightly doped epitaxial surface layer 11 in which is provided N-type retrograde implanted well 12 formed at its upper surface in which the transfer device 14 is formed. A control gate electrode 16 is responsive to signals from word line circuitry to couple data signals applied to the bit or data line diffused region 18 to the diffused storage node region 20. A deep trench 22 is provided in which a storage capacitor is formed. A polysilicon storage node plate 24 is formed in the trench and isolated from the substrate 10 by a thin storage node dielectric, not shown. A conductive strap 26 connects the diffused node 20 to the plate 24.

Manufacturing experience has shown that the SPT DRAM cell described is not suitable for extension to greater than 16 Mbit applications due in part to the performance limitations of P-array transfer devices and the existence of a parasitic device formed by the diffused storage node 20, the polysilicon plate 24 and the substrate 10. Simple conversion to N-type transfer devices is not practical and reduction of electrical stress on the capacitor dielectric by using Vdd/2 reference node biasing is not possible. The subject invention addresses the unsolved problems of the prior art by providing a solution to barriers presented in extending the manufacturability of the simple SPT cell to 64 Mb DRAM and beyond.

SUMMARY OF THE INVENTION

An object of the invention is to provide a double grid SPT cell in which the density limitations of the prior art are removed.

Another object of the invention is to provide a Substrate Plate Trench DRAM design which does not increase process complexity at the expense of product yield.

Yet another object of the invention is to provide an SPT DRAM cell which has a minimum impact on existing processing technologies.

The present invention relates to methods for providing a cell design in which all of the historically limiting parameters of DRAM cells are dealt with in a unifying manner to provide a near optimum design in which charge leakage factors are minimized and device bias conditions are optimized. The invention includes a substrate plate trench DRAM cell array in which a buried region is used to electrically and physically isolate regions in a semiconductor substrate such that the cell transfer device can be operated independently from other support devices formed in the substrate. Sub-half micron feature sizes are used cooperatively with previously known techniques to provide a simple buried isolation layer.

These and other objects and features of the invention will become more fully apparent from the several drawings and description of the preferred embodiment.

DESCRIPTION OF THE DRAWINGS

FIG. 1 is a simplified schematic sectional view of a substrate plate trench (SPT) DRAM cell of the prior art illustrating the basic electrical connections of the cell.

FIG. 2 is a simplified schematic sectional view of the double grid substrate plate trench (SPT) DRAM cell of the invention illustrating the basic electrical connections of the cell.

FIGS. 3–10 are schematic sectional views of the array of the invention illustrating the array at various steps in a preferred manufacturing process.

DESCRIPTION OF THE PREFERRED EMBODIMENT

Referring to FIG. 2, there is shown the basic elements of the double well Substrate Plate Trench DRAM cell of the invention. The cell is an improvement of the prior art SPT DRAM cell as described by Lu et al. in U.S. Pat. No. 4,688,063 and as modified by Kenney in U.S. Pat. No. 4,801,988, both of which are incorporated herein by reference. The cell includes the following major features. A substrate 10 of P+ type

Figure 1.A.1 (Continued)

semiconductor material has an upper epitaxial layer of P- type material, as in the prior art. Using bipolar subcollector fabrication technology, a first epitaxial layer 11 can be formed, followed by the localized formation of an N-type buried layer 32 and an additional P-type epitaxial layer 13. Surface diffused reach through regions 34 connect to the perimeter of the buried layer 12 such that the substrate region 12 in which the DRAM cells are formed becomes both physically and electrically isolated from the semiconductor substrate 10. A P-type isolated region 12 is formed at its upper surface into which N-channel transfer devices 14 are formed. A control gate electrode 16 of device 14 is responsive to a word access line of the DRAM array support circuits, not shown, to couple data between data or bit line diffused N-type region 18 and diffused N-type storage node region 20 through the channel region formed in region 12. In a manner similar to the prior art, a storage capacitor is formed in a deep trench 22 adjacent to the storage node 20 and includes a signal storage node 20 formed by conductive N-type polysilicon electrode 24 isolated from substrate 10 by a thin dielectric layer. Diffused surface storage node 20 and signal storage node 24 in the trench 22 are connected by a conductive strap 26. At the top of the storage trench a thick insulating collar 28 is provided to increase the threshold voltage of the vertical parasitic FET formed by the diffused storage node 20 and the substrate within the P-region 12. In order to eliminate the trench gated-induced diode leakage mechanism, the collar 28 must extend down the sidewall of trench 22 to a point below the lowermost pn junction of N-region 32. Local surface isolation 30 is also provided, as is well known in the prior art.

It is an important aspect of the invention to provide arrays of cells as described in connection with FIG. 2 in a simple easy to manufacture processing sequence. Because the basic SPT cell is effectively placed inside a P-well-which is contained in an N-well, the additional processing steps necessary to fabricate the invention are minimized. Typically, the P-well 12 can be biased at about −1 volt, the P+ substrate 10 at a level between ground and Vdd and the N-well 32 at a potential greater than or equal to the substrate potential Vsub.

The following key points describe key relationships available to the cell of the invention due to the ability to independently bias the several pn junction-isolated regions:

1. By biasing regions 32/34 positively with respect to region 12, the substrate region of the transfer device 14, vertical sidewall trench parasitic FET subthreshold leakage and other leakage mechanisms along the trench sidewall will be collected on the electrode Vbn of N-type regions 32/34, instead of the storage node, region 20, of the transfer device improving retention time over the PRIOR ART cell (FIG.I);

2. By biasing regions 32/34 positively with respect to region 12, minority carrier generation in region 12 and injection from data line diffused region 18 will be collected on the electrode Vbn of N-type regions 32/34, instead of region 20, improving retention time over the PRIOR ART (FIG.I);

3. By biasing regions 32/34 positively with respect to semiconductor substrate region 10, in the case where deep collar structure region 28 is above or below N+ region 34 lower edge, leakage current along the trench dielectric sidewall of trench capacitor structure region 22 (generated in region 10) and carrier generation in the bulk of region 10 is collected on the electrode Vbn of region 34, instead of region 12, improving well voltage drops over PRIOR ART (FIG. 1); and

4. By isolation of region 12 from semiconductor substrate region 10 using N-type regions 32/34, independent bias conditions allow for different back-bias conditions for the n-channel MOSFET devices in region 12 and in peripheral surface P-type region 13 improving design point flexibility not possible in PRIOR ART (FIG. 1).

If one chooses to fabricate N-channel support FETs directly in the surface of the epitaxial layer 13, then a Vsub of 0 volts is advantageous. Additional advantages over the prior SPT cells which might not be readily apparent include the use of N-channel transfer devices providing faster I/O operations of the DRAM cell, providing isolation of all of the array transfer devices by the double pn junctions between regions 12 and 32 and 32 and 11 and enabling the substrate to be biased to reduce the stress across the storage node dielectric.

Reference is now made to FIGS. 3–10 which describe the preferred process sequence used to fabricate the double grid substrate plate trench DRAM cell array.

Referring to FIG. 3, starting with a heavily doped P+ type semiconductor wafer 10, a lightly doped p-epitaxial layer 11 having a thickness of about 0.25 micron is provided. Next, heavily doped N-type regions 32, preferably using arsenic as the dopant impurity, are formed by a simultaneous out-diffusion and epitaxial growth step to form lightly doped layer 13 having a thickness of about 2.5 microns. On the upper surface of epitaxial layer 13 an oxide/nitride layer 50 having a thickness of about 175 nm is formed on substrate 10 to act as an etch/polish stop to be used in subsequent steps. A relatively thick, about 500 nm, layer 52 of oxide is deposited by a conventional CVD TEOS technique to act as an etch mask for trenches 22. A photolithographic mask is formed using a high resolution photoresist and is used to define the pattern of trenches 22 which are to be etched in substrate 10. The mask pattern is transferred to the thick oxide layer 52 and the oxide/nitride layer 50 by a dry plasma etching process using oxygen and carbon tetrafluoride (CF4) as the active agents. After stripping the photoresist, trenches 22′ are etched to a depth of about 5.0 microns using a anisotropic RIE process to provide the structure shown in FIG. 3. It should be noted that the bottom of the etched trenches are not shown as a matter of convenience in describing the invention.

Next, as shown in FIG. 4, the trench capacitor structure is formed by thermally oxidizing the now exposed silicon sidewalls and bottom of the trenches to a thickness of about 4 nm. Then, about 7 nm of silicon nitride is conformally deposited. The nitride layer is then oxidized to form about 1.5 nm silicon dioxide to complete the ONO cell node dielectric. The trenches are then filled by conformally depositing polysilicon doped to at least 1E19 atoms/cubic centimeter to a thickness of about 900 nm above the surface of the substrate. A thermal anneal step at about 1000 degrees C. in nitrogen heals any inadvertent seams formed in the polysilicon in the deep trenches 22. A polysilicon RIE process selective to silicon dioxide and silicon nitride then removes all of the polysilicon on the planar areas of the substrate and etches into the polysilicon at the tops of the trenches to a level of about 1.5 microns below the surface of the substrate leaving doped polysilicon 55 in the bottom of the trenches. The trench collar 28 is then formed on the sidewalls of the exposed trench tops by

Figure 1.A.1 (Continued)

conformally depositing CVD of about 90 nm of silicon dioxide and then anisotropically etching the oxide from planar areas, including the bottoms of the trenches, to leave a collar 28 on the upper sidewalls of the recessed trenches in a manner similar to that in U.S. Pat. No. 4,801,988. The resulting structure is shown in FIG. 4.

Referring now to FIG. 5, the trenches are again filled with arsenic doped polysilicon to a surface thickness of about 600 nm and annealed, as above. Then all of the thus deposited polysilicon formed on the back side of the substrate is removed preferably by a planarizing process such as chemical-mechanical polishing in order to reduce any undesirable stress caused by this nonfunctional layer. The front, or trench containing side, of the substrate is then planarized to remove the last deposited 600 nm of polysilicon from all planar surfaces. In order to achieve superior planarity, it is preferable to use a chemical-mechanical polishing technique. Such techniques are described in greater detail in U.S. Pat. No. 4,994,836 to Beyer et al. and U.S. Pat. No. 4,789,648 to Chow et al. Next, polysilicon 55' in the top of the trenches is recessed about 50 nm to 100 nm below the substrate surface in order to prevent subsequently applied polysilicon word lines from shorting to the signal storage node of the trench capacitors. The resulting structure is shown in FIG. 5.

Next the local isolation in the form of shallow trench isolation (STI) is formed, shown in FIG. 6. An STI mask is applied to the substrate and defines all of the regions where STI is desired. The exposed oxide/nitride etch stop layer 50 is etched to expose the silicon substrate surface and overlapped polysilicon filled trench tops. Preferably in the same processing chamber, the exposed substrate, trench collar and polysilicon are etched to a depth of about 350 nm. A LPCVD TEOS oxide layer 56 of about 630 nm is then conformally deposited over the entire substrate as shown in FIG. 6.

Next, as shown in FIG. 7, STI oxide layer 56 is planarized, preferably by a combination of RIE etch-back and chemical-mechanical polishing as described in copending application Ser. No. 07/427,153 filed Oct. 25, 1989 entitled "Forming Wide Dielectric Filled Trenches in Semiconductors" by Kerbaugh et al. Next, any remaining oxide/nitride layer 50 is removed by hot phosphoric acid and buffered HF. At this point, a sacrificial oxide may be grown on the exposed substrate surfaces, as these will become the active device areas for N- and P-channel devices of the CMOS process into which the array of the invention is integrated.

Next, N-wells for P-channel devices and for providing the buried N-type region 34 are formed by the conventional use of an N-well mask which covers all of the substrate except where N-wells are desired. It will be recognized that additional processing steps may be used to form the reach through region 34 as a separate operation, if desired. After formation of the N-well mask, the substrate is exposed to a plurality of ion implantation steps to form retrograde N-wells 34. Phosphorus ions are implanted at about 900 KEV with a dose of about 5E13 atoms per square centimeter to form the higher concentration deepest portion of the well, at about 500 KEV with a dose of about 2.3E13 atoms per square centimeter to form the bulk of the well, and at about 150 KEV with a dose of about 1.9E12 atoms per square centimeter to control punch through. If desired an additional N-well mask can be used at this point to selectively implant arsenic at about 80 KEV with a dose of about 1.3E12 atoms per square centimeter to control the threshold voltage of the P-channel FETs formed in selective N-wells. Additional implant masks and implants may also be used to further tailor specific device threshold voltages.

Following the formation of the N-wells, a conventional P-well mask is formed in a similar manner in order to mask the substrate against boron ions used to form the P-wells 58, also shown in FIG. 7. To form the P-wells, boron ions are implanted at about 200 KEV with a dose of about 8E12 atoms per square centimeter to form the bulk of the well, at about 80 KEV with a dose of about 1.6E12 atoms per square centimeter to control punch through region and at about 7.3 KEV with a dose of about 3.7E12 atoms to control the threshold voltage of the N-channel FETs used in the array and support circuitry of the DRAM. Thus, the structure in FIG. 7 results. It will be clear from the above description, that the N-well regions 32 make physical contact with the substrate epitaxial layer 11 and the P-wells 58 in order to physically and electrically isolate the P-region 58 and 11, including P+ substrate 10. Unlike the prior art, there is no need to ensure that the substrate regions of the transfer devices can be electrically coupled to the semiconductor substrate 10.

Next, the gate stack structure comprising the gate insulator, a conductive gate and a silicon nitride cap is formed as shown in FIG. 8. After stripping the sacrificial oxide referred to above, a gate insulator layer 60 is formed by growing about 10 nm of silicon dioxide on the exposed silicon surfaces of the substrate. A polysilicon layer 62 of about 200 nm is deposited and doped by ion implanting phosphorus at about 25 KEV with a dose of about 6E15 atoms per square centimeter. This is followed by deposition of about 100 nm of titanium silicide layer 64 by sputtering to reduce the resistivity of the wordlines. The gate stack is completed by depositing a layer 66 of silicon dioxide and a layer 68 of silicon nitride of about 80 nm resulting in the structure shown in FIG. 8.

As shown in FIG. 9, the multilayered gate stack is selectively etched to define the first level of interconnect and the gate electrodes for the CMOS FET devices to be formed on the planarized substrate. The exposed polysilicon is oxidized slightly at about 1050 degrees C. Next, a blocking mask is used to protect all of the device regions except where N-channel FETs are to be formed. A lightly doped N-type region 70 is then formed by implanting phosphorus at about 30 KEV with a dose of about 1E14 atoms per square centimeter. After removal of the blocking mask, a sidewall spacer 72 is formed by depositing about 45 nm of CVD silicon nitride followed by an anisotropic RIE of nitride present on planar surfaces. Next, the doped regions are silicided in a conventional manner by evaporating about 20 nm of cobalt, annealing at about 750 degrees C. and removing unreacted cobalt in dilute nitric acid.

The devices and DRAM structure are completed with the following steps illustrated with reference to FIG. 10. A silicon nitride sidewall 74 is formed by depositing a layer of silicon nitride of about 15 nm followed by an anisotropic RIE step. The N+-type diffusions 76 for NFET devices are formed, after placing a blocking mask to protect PFET regions, by implanting arsenic at about 50 KEV with a dose of about 5E15 followed by a drive-in step in nitrogen at about 900 degrees C. The P+-type diffusions for PFET devices are formed, after placing a blocking mask to protect NFET regions, by implanting boron at about 10 KEV

5,384,474

with a dose of about 5E15 atoms per square centimeter. A polysilicon surface strap 26 is formed to connect the N-type diffusions forming the storage node region 20 to the polysilicon 55' at the tops of the deep trenches is formed by the use of a blocking mask exposing the storage node regions, selectively etching the oxide at the trench top, depositing N-type polysilicon and planarizing by a chemical-mechanical polishing process to leave polysilicon straps 26. Interconnects 78 of titanium nitride and tungsten are formed as borderless contacts and an interlevel passivating layer 80 of phosphorus-doped glass is deposited and planarized, again by chemical-mechanical polishing techniques. The DRAM is completed by providing a number of additional planarized interconnect levels, as required by the complexity of the circuits to be interconnected.

While the invention has been described in terms of a single preferred embodiment, those skilled in the art will recognize that many of the steps described above can be altered and that dopant species and types as well as other material substitutions can be freely made without departing from the spirit and scope of the invention.

What is claimed is:

1. A dynamic random access memory device comprising:
   a semiconductor substrate having a first region of a first conductivity type,
   at least one array of dynamic memory cells, each cell comprising an access transistor coupled to a storage capacitor, the transistor of each memory cell being formed within a second region of said first conductivity type of said semiconductor substrate, each access transistor having a control electrode, a data line contact region, a storage node region, and a channel region;
   a plurality of signal storage capacitors formed in a plurality of trenches in said substrate, each capacitor including a signal storage node and a reference voltage node separated by a dielectric insulator, at least a portion of the reference voltage node being formed in said first region of said substrate and the signal storage node of each capacitor being formed in one of said trenches and connected to a corresponding storage node region of one of said access transistors;
   means for physically and electrically isolating all of the channel region of said access transistors within said one array from said first region of said substrate, said isolating means comprising a third region of opposite conductivity type to said first region, a first portion of said third region being formed laterally between said first and second regions, said first portion of said third region intersecting all of said trenches, said third region further comprising a second subregion extending substantially around said array and extending between said reference voltage node region and the surface of said substrate;
   means for biasing said first, second and third regions of said substrate to first, second and third different voltages; and
   the signal storage capacitor includes an insulating layer at the top of each of said trenches having a thickness greater than the thickness at the bottom of said trenches.

2. The dynamic random access memory device of claim 1 wherein the third region has a impurity doping level which varies with depth from the surface of the substrate.

3. The dynamic random access memory device of claim 1 wherein the impurity type of said first region is P-type.

4. The dynamic random access memory cell of claim 1 wherein the thicker portion of the insulating layer in said trenches extends down to at least the level of the top of said third region.

5. The dynamic random access memory cell of claim 4 wherein the thicker portion of the insulating layer in said trenches extends down to at least the level of the bottom of said third region.

6. The dynamic random access memory device of claim 4 wherein the dopant impurity of said first region is boron and the dopant impurity in the third region is N-type.

7. The dynamic random access memory device of claim 5 wherein the access transistors are N-channel FET devices.

* * * * *

Figure 1.A.1 (Continued)

United States Patent [19]
Park et al.

[11] Patent Number: 5,521,115
[45] Date of Patent: May 28, 1996

[54] **METHOD OF MAKING DOUBLE GRID SUBSTRATE PLATE DRAM CELL ARRAY**

[75] Inventors: **Jong W. Park**, Essex Junction; **Steven H. Voldman**, Burlington, both of Vt.

[73] Assignee: **International Business Machines Corporation**, Armonk, N.Y.

[21] Appl. No.: **316,693**

[22] Filed: **Sep. 30, 1994**

Related U.S. Application Data

[62] Division of Ser. No. 149,146, Nov. 5, 1993, Pat. No. 5,384,474, which is a continuation of Ser. No. 819,159, Jan. 9, 1992, abandoned.

[51] Int. Cl.^{6} H01L 21/8242
[52] U.S. Cl. 437/60; 437/64; 437/203
[58] Field of Search .... 437/47, 60, 64, 437/203, 919; 148/DIG. 14, DIG. 50

[56] **References Cited**

U.S. PATENT DOCUMENTS

4,688,063	8/1987	Lu et al.	357/23.6
4,794,434	12/1988	Pelley	357/23.6
4,801,988	1/1989	Kenney	357/23.6
4,829,017	5/1989	Malhi	437/47
4,873,560	10/1989	Sunami et al.	357/23.6
4,912,054	3/1990	Tomassetti	437/31
4,918,502	4/1990	Kaga et al.	357/23.6
4,944,682	7/1990	Cronin et al.	437/192
5,348,905	9/1994	Kenney	437/60
5,362,663	11/1994	Bronner et al.	437/60

FOREIGN PATENT DOCUMENTS

0283964	9/1988	European Pat. Off.
2215913	9/1989	United Kingdom

OTHER PUBLICATIONS

"Trench and Compact Structures for DRAMs" by P. Chatterjee et al, International Electron Devices Meeting 1986, Technical Digest paper 6.1, pp. 128–131.

"CMOS Semiconductor Memory Structural Modification to Allow Increased Memory Charge" Anonymous, IBM Technical Disclosure Bulletin, vol. 31, No. 11, Apr. 1989, pp. 162–165.

"A 45–ns 16Mb DRAM with Triple–Well Structure" by S. Fujii et al., IEEE Journal of Solid–State Circuits, vol. 24, No. 5, Oct. 1989, pp. 1170–1175.

"New Well Structure for Deep Sub–micron CMOS/BiCMOS Using Thin Epitaxy over Buried Layer and Trench Isolation" by Y. Okazaki et al., 1990 Symposium on VLSI Technology, Digest of Technical Papers, paper 6C–4, pp. 83–84.

Primary Examiner—Chandra Chaudhari
Attorney, Agent, or Firm—Howard J. Walter, Jr.

[57] **ABSTRACT**

A high density substrate plate trench DRAM cell memory device and process are described in which a buried region is formed adjacent to deep trench capacitors such that the substrate region of DRAM transfer FETs can be electrically isolated from other FETs on a semiconductor substrate. The buried region is contacted along its perimeter by a reach through region to complete the isolation. The combined regions reduce charge loss due to better control of device parasitics.

3 Claims, 5 Drawing Sheets

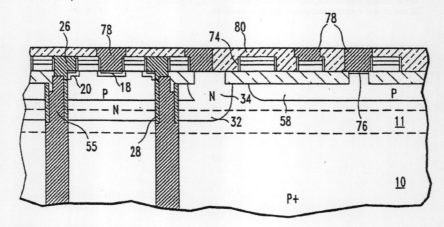

Figure 1.A.2 US Patent No. 5,521,115.

第1章 介绍 | 27

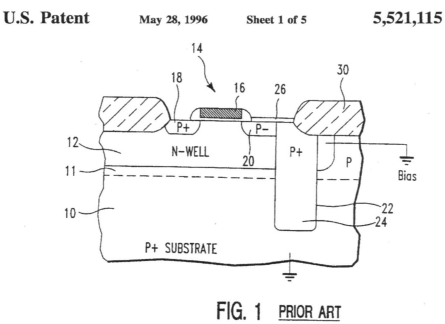

Figure 1.A.2 *(Continued)*

U.S. Patent **May 28, 1996** **Sheet 2 of 5** **5,521,115**

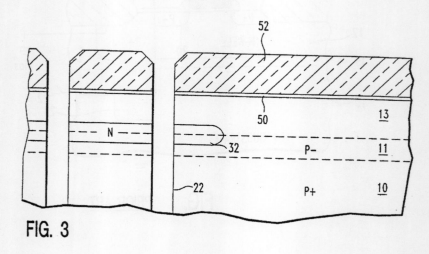

FIG. 3

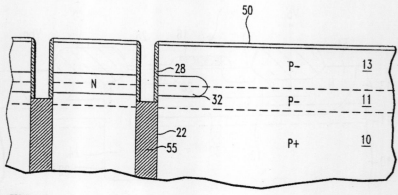

FIG. 4

Figure 1.A.2 (*Continued*)

U.S. Patent　May 28, 1996　Sheet 3 of 5　5,521,115

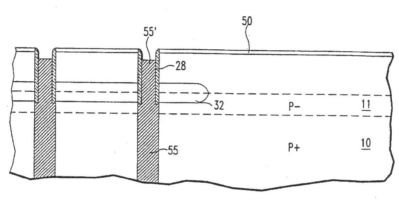

FIG. 5

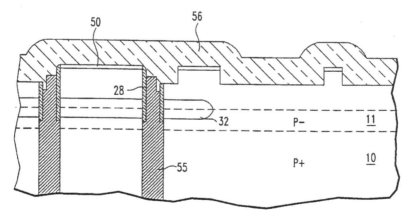

FIG. 6

Figure 1.A.2 (*Continued*)

U.S. Patent May 28, 1996 Sheet 4 of 5 **5,521,115**

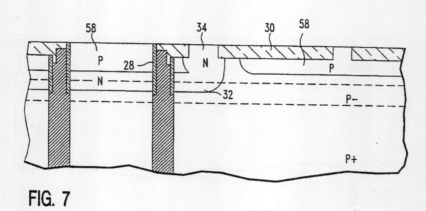

FIG. 7

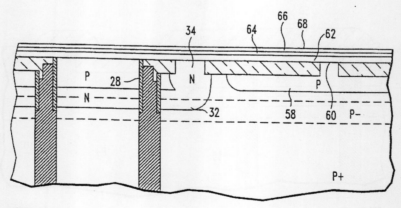

FIG. 8

Figure 1.A.2 *(Continued)*

U.S. Patent **May 28, 1996** **Sheet 5 of 5** **5,521,115**

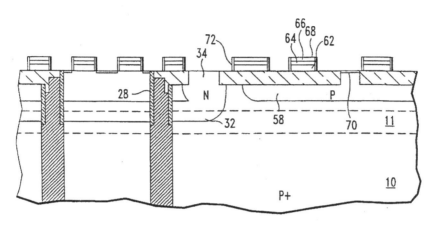

FIG. 9

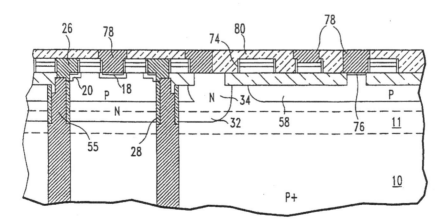

FIG. 10

Figure 1.A.2 *(Continued)*

5,521,115

METHOD OF MAKING DOUBLE GRID SUBSTRATE PLATE DRAM CELL ARRAY

This is a divisional of application Ser. No. 08/149,146 now U.S. Pat. No. 5,384,474 filed on Nov. 5, 1993 which is a continuation of Ser. No. 07/819,159 filed on Jan. 9, 1992, now abandoned.

RELATED APPLICATIONS

This application is related to co-pending application Ser. Nos. 07/819,148 and 07/818,668 filed concurrently herewith and entitled "Diffused Buried Plate Trench DRAM Cell Array" by D. M. Kenney and "Double Well Substrate Plate Trench DRAM Cell Array" by G. B. Bronner et al, now U.S. Pat. Nos. 5,264,716 and 5,250,829, respectively.

BACKGROUND OF THE INVENTION

1. Field of the Invention

This invention relates to semiconductor memory devices and particularly to high density dynamic random access memory cells and methods for their manufacture in submicron technologies.

2. Description of the Prior Art

Designers of technologies for producing semiconductor devices have been continually pressured to increase effective device densities in order to remain cost and performance competitive. As a result, VLSI and ULSI technologies have entered the sub-micron realm of structural dimensions and now are designing technologies in the deep submicron feature size range. In the foreseeable future absolute atomic physical limits will be reached in the conventional two-dimensional design approach to semiconductor device design. Traditionally, Dynamic Random Access Memory (DRAM) designers have met the severest of challenges in advancing technologies by pushing the limits of feature size resolution with each generation of DRAM. For example, designers of 64K bit DRAMs were perplexed to learn that a practical physical limit to charge capacity of storage capacitors in planar cell layouts had already been reached due to the minimum charge capacity required to allow reliable data signal sensing in the presence of naturally occurring atomic particle radiation inherently present in fabrication materials and the operating environment. Storage capacitors in the range of about 50 femtofarads were considered to be a physical limit. From a practical view, this limitation prevented a continuation of the scaling of DRAM dimensions and voltages initiated in the early 1980s. Reduction in the surface area of semiconductor substrate utilized by the DRAM storage capacitor has been severely restricted. Due to deceases in the thickness of reliable capacitor dielectric materials, existing 1 Megabit (1 Mb) DRAM technologies continue to enjoy the freedom of planar, two-dimensional device and circuit design. Beginning with 4 Mb DRAMs, the world of three-dimensional design has been utilized to the extent that the simple single device/capacitor memory cell has been altered to provide the capacitor in a vertical dimension. In such designs, the capacitor has been formed in a trench formed in the surface of the semiconductor substrate. In yet denser designs, other forms of three-dimensional capacitors have been proposed, such as stacking the plates of the capacitor above the transfer device. Such designs, however, present difficulties in forming the interconnections to the required word access and data bit lines to the DRAM memory cell. Additional designs have been proposed in which the transfer device and its associated capacitor are both formed within a trench of preferably minimum feature size. Currently, insurmountable processing difficulties make such designs impractical for product manufacturing processes.

A large number of proposals for 16 Mb and greater density DRAM cell designs have avoided continuing development of trench cell technology because of the existence of charge leakage mechanisms known to be present in trench capacitor structures. As these leakage mechanisms have become known, extensions of trench DRAM cells designs have been used successfully in 16 Mb designs.

The following references describe various aspects of prior art techniques used in DRAM and other semiconductor technologies.

The article "Trench and Compact Structures for DRAMs" by P. Chatterjee et al., International Electron Devices Meeting 1986, Technical Digest paper 6.1, pp. 128–131, describes variations in trench cell designs through 16 Mb DRAM designs, including the Substrate Plate Trench (SPT) cell described in more detail in U.S. Pat. No. 4,688,063 issued Aug. 18, 1987 to Lu et al. and assigned to the assignee of the instant invention. The SPT cell uses a highly conductive substrate as the DRAM cell plate. The storage node of each cell is formed in a deep trench in the substrate. U.S. Pat. No. 4,801,988 issued Jan. 31, 1989 to Kenney and assigned to the assignee of the instant invention, describes an improved SPT cell which includes a thick isolation region formed within the trench to enable higher density packing of DRAM cells. The article "CMOS Semiconductor Memory Structural Modification to Allow Increased Memory Charge" anonymous, IBM Technical Disclosure Bulletin, Vol. 31, No. 11, April 1989, pp.162–5, teaches a method of isolating the substrate plate of an SPT cell from support devices by providing a buried region under support devices in order to allow the plate reference voltage to be separately biased at an optimum Vdd/2 volts.

U.S. Pat. No. 4,912,054 issued Mar. 27, 1990 to Tomassetti describes methods of isolating bipolar-CMOS circuit devices through the use of various epitaxial layers as commonly found in bipolar device technologies. The article "A 45-ns 16-Mbit DRAM with Triple-Well Structure" by S. Fujii et al., IEEE Journal of Solid-State Circuits, Vol. 24, No. 5, October 1989., pp. 1170–1175, describes techniques for isolating various different functional device types in which the entire array of trench DRAM cells is formed within a surface implanted P-well.

U.S. Pat. No. 4,829,017 issued May 9, 1989 to Malhi describes a method of forming a buried doped layer in a substrate by forming a shallow trench, protecting its sidewalls, further extending the trench and finally doping the walls of the extended trench to form a continuous doped region useful as the storage node of a trench DRAM.

The article "New Well Structure for Deep Sub-micron CMOS/BiCMOS Using Thin Epitaxy over Buried Layer and Trench Isolation" by Y. Okazaki et al., 1990 Symposium on VLSI Technology, Digest of Technical Papers, paper 6C-4, pp. 83–4, describes the use of buried epitaxial layers to isolate surface devices from the substrate.

The following references relate specifically to variations in SPT DRAM cells in which a buried region of opposite conductivity type from the substrate is used a one plate of the DRAM storage capacitor. U.S. Pat. No. 4,918,502 issued Apr. 17, 1990 to Kaga et al. describes a buried plate trench DRAM cell in which the storage node of the cell and a sheath plate are formed in a single trench. At the bottom of the trench a diffusion of opposite type from the substrate is

5,521,115

3

formed such that the diffusions of adjacent cells interconnect forming a grid-like structure. One or more trenches-not associated with a DRAM cell is formed to act as a reach through to enable the doped region to be biased at a suitable reference voltage. FIG. 12, thereof, clearly illustrates the grid-like aspect of the buried region. European published application 0 283 964, published Sep. 28, 1988 describes a buried plate SPT DRAM cell in which an out-diffused region from the DRAM trenches, similar to that in Kaga et al., in which the diffused region forms the plate of the SPT cell. As in Kaga et al. a grid-like region is formed and is contacted by a non-cell trench. U.S. Pat. No. 4,873,560 issued Oct. 10, 1989 to Sunami et al, describes yet another buried plate SPT cell in which the access transistor is formed in the cell trench. FIG. 30, thereof, and its related text, describes the importance of maintaining the grid-like structure of the buried region in order to enable proper operation of the cell transfer device. Sunami et al, further cautions that in the event that opening in the grid-like buried region should be "filled by the depletion layer" isolating the surface devices from the substrate a separate connection can be made to the "isolated" surface region in order to bias it to the same potential as the substrate. UK Patent Application GB 2 215 913 A, published Sep. 27, 1989 describes yet another variation in the buried SPT DRAM cell design in which the dopant for the buried region is provided by ion implantation into the sidewalls of the deep trench of the DRAM cell. Finally, U.S. Pat. No. 4,794,434 issued Dec. 27, 1988 to Pelley, describes a buried plate SPT DRAM cell formed using bipolar device processing methods in which the buried plate region is formed from a buried sub-collector structure normally part of a bipolar transistor.

While the above cited references illustrate the diverse and concentrated efforts made by DRAM designers in attempting to overcome the inherent barriers in continuing to reduce the size, and increase the density, of DRAM cells, none provide the capability to carry DRAM technology into the sub-0.5 micron feature size range, a feat which must be achieved in order to continue the two decade "tradition" of providing ever increasing density of DRAM technology. DRAM designers have turned to the process-complicating use of "stacked capacitor" DRAM cells knowing that the addition of processing steps decreases the manufacturability of a specific design.

Referring to FIG. 1, there is shown a schematic cross-sectional view of the basic Substrate Plate Trench (SPT) DRAM cell described in U.S. Pat. No. 4,688,063 to Lu et al. entitled "Dynamic RAM Cell with MOS Trench Capacitor in CMOS". A P+ type semiconductor substrate 10 is provided with a lightly doped epitaxial surface layer 11 in which is provided N-type retrograde implanted well 12 formed at its upper surface in which the transfer device 14 is formed. A control gate electrode 16 is responsive to signals from word line circuitry to couple data signals applied to the bit or data line diffused region 18 to the diffused storage node region 20. A deep trench 22 is provided in which a storage capacitor is formed. A polysilicon storage node plate 24 is formed in the trench and isolated from the substrate 10 by a thin storage node dielectric, not shown. A conductive strap 26 connects the diffused node 20 to the plate 24.

Manufacturing experience has shown that the SPT DRAM cell described is not suitable for extension to greater than 16 Mbit applications due in part to the performance limitations of P-array transfer devices and the existence of a parasitic device formed by the diffused storage node 20, the polysilicon plate 24 and the substrate 10. Simple conversion to N-type transfer devices is not practical and reduction of

4

electrical stress on the capacitor dielectric by using Vdd/2 reference node biasing is not possible. The subject invention addresses the unsolved problems of the prior art by providing a solution to barriers presented in extending the manufacturability of the simple SPT cell to 64 Mb DRAM and beyond.

SUMMARY OF THE INVENTION

An object of the invention is to provide a double grid SPT cell in which the density limitations of the prior art are removed.

Another object of the the invention is to provide a Substrate Plate Trench DRAM design which does not increase process complexity at the expense of product yield.

Yet another object of the invention is to provide an SPT DRAM cell which has a minimum impact on existing processing technologies.

The present invention relates to methods for providing a cell design in which all of the historically limiting parameters of DRAM cells are dealt with in a unifying manner to provide a near optimum design in which charge leakage factors are minimized and device bias conditions are optimized. The invention includes a substrate plate trench DRAM cell array in which a buried region is used to electrically and physically isolate regions in a semiconductor substrate such that the cell transfer device can be operated independently from other support devices formed in the substrate. Sub-half micron feature sizes are used cooperatively with previously known techniques to provide a simple buried isolation layer.

These and other objects and features of the invention will become more fully apparent from the several drawings and description of the preferred embodiment.

DESCRIPTION OF THE DRAWINGS

FIG. 1 is a simplified schematic sectional view of a substrate plate trench (SPT) DRAM cell of the prior art illustrating the basic electrical connections of the cell.

FIG. 2 is a simplified schematic sectional view of the double grid substrate plate trench (SPT) DRAM cell of the invention illustrating the basic electrical connections of the cell.

FIGS. 3–10 are schematic sectional views of the array of the invention illustrating the array at various steps in a preferred manufacturing process.

DESCRIPTION OF THE PREFERRED EMBODIMENT

Referring to FIG. 2, there is shown the basic elements of the double well Substrate Plate Trench DRAM cell of the invention. The cell is an improvement of the prior art SPT DRAM cell as described by Lu et al. in U.S. Pat. No. 4,688,063 and as modified by Kenney in U.S. Pat. No. 4,801,988, both of which are incorporated herein by reference. The cell includes the following major features. A substrate 10 of P+ type semiconductor material has an upper epitaxial layer of P– type material, as in the prior art. Using bipolar sub-collector fabrication technology, a first epitaxial layer 11 can be formed, followed by the localized formation of an N-type buried layer 32 and an additional P-type epitaxial layer 13. Surface diffused reach through regions 34 connect to the perimeter of the buried layer 12 such that the substrate region 12 in which the DRAM cells are formed becomes both physically and electrically isolated from the

Figure 1.A.2 (*Continued*)

semiconductor substrate **10**. A P-type isolated region **12** is formed at its upper surface into which N-channel transfer devices **14** are formed. A control gate electrode **16** of device **14** is responsive to a word access line of the DRAM array support circuits, not shown, to couple data between data or bit line diffused N-type region **18** and diffused N-type storage node region **20** through the channel formed in region **12**. In a manner similar to the prior art, a storage capacitor is formed in a deep trench **22** adjacent to the storage node **20** and includes a signal storage node formed by conductive N-type polysilicon electrode **24** isolated from substrate **10** by a thin dielectric layer. Diffused surface storage node **20** and signal storage node **24** in the trench **22** are connected by a conductive strap **26**. At the top of the storage trench a thick insulating collar **28** is provided to increase the threshold voltage of the vertical parasitic FET formed by the diffused storage node **20** and the substrate within the P-region **12**. In order to eliminate the trench gated-induced diode leakage mechanism, the collar **28** must extend down the sidewall of trench **22** to a point below the lowermost pn junction of N-region **32**. Local surface isolation **30** is also provided, as is well known in the prior art.

It is an important aspect of the invention to provide arrays of cells as described in connection with FIG. 2 in a simple easy to manufacture processing sequence. Because the basic SPT cell is effectively placed inside a P-well which is contained in an N-well, the additional processing steps necessary to fabricate the invention are minimized. Typically, the P-well **12** can be biased at about −1 volt, the P+ substrate **10** at a level between ground and Vdd and the N-well **32** at a potential greater than or equal to the substrate potential Vsub.

The following key points describe key relationships available to the cell of the invention due to the ability to independently bias the several pn junction-isolated regions:

1. By biasing regions **32/34** positively with respect to region **12**, the substrate region of the transfer device **14**, vertical sidewall trench parasitic FET subthreshold leakage and other leakage mechanisms along the trench sidewall will be collected on the electrode Vbn of N-type regions **32/34**, instead of the storage node, region **20**, of the transfer device improving retention time over the PRIOR ART cell (FIG. 1);
2. By biasing regions **32/34** positively with respect to region **12**, minority carrier generation in region **12** and injection from data line diffused region **18** will be collected on the electrode Vbn of N-type regions **32/34**, instead of region **20**, improving retention time over the PRIOR ART (FIG. 1);
3. By biasing regions **32/34** positively with respect to semiconductor substrate region **10**, in the case where deep collar structure region **28** is above or below N+ region **34** lower edge, leakage current along the trench dielectric sidewall of trench capacitor structure region **22** (generated in region **10**) and carrier generation in the bulk of region **10** is collected on the electrode Vbn of region **34**, instead of region **12**, improving well voltage drops over PRIOR ART (FIG. 1); and
4. By isolation of region **12** from semiconductor substrate region **10** using N-type regions **32/34**, independent bias conditions allow for different back-bias conditions for the n-channel MOSFET devices in region **12** and in peripheral surface P-type region **13** improving design point flexibility not possible in PRIOR ART (FIG. 1).

If one chooses to fabricate N-channel support FETs directly in the surface of the epitaxial layer **13**, then a Vsub of 0 volts is advantageous. Additional advantages over the prior SPT cells which might not be readily apparent include the use of N-channel transfer devices providing faster I/O operations of the DRAM cell, providing isolation of all of the array transfer devices by the double pn junctions between regions **12** and **32** and **32** and **11** and enabling the substrate to be biased to reduce the stress across the storage node dielectric.

Reference is now made to FIGS. 3–10 which describe the preferred process sequence used to fabricate the double grid substrate plate trench DRAM cell array.

Referring to FIG. 3, starting with a heavily doped P+ type semiconductor wafer **10**, a lightly doped P− epitaxial layer **11** having a thickness of about 0.25 micron is provided. Next, heavily doped N-type regions **32**, preferably using arsenic as the dopant impurity, are formed by a simultaneous out-diffusion and epitaxial growth step to form lightly doped layer **13** having a thickness of about 2.5 microns. On the upper surface of epitaxial layer **13** an oxide/nitride layer **50** having a thickness of about 175 nm is formed on substrate **10** to act as an etch/polish stop to be used in subsequent steps. A relatively thick, about 500 nm, layer **52** of oxide is deposited by a conventional CVD TEOS technique to act as an etch mask for trenches **22**. A photolithographic mask is formed using a high resolution photoresist and is used to define the pattern of trenches **22** which are to be etched in substrate **10**. The mask pattern is transferred to the thick oxide layer **52** and the oxide/nitride layer **50** by a dry plasma etching process using oxygen and carbon tetrafluoride (CF4) as the active agents. After stripping the photoresist, trenches **22'** are etched to a depth of about 5.0 microns using a anisotropic RIE process to provide the structure shown in FIG. 3. It should be noted that the bottom of the etched trenches are not shown as a matter of convenience in describing the invention.

Next, as shown in FIG. 4, the trench capacitor structure is formed by thermally oxidizing the now exposed silicon sidewalls and bottom of the trenches to a thickness of about 4 nm. Then, about 7 nm of silicon nitride is conformally deposited. The nitride layer is then oxidized to form about 1.5 nm silicon dioxide to complete the ONO cell node dielectric. The trenches are then filled by conformally depositing polysilicon doped to at least 1E19 atoms/cubic centimeter to a thickness of about 900 nm above the surface of the substrate. A thermal anneal step at about 1000 degrees C. in nitrogen heals any inadvertent seams formed in the polysilicon in the deep trenches **22**. A polysilicon RIE process selective to silicon dioxide and silicon nitride then removes all of the polysilicon on the planar areas of the substrate and etches into the polysilicon at the tops of the trenches to a level of about 1.5 microns below the surface of the substrate leaving doped polysilicon **55** in the bottom of the trenches. The trench collar **28** is then formed on the sidewalls of the exposed trench tops by conformally depositing CVD of about 90 nm of silicon dioxide and then anisotropically etching the oxide from planar areas, including the bottoms of the trenches, to leave a collar **28** on the upper sidewalls of the recessed trenches in a manner similar to that in U.S. Pat. No. 4,801,988. The resulting structure is shown in FIG. 4.

Referring now to FIG. 5, the trenches are again filled with arsenic doped polysilicon to a surface thickness of about 600 nm and annealed, as above. Then all of the thus deposited polysilicon formed on the back side of the substrate is removed preferably by a planarizing process such as chemical-mechanical polishing in order to reduce any undesirable stress caused by this non-functional layer. The front, or

Figure 1.A.2 (*Continued*)

5,521,115

7

trench containing side, of the substrate is then planarized to remove the last deposited 600 nm of polysilicon from all planar surfaces. In order to achieve superior planarity, it is preferable to use a chemical-mechanical polishing technique. Such techniques are described in greater detail in U.S. Pat. No. 4,994,836 to Beyer et al. and U.S. Pat. No. 4,789,648 to Chow et al. Next, polysilicon **55'** in the top of the trenches is recessed about 50 nm to 100 nm below the substrate surface in order to prevent subsequently applied polysilicon word lines from shorting to the signal storage node of the trench capacitors. The resulting structure is shown in FIG. **5**.

Next the local isolation in the form of shallow trench isolation (STI) is formed, shown in FIG. **6**. An STI mask is applied to the substrate and defines all of the regions where STI is desired. The exposed oxide/nitride etch stop layer **50** is etched to expose the silicon substrate surface and overlapped polysilicon filled trench tops. Preferably in the same processing chamber, the exposed substrate, trench collar and polysilicon are etched to a depth of about 350 nm. A LPCVD TEOS oxide layer **56** of about 630 nm is then conformally deposited over the entire substrate as shown in FIG. **6**.

Next, as shown in FIG. **7**, STI oxide layer **56** is planarized, preferably by a combination of RIE etch-back and chemical-mechanical polishing as described in co-pending application Ser. No. 07/427,153 filed Oct. 25, 1989 entitled "Forming Wide Dielectric Filled Trenches in Semiconductors" by Kerbaugh et al. Next, any remaining oxide/nitride layer **50** is removed by hot phosphoric acid and buffered HF. At this point, a sacrificial oxide may be grown on the exposed substrate surfaces, as these will become the active device areas for N- and P-channel devices of the CMOS process into which the array of the invention is integrated.

Next, N-wells for P-channel devices and for providing the buried N-type region **34** are formed by the conventional use of an N-well mask which covers all of the substrate except where N-wells are desired. It will be recognized that additional processing steps may be used to form the reach through region **34** as a separate operation, if desired. After formation of the N-well mask, the substrate is exposed to a plurality of ion implantation steps to form retrograde N-wells **34**. Phosphorus ions are implanted at about 900 KEV with a dose of about 5E13 atoms per square centimeter to form the higher concentration deepest portion of the well, at about 500 KEV with a dose of about 2.3E13 atoms per square centimeter to form the bulk of the well, and at about 150 KEV with a dose of about 1.9E12 atoms per square centimeter to control punch through. If desired an additional N-well mask can be used at this point to selectively implant arsenic at about 80 KEV with a dose of about 1.3E12 atoms per square centimeter to control the threshold voltage of the P-channel FETs formed in selective N-wells. Additional implant masks and implants may also be used to further tailor specific device threshold voltages.

Following the formation of the N-wells, a conventional P-well mask is formed in a similar manner in order to mask the substrate against boron ions used to form the P-wells **58**, also shown in FIG. **7**. To form the P-wells, boron ions are implanted at about 200 KEV with a dose of about 8E12 atoms per square centimeter to form the bulk of the well, at about 80 KEV with a dose of about 1.6E12 atoms per square centimeter to control punch through region and at about 7.3 KEV with a dose of about 3.7E12 atoms to control the threshold voltage of the N-channel FETs used in the array and support circuitry of the DRAM. Thus, the structure in FIG. **7** results. It will be clear from the above description, that the N-well regions **32** make physical contact with the substrate epitaxial layer **11** and the P-wells **58** in order to physically and electrically isolate the P-region **58** and **11**, including P+ substrate **10**. Unlike the prior art, there is no need to ensure that the substrate regions of the transfer devices can be electrically coupled to the semiconductor substrate **10**.

8

Next, the gate stack structure comprising the gate insulator, a conductive gate and a silicon nitride cap is formed as shown in FIG. **8**. After stripping the sacrificial oxide referred to above, a gate insulator layer **60** is formed by growing about 10 nm of silicon dioxide on the exposed silicon surfaces of the substrate. A polysilicon layer **62** of about 200 nm is deposited and doped by ion implanting phosphorus at about 25 KEV with a dose of about 6E15 atoms per square centimeter. This is followed by deposition of about 100 nm of titanium silicide layer **64** by sputtering to reduce the resistivity of the wordlines. The gate stack is completed by depositing a layer **66** of silicon dioxide and a layer **68** of silicon nitride of about 80 nm resulting in the structure shown in FIG. **8**.

As shown in FIG. **9**, the multilayered gate stack is selectively etched to define the first level of interconnect and the gate electrodes for the CMOS FET devices to be formed on the planarized substrate. The exposed polysilicon is oxidized slightly at about 1050 degrees C. Next, a blocking mask is used to protect all of the device regions except where N-channel FETs are to be formed. A lightly doped N-type region **70** is then formed by implanting phosphorus at about 30 KEV with a dose of about 1E14 atoms per square centimeter. After removal of the blocking mask, a sidewall spacer **72** is formed by depositing about 45 nm of CVD silicon nitride followed by an anisotropic RIE of nitride present on planar surfaces. Next, the doped regions are silicided in a conventional manner by evaporating about 20 nm of cobalt, annealing at about 750 degrees C. and removing unreacted cobalt in dilute nitric acid.

The devices and DRAM structure are completed with the following steps illustrated with reference to FIG. **10**. A silicon nitride sidewall **74** is formed by depositing a layer of silicon nitride of about 15 nm followed by an anisotropic RIE step. The N+-type diffusions **76** for NFET devices are formed, after placing a blocking mask to protect PFET regions, by implanting arsenic at about 50 KEV with a dose of about 5E15 followed by a drive-in step in nitrogen at about 900 degrees C. The P+-type diffusions for PFET devices are formed, after placing a blocking mask to protect NFET regions, by implanting boron at about 10 KEV with a dose of about 5E15 atoms per square centimeter. A polysilicon surface strap **26** is formed to connect the N-type diffusions forming the storage node region **20** to the polysilicon **55'** at the tops of the deep trenches is formed by the use of a blocking mask exposing the storage node regions, selectively etching the oxide at the trench top, depositing N-type polysilicon and planarizing by a chemical-mechanical polishing process to leave polysilicon straps **26**. Interconnects **78** of titanium nitride and tungsten are formed as borderless contacts and an interlevel passivating layer **80** of phosphorus-doped glass is deposited and planarized, again by chemical-mechanical polishing techniques. The DRAM is completed by providing a number of additional planarized interconnect levels, as required by the complexity of the circuits to be interconnected.

While the invention has been described in terms of a single preferred embodiment, those skilled in the art will recognize that many of the steps described above can be altered and that dopant species and types as well as other material substitutions can be freely made without departing from the spirit and scope of the invention.

Figure 1.A.2 (*Continued*)

5,521,115

What is claimed is:

1. The method of making a dynamic random access memory device comprising the steps of:
  providing a semiconductor substrate of a first conductivity type;
  forming a buried region of second conductivity type over at least a portion of the area of said substrate;
  forming an array of trenches in a matrix pattern in the top surface of said substrate, each trench penetrating into said substrate and substantially into said buried region;
  forming a dielectric layer of a predetermined thickness on the inside surfaces of said trenches and filling said trenches with a conductive electrode material;
  ion implanting and diffusing around the perimeter of said array of trenches a diffused region of said second conductivity type having a depth so as to physically and electrically isolate a portion of the substrate within the matrix pattern above said buried region; and
  forming within the isolated portion of the matrix pattern a plurality of semiconductor devices for coupling signals to and from said conductive electrode material within at least some of said trenches.

2. The method of making a dynamic random access memory device of claim **1** wherein the step of forming a dielectric layer on the inside surfaces of said trenches includes a step of forming a second dielectric layer thicker than said first dielectric layer to form a collar at the top of said trenches.

3. The method of making a dynamic random access memory device of claim **1** wherein the source of dopant material used to form the buried region is N-type.

* * * * *

Figure 1.A.2 (*Continued*)

第 2 章 发　　明

2.1　引言

本章将讨论如何成为一名发明人[1,2]，如何增加对创作过程的关注，以及如何成为一名多产的发明人。本章还讨论了如何学习撰写专利和发明[3,4]，以及产生发明的方法等内容[5-28]。

2.2　如何成为发明人

在本节中，将讨论成为发明人的步骤和过程。

2.2.1　做什么才是有创造性的？

在成为发明人的步骤中，第一步是了解你所做的事情，什么是具有创造性的。我们或许不想承认已经做过的事情中含有部分创作过程（如图 2-1 所示），但是我们许多人已经做了以下事情[1,2]：

- 画图
- 绘画
- 演奏乐器
- 唱歌

- 计算机艺术
- 制图
- 软件
- 工程工作
- 设计
- 写作

图 2-1　创造性过程

所有这些都是创造性的过程，可以帮助您成为发明家。就我而言，我曾经在画布上用丙烯画画，甚至在有机玻璃的后面画画。我曾经玩过小号和长号。但我不会唱歌！我还使用过诸如 CorelDraw 等计算机图形工具处理专利技术。同时我写书，这也是我创作过程的一部分。当然，我也写专利[1, 2]。

2.2.2　你如何思考？

人们会有不同的想法（如图 2-2 所示）。了解自己的优势并充分利用这些优势是非常重要的。以下是人们思考的一些不同方式[1,2]：

- 有些人是善于分析的。
- 有些人是注意细节的。
- 有些人是形象化的。
- 有些人会在他们手工作业时思考。
- 有些人是机械呆板的。

你的优势是什么？

就我而言，我的思维模式取决于我所处的环境。我会以形象化为导向，并且从发明创作到写作专利都会运用这一技能。我是左撇子，这意味着我更倾向于用"右脑"[1,2]。

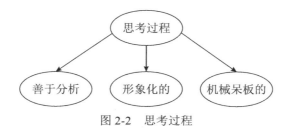

图 2-2　思考过程

2.2.3　你在哪里思考？

一个重要的概念是了解你个人在哪里思考以及如何思考。这是让你用自己的时间来思考发明的关键。

不同的人在不同的地方思考。有些人发现他们在办公室的办公桌上会尽全力思考。而对于其他人来说，今天的办公场所，有太多干扰有效思维的分心事[1,2]。

当我在马来西亚、新加坡和斯里兰卡教授发明课程时，我会要求与会者说说思考的最佳场所。每一堂课上我都会问这个问题，总是会有一位经理说，他会坐在马桶上尽全力思考。当我问他为什么时，他回答说，这是他唯一可以获得的 5 到 10 分钟的个人时间。

2.2.4　你什么时候思考？

有些人在早上清醒的时候会尽全力思考，这可能是在家里喝咖啡时或家人醒来之前。有时候，当你在睡觉时，你的头脑正在考虑事情，最好是在早上记录下来。我以前早上醒来后，会在早上 5 点或 6 点去吃早餐，并且在咖啡到来之前拿出一张餐巾纸，潦草地写下我的发明思想。当我上班时，我会开始记录我的想法并提交这些发明构想。

有些人醒来后会在洗澡时思考。这时你的思路清晰，文思泉涌。还有些人喜欢听音乐时思考。

当我在麻省理工学院的时候，有一个天才小孩会整天走路。当他想思考的时候，他就开始走路。

有时候，我最好的想法也是在慢跑或跑步时产生的。对我而言，最好的方式是开始思考一个主题，然后进行一个小时的慢跑。这让我专注于思考。

所以，重要的是要明白你在哪里思考能得出最好的想法，然后就在精神上或身体上去那个地方[1, 2]。

2.2.5 捕捉你的想法和发明

发明过程的一个重要部分是捕捉你的想法,并记录下来。这一点几乎每个人都忘记了。捕捉你的想法对发明很重要 [1, 2]:

- 将它存储在发明笔记本中。
- 将它存储在记事本上。
- 将它存储在手机上。
- 将文档输入到电子发明数据库。

从历史上看,企业习惯于拥有发明交底书笔记本,这些发明交底书笔记本是属于受控文件的公司财产。头版上有一个号码,还有你的名字和日期。每个页面都被编号,并且有一个签名行、一个见证行和一个日期。这些笔记本被用来记录员工的想法和策略。对于发明来说,签名、日期和见证很重要。在"First to Invent"系统下,这是一份法律文件,用于证明谁是第一个发明人。有些公司还把这种方法用于文件记录。

在一些公司,它们现在已经转向电子在线数据库。例如,IBM 改用了一个可以记录日期、时间和发明者的在线数据库系统。它也被用于追踪、调度和记录,向该系统提交发明同时也完成了启动新发明的法律文件。

因此,记录是发明过程的重要组成部分,所以您不会失去自己的想法,并从法律上保护它。

2.2.6 时间

时间对发明过程至关重要(如图 2-3 所示)。你的想法提出的时间会影响该想法能否被接受。

图 2-3 时间

如果一项发明太过遥远,可能不会被接受,因为存在价值和可操作性的问题。问题是,大多数人无法意识到太超前发明的价值。相反,如果发明过于接近评价当前价值,就会有人发明类似的想法，或者已经将它们提交给专利局。因此，会有一个"时间窗口"，对于发布和提交想法至关重要[1,2]。

所以，问题是何时是发布发明的理想时机。以我自己的工作类型和经验表明，发布具有价值的发明的最佳时机是三到五年。对于更多革命性的想法，可能有一个更大的窗口（例如10年）。根据我的经验，当我提出未来10年的发明时，它们没有被接受或看重。专利审查委员会看不到未来5年的事情。讽刺的是，10年后提出了相同的想法就会被接受。

因此，发布发明构想的时机对于接受发明构想至关重要。

2.3 研究发明人

研究多产发明人的工作习惯和目标有助于找到成为多产发明人的可能途径。人们会发现作家、艺术家和发明人都有一定的工作习惯、道德规范以及形成富有成效过程的习惯。

2.3.1 研究多产发明人

成为发明人，甚至是多产发明人的一个步骤是研究其他发明人（见图2-4）。为了研究发明人，可以按照如下步骤[1-3]进行，如图2-4所示。

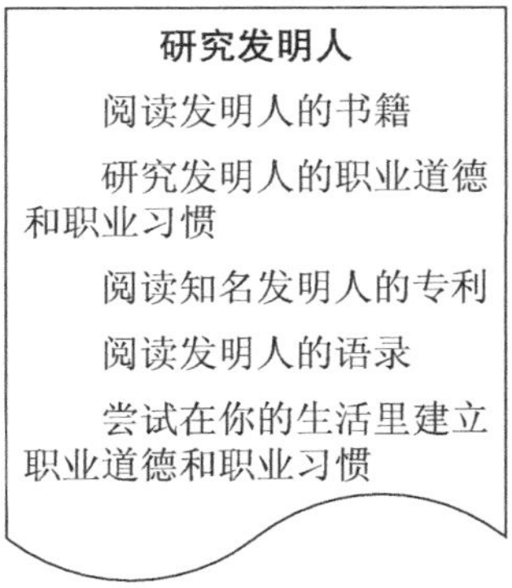

研究发明人
阅读发明人的书籍
研究发明人的职业道德和职业习惯
阅读知名发明人的专利
阅读发明人的语录
尝试在你的生活里建立职业道德和职业习惯

图2-4 研究发明人

阅读有关发明人的书籍是很有价值的。你可以学习他们的工作习惯、思想和想法。

2.3.2 研究发明人的习惯

我从托马斯·阿尔瓦·爱迪生的工作习惯和发明过程中学到了很多关于他的知识。

我去了他在佛罗里达州的家，也看到了他的工作环境。托马斯·阿尔瓦·爱迪生是一个很好的例子，因为他建立了许多发明方面的工作纪律和目标。以他为例，从他睡觉的地方以及他每天睡几个小时可以看出他对发明的渴望。他设定了每周、每年要创造多少个发明的目标，以及有多少重要发明的计划表[1-3]。

在麻省理工学院，作为一名等离子物理实验室的实验人员，我开始养成了一些托马斯·阿尔瓦·爱迪生的工作习惯，以提高我的睡眠质量、工作效率和生产力。爱迪生曾经以四小时为周期轮班睡觉。我也养成了类似的工作和休息的习惯，把睡眠周期与发明过程有益的结合起来。

2.3.3 研究发明人的目标

在研究发明人的过程中，人们会发现许多人都有目标。如今，在现代企业中，大多数企业设立的研发部门都有发明目标。一些发明人也实践了这个过程。

托马斯·阿尔瓦·爱迪生制订了有关他多长时间有一个发明的时间表。他的目标是每周有一项发明。他还有一个目标是每六个月有一个重要发明。所以，他并没有抛弃那些不那么重要的发明，而是清楚自己想在特定时期内达到一个重要的目标[1-3]。

2.4 创作周期

许多多产个体都有一个创作周期来帮助它们保持多产。这将在 2.5 节中讨论。

2.4.1 什么是创作周期？

许多多产的人发现，他们需要激发"创作周期"来启动并维持创作过程（如图 2-5 所示）。作家、演员、喜剧演员、发明家和其他专业人士需要这个周期以保持其创作过程的持续[1,2]。

图 2-5　创作周期

当作家再次开始写作时，他们会得"书写痉挛症"。作家，像发明家一样，会有技巧来避免这个问题。一些作家每天将日常职业准则与目标结合起来。一些多产的作家从早上 5 点到上午 10 点工作，然后一天中的其余时间则停止工作。有些作者每天有一页的目标；实际上，作者甚至不关心他 / 她是否使用在那天写出的那些材料。为什么？因为他 / 她的目标是保持每日写作练习。

每个人都有适合他们自己的创作周期。就我而言，这个过程包括实验工作、实现发明、撰写专利，并重复这个过程。有时候，技术写作会产生更多的发明思想。在我撰写专利或进行画图的过程中，很多时候都会出现新的发明[1-3]。

2.4.2 激发创作周期

为了激发创作周期，不同的人使用不同的方法。一些方法包括以下内容：

- 写文字
- 整理笔记
- 绘制图片

以我为例，我做了以上所有事情。有时候，我需要写主题文本。在其他情况下，我需要画图。可以在纸上或在电子设备上。此外，我在维恩图中绘制主题，其中维恩图中的每个形状都是不同的主题或学科。维恩图中的重叠区域是一门交叉学科。因此，通过定义"已知"学科的界限，就可以找到它们之间没有现有发明的领域。通过将第一区域与第二区域相结合，就有了发明的机会[1-3]。

2.5 左脑和右脑

大脑的不同区域驱动不同的功能。这些功能还与您是左撇子还是右撇子有关。大脑的右侧驱动直觉、可视化和创造力，而左侧则驱动逻辑思维和分析过程。

由于文化或宗教信仰，许多社会地区不允许人们使用左手。在许多文化中，要求人们使用右手。而在一些文化中，人的左手将被折断，以防止用左手写作。

问题在于，如果限制左手的使用，左撇子的发展过程将会受到制约。

专利撰写和发明会考虑到分析、逻辑以及创意过程。

2.6 跳出盒子去思考

今天，想法会受到过去、现在或者已知因素的约束和影响，并且很难摆脱"盒子"（如图2-6所示）。有时，发明是已知想法或发明的一个小延伸。一些公司的文化以"增量发明"而闻名，这为已知的发明提供了改进。

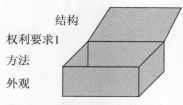

图 2-6　跳出盒子去思考

许多发明都是已知思想的演变。有时，这会受到不同事情的风险的影响，或者受到完成新想法并产生新变化的时间表和时间的影响。

在某些情况下，它们是革命性的。当人们提出革命性的概念时，许多人认为这是"跳出盒子去思考"。我参与了革命性的技术，有时它们成功，有时却失败。因此，风险和时间是革命性变化的因素。一些情况下，革命性的变化需要开发资金、新设备和新工具。

在IBM时，我曾经听到一个开发团队的高级开发经理说："在开发过程中，变得疯狂；当你接近制造业时，变得保守。"这是一种在演进发明与革命发明之间创造平衡的做法。一位工程师曾对我说过，"跳出盒子去思考。"我问他，"什么是盒子？"

2.7 横向思维与批判性思维

有两种与思维相关的类别。第一类称为横向思维，第二类称为批判性思维[1-3,5-8]。

批判性思维是指判断陈述的真伪，以及在思维过程中寻找错误的思考方式。

横向思维则与陈述和想法的转向价值有关。当他/她想从一个已知的想法转向创造一个新想法时，一个人会使用横向思维。

爱德华·德博诺说："我们解决问题的方式，可能并不是通过消除原因，而是通过寻找前进的方法，即使原因仍然存在"[1,2]。

2.8 在学科之间的界限发现发明

教育和学习分为学科和课程。工程分为电气、机械、民用和其他领域。科学分为物理学、化学和生物学。

在已建立学科的领域很难做出发明和贡献，而在新领域就较容易产生发明。在"跨学科"领域则更容易做出发明和贡献。

既定学科的界限之间是创新和发明的领域。新的学科也是机会领域[1,2]（如图 2-7 所示）。

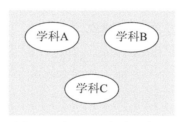

图 2-7 在界限之间发现发明

以我为例，我在布法罗大学的工程科学习交叉学科，其中包括电气、机械和热科学。在一个共同的课程中，我的同事专注于能源、生物医学、航空航天和核工程。因此，在我的教育培训中，跨学科的某一领域就被整合起来，以便在这些新领域开展事业和发展。

因此，学科之间的界限是创新和发明的机会！

2.9 结构化发明 TRIZ

这里存在解决问题和进行发明的结构化方法。一种流行的方法称为 TRIZ（Teoriya Resheniya Izobetaltelskikh Zadatch，TRIZ）[5-8]。该方法由 Genrikh Saulovich Altshuller 于 1948 年开发。TRIZ 意思是解决创造性问题的理论，或者也可以说是创造性解决问题的理论。

这个过程指的是通过消除矛盾来进行发明。Genrikh Altshuller 表示："发明是在某些原则的帮助下消除矛盾。"

Altshuller 的理念是开发一种发明的方法，他认为，必须要浏览大量的发明，确定其背后的矛盾，并且确定发明人所使用的原则。Altshuller 的方法专注于发明的移除概念。

2.10 专利示例

在本节中，给出了一些"跳出盒子去思考"的专利示例。美国专利商标局（USPTO）将这一部分的专利强调为"自然发明"，以应用于磁盘驱动器操作来保护磁记录头。

2.11 总结

第 2 章讨论了如何成为一名发明人，并找到自己内心的自我去创造。著名发明家的例子也将说明他们是如何工作和思考的，从而变得更有创造力和多产。此外，本章讨论还包括创作周期、跳出盒子去思考和 TRIZ。

第 3 章将开启关于专利和专利法的讨论。首先讨论专利搜索引擎，包括专利网站和 Google 专利。还将讨论专利语言，以便人们了解专利流程的语言。在第 3 章中，会讨论专利交底书的结构。

问题

1. 你有没有发明过？你第一次发明的年龄是多大？
2. 你有没有建立或绘制一项发明？
3. 你有没有构造过一种新颖的设计？
4. 当你还是个孩子的时候你做了什么？列出所有包含创造性过程的游戏类型。
5. 你有没有专利？发明过程是什么？
6. 你如何思考？
7. 你在哪里思考？
8. 你什么时候思考？
9. 你把想法写下来了吗？
10. 你有没有研究过发明人？
11. 你曾与别人合作过一个发明吗？
12. 你如何在产品中找到发明？
13. 你有过包含发明的产品吗？
14. 你如何从发明中提取新颖性？你如何定义新颖性？

15. 你有创作周期吗？它是如何工作的？
16. 你认为你是"在盒子里思考"还是"跳出盒子去思考"？
17. "在盒子里"思考的优点是什么？
18. 你研究过各种系统的发明方法吗？有什么方法？
19. 你如何开始发明过程？
20. 你是在边界之间还是在学科之内更容易产生想法？解释其优点和缺点。

案例研究

案例研究 A

列出所有涉及创作的活动。

案例研究 B

列出你在什么地方思考。根据你最想思考的地方排列这些项目。

案例研究 C

你的创作周期是什么？用图表解释你的思维过程。

参考文献

1. Voldman, S. (2014). Short Course, *Innovating, Inventing, and Patenting*, Dr. Steven H. Voldman LLC, Ministry of Science Technology and Innovation (MOSTI), Putrajaya, Malaysia, May 2014.

2. Voldman, S. (2015). Short Course, *Innovating, Inventing, and Patenting*, Dr. Steven H. Voldman LLC, FITIS, Sri Lanka, February 2015.

3. Voldman, S. (2016). Short Course, *Writing and Generating Patents*, Dr. Steven H. Voldman LLC, FITIS, Sri Lanka, February 2016.

4. Amernick, B.A. 1991. *Patent Law for the Nonlawyer: A Guide for the Engineer, Technologist, and Manager*, 2e. Von Nostrand Reinhold. ISBN: 13 978-0442001773.

5. TRIZ Journal. https://triz-journal.com (accessed 19 December 2017).

6. Ilevbare, I.M., Probert, D., and Phaal, R. (2013). A review of TRIZ and its benefits and challenges in practice. Technovation 33 (2-3): 30-37.

7. Rantanen, K. and Domb, E. (2008). Simplified TRIZ, 2e. New York: Auerbach Publications, Taylor and Francis Group.

8. Creating Minds. TRIZ 40 design principles. https://www.triz40.com (accessed 19 December 2017).

9. U.S. Patent Office (USPTO). https://www.uspto.gov (accessed 19 December 2017).

10. European Patent Office (EPO). https://www.epo.org (accessed 19 December 2017).

11. Japan Patent Office. https://www.jpo.go.jp (accessed 19 December 2017).

12. Malaysian Patent Office (MyIPO). https://www.myipo.gov/my (accessed 19 December 2017).

13. World Intellectual Property Organization (WIPO). https://www.wipo.int (accessed 19 December 2017).

14. Organisation Africaine de la Propriete Intellectuelle (OAPI). https://www.oapi.int (accessed 19 December 2017).

15. African Regional Intellectual Property Organization (ARIPO). https://www.airpo.org (accessed 19 December 2017).

16. U.S. Department of Commerce (2000). *Patents and How to Get One: A Practical Handbook*. U.S. Department of Commerce, Courier Corporation.

17. Mueller, J.M. (2016). Patent Law, 5e. Wolter Kluwer Publications.

18. Stim, R. (2016). *Patent, Copyright, and Trademark: An Intellectual Property Desk Reference*. Nolo. ISBN: 978-1-4133-2221-7.

19. Slusky, R. (2013). *Invention Analysis and Claiming: A Patent Lawyer's Guide*, 2e. ABA Book Publishing. ISBN: 13 978-1614385615.

20. Rosenberg, M. (2016). *Essentials of Patent Claim Drafting*, LexisNexis IP Law and Strategy Series. OUP.

21. Adams, D.O. (2015). Patents Demystified: An Insider's Guide to Protecting Ideas and Invention. American Bar Association. ISBN: 13 978-163425679.

22. Pressman, D. and Tuytschaevers, T. (2016). *Patent It Yourself: Your Step-by-Step Guide to filing at the U.S. Patent Office*. Nolo.

23. Charmsson, H.J.A. and Buchaca, J. (2008). *Patents, Copyrights, and Trademarks for Dummies*. Wiley.

24. Stim, R. and Pressman, D. (2015). *Patent Pending in 24 Hours*, 7e. Nolo. ISBN: 978-1-4133-2201-9.

25. Lo, J. and Pressman, D. (2015). *How to Make Patent Drawings*, 7e. Nolo. ISBN: 978-1-4133-2156-2.

26. Durham, A.L. (2013). *Patent Law Essentials: A Concise Guide*, 4e. Oxford: Praeger, ABC-CLIO, LLC. ISBN: 13 978-1440828782.

27. DeMatteis, B., Gibb, A., and Neustal, M. (2006). *The Patent Writer: How to Write Successful Patent Applications*. Garden City Park, NY: Patents for Commerce, Square One Publishers.

28. Sutton, E. *Software Patents: A Practical Perspective*, 2016. CreateSpace Independent Publishing Platform.

附录 2.A

United States Patent [19]
Arya et al.

[11] Patent Number: 5,710,682
[45] Date of Patent: Jan. 20, 1998

[54] ELECTROSTATIC DISCHARGE PROTECTION SYSTEM FOR MR HEADS

[75] Inventors: **Satya P. Arya**, San Jose; **Timothy Scott Hughbanks**, Morgan Hill, both of Calif.; **Steven Howard Voldman**, South Burlington, Vt.; **Albert John Wallash**, Morgan Hill, Calif.

[73] Assignee: **International Business Machines Corporation**, Armonk, N.Y.

[21] Appl. No.: **799,259**
[22] Filed: **Feb. 13, 1997**

Related U.S. Application Data

[63] Continuation of Ser. No. 706,107, Aug. 30, 1996, abandoned, which is a continuation of Ser. No. 613,928, Mar. 11, 1996, Pat. No. 5,644,454.

[51] Int. Cl.^{6} G11B 5/48; G11B 21/16
[52] U.S. Cl. **360/106**; 360/113
[58] Field of Search 360/104, 106, 360/107, 109, 113; 29/603.04, 603.07

[56] **References Cited**

U.S. PATENT DOCUMENTS

4,800,454	1/1989	Schwarz et al.	360/123
5,247,413	9/1993	Shibata et al.	360/113
5,270,882	12/1993	Jove et al.	360/67
5,272,582	12/1993	Shibata et al.	360/113
5,465,186	11/1995	Bajorek et al.	360/113
5,491,597	2/1996	Bennin et al.	360/105
5,491,605	2/1996	Hughbanks et al.	360/113
5,508,857	4/1996	Horita	360/105
5,526,206	6/1996	Shimizu	360/105

FOREIGN PATENT DOCUMENTS

0 556 891 A1	8/1993	European Pat. Off.	
6-162447	6/1994	Japan	

OTHER PUBLICATIONS

IBM Tech. Disclosure Bul., Greaves et al, "Electro Static Discharge Protection....", vol. 36, No. 12, Dec. 1993, pp. 271–272.

Primary Examiner—Stuart S. Levy
Assistant Examiner—David L. Ometz
Attorney, Agent, or Firm—Baker, Maxham, Jester & Meador

[57] **ABSTRACT**

An MR head receives ESD protection from a mechanism that automatically and releasably shorts the MR head whenever a suspension assembly on which the head is mounted is not installed in an HDA. The suspension assembly includes a flexure underlying a load beam, which is connected to an actuator arm. The MR head is mounted to a distal end of the flexure, leads from components of the MR head being brought out in the form of MR wire leads running along the load beam and the support arm to a nearby terminal connecting side tab. The conductors are separated and exposed at a designated point along the flexure to provide a contact region. A shorting bar, which comprises an electrically conductive member attached to the actuator arm, automatically connects the MR wire leads at the contact region when absence of support for the MR head permits the load beam to bend sufficiently toward the shorting bar. Thus, when the assembly is removed from installation in an HDA, the flexure is permitted to move toward the shorting bar, bringing the contact region and the shorting bar in electrical contact to short the MR wired leads and thereby disable the MR sensor. When the assembly is installed in an HDA, the MR head is supported by an air bearing or the disk itself, depending upon whether the disk is rotating or stopped, respectively. In either case, the load beam is not permitted to droop and the shorting bar cannot contact the conductors, thus activating the MR sensor. Temporary ESD protection mechanisms are also provided, these being removable prior to operation of the HDA by breaking and removing various temporary shorting mechanisms.

16 Claims, 4 Drawing Sheets

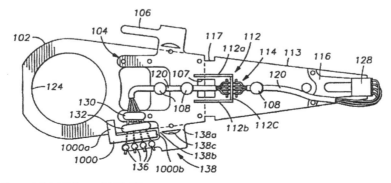

Figure 2.A.1 US Patent 5,710,682.

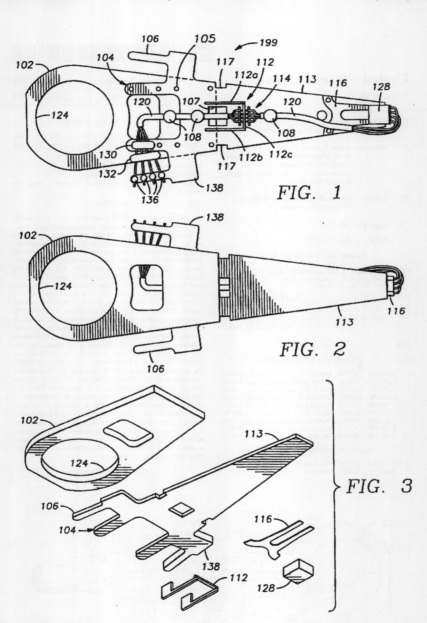

Figure 2.A.1 *(Continued)*

第 2 章 发 明 | 51

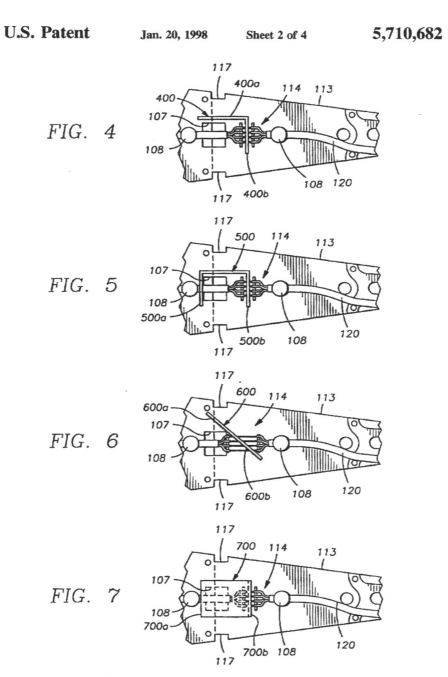

Figure 2.A.1 (*Continued*)

U.S. Patent Jan. 20, 1998 Sheet 3 of 4 **5,710,682**

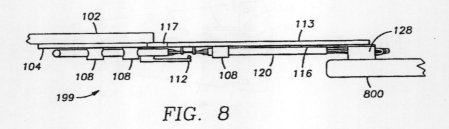

FIG. 8

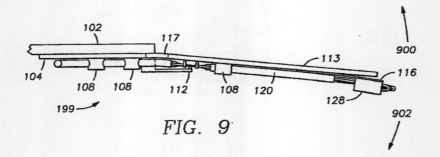

FIG. 9

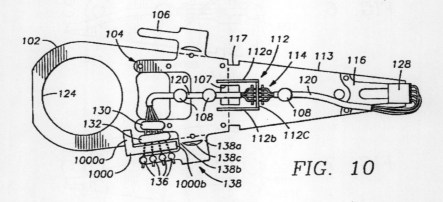

FIG. 10

Figure 2.A.1 *(Continued)*

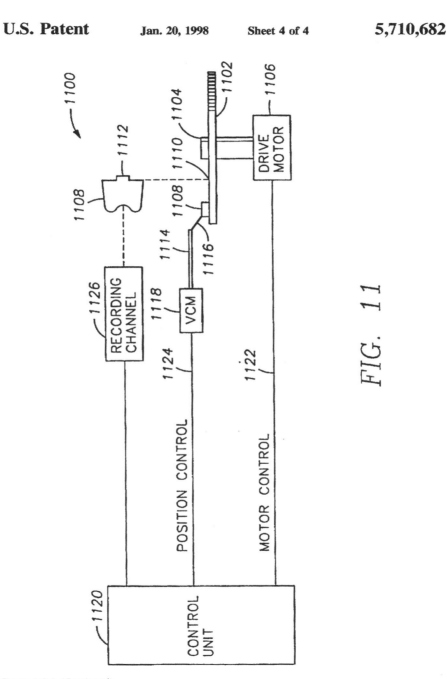

Figure 2.A.1 *(Continued)*

5,710,682

ELECTROSTATIC DISCHARGE PROTECTION SYSTEM FOR MR HEADS

This application is a continuation of application No. 08/706,107, filed Aug. 30, 1996, now abandoned, which is a continuation of application No. 08/613,928 filed Mar. 11, 1996, now U.S. Pat. No. 5,644,454.

BACKGROUND OF THE INVENTION

1. Field of the Invention

The present invention relates to the protection of magnetoresistive ("MR") recording heads from damage caused by electrostatic discharge ("ESD"). More particularly, the invention concerns a method and apparatus for providing ESD protection by shorting MR heads during selected times.

2. Description of the Related Art

Many modern disk drives employ MR recording heads, also called "MR heads", "MR sensors", or "MR elements". MR heads provide improved performance in a number of important respects. However, compared to previous generation thin-film recording heads, MR heads are typically about 100 times more sensitive to damage caused by ESD. Some tests have estimated 14,000 volts as the failure voltage for one model of IBM's thin-film inductive recording head. This estimation was made using the known Human Body Model of failure analysis, known as "HBM". In contrast, tests of conventional MR heads indicate an HBM failure voltage of only 150 volts. These tests simply reflect one known characteristic of MR heads—their high susceptibility to damage from ESD.

During operation of a magnetic storage drive, ESD is typically a relatively insignificant problem. The storage drive is usually encased within a computer, where it is protected from static discharge, particle contaminants, human interference, and other damage. In contrast, during the manufacture of magnetic storage drives, ESD can be a significant and perplexing problem, significantly reducing the effective yield of manufacturing operations.

As a result, engineers are continually seeking effective ways to prevent ESD damage during manufacturing operations. Traditionally, one of the best ways to reduce yield losses from ESD damage is to short the leads of an MR head together. This provides an electrical discharge path around the MR element, rather than through it. Experiments conducted by the inventors using a conventional MR head have shown that spanning the MR sensor with a 1 ohm connection increases the HBM failure voltage from 150 to 2000 volts.

Although the MR sensor is protected from ESD when its leads are shorted, this effectively renders the MR sensor inoperative. Therefore, to activate the MR sensor for manufacturing tests and the like, the shorted leads must be removed, disabled, or otherwise electrically disconnected. Likewise, after such tests, the interconnection between the leads must be reconnected to protect the MR sensor again. Manually shorting the leads in this manner, however, fails to provide a sufficiently convenient mechanism for protecting the MR head.

SUMMARY OF THE INVENTION

Broadly, the present invention provides ESD protection for an MR recording head by selectively activating a mechanism that shorts and consequently disables the MR head. The invention includes both method and apparatus aspects. According to the apparatus aspect of the invention, an MR head is provided on a slider mounted to an integrated suspension assembly. The suspension assembly includes a flexure underlying a load beam, the load beam being connected to an actuator arm. The MR sensor leads are brought out in MR lead wires running along the load beam and the actuator arm to nearby terminal connecting side tabs. The MR lead wires are separated and exposed along the load beam to provide a contact region designed for repeatable electrical contact with a rigid shorting bar, which may be embodied in an electrically conductive member attached to the actuator arm.

The shorting bar and contact region MR lead wires are configured such that the shorting bar automatically interconnects the wires at the contact region when lack of support for the MR head permits the load beam to bend toward the shorting bar. As a result, when the suspension assembly is removed from its head disk assembly ("HDA"), the load beam bends toward the shorting bar, bringing the contact region in electrical contact with the shorting bar to short the MR lead wires and thereby disable the MR sensor. When the suspension assembly has been installed into a disk drive, the MR head is supported by an air bearing (when the disk is rotating) or by the disk itself (when the disk is stopped). In either case, the load beam is not permitted to droop and the shorting bar cannot touch the MR lead wires; as a result, activating the MR sensor is activated.

In another embodiment, where the shorting bar is electrically grounded, the MR sensor is grounded whenever the shorting bar meets the MR lead wires.

In a different embodiment, a one-time temporary shorting mechanism is provided to protect the MR head. Here, the MR lead wires are brought out to the terminal connecting side tabs and stripped of insulation in this region. Each MR head wire is adhered at this point to the side tab. A conductive finger overlies the uninsulated MR lead wires at this region, shorting the wires together. After installation of the integrated suspension assembly, the conductive finger and side tab are broken off of the support arm, and the MR lead wires are connected to a disk controller or another electronic module via a ribbon cable or another appropriate means.

In another embodiment, a one-time temporary shorting mechanism is provided to protect the MR head. Here, the MR lead wires are brought out to an electrically conductive side tab and stripped of insulation in this region Uninsulated portions of each MR head wire are adhered to the side tab at this point using conductive adhesive, thereby shorting all MR lead wires together. The conductive side tab may be grounded, if desired. After installation of the integrated suspension assembly, the side tab is separated from the support arm and the MR lead wires, and the wires are connected to a disk controller or another electronic module via a ribbon cable, flex cable, or another suitable means.

In contrast to the apparatus aspect of the invention, a different aspect of the invention concerns a method for automatically activating and deactivating an MR recording head by selectively shorting its leads to protect the head from ESD-induced damage.

The present invention provides a number of distinct advantages. For example, the invention provides a simple and inexpensive method and apparatus for protecting an MR sensor from ESD. Also, the invention is beneficial since it may be implemented to provide ESD protection that is automatically activated when needed.

BRIEF DESCRIPTION OF THE DRAWINGS

The nature, objects, and advantages of the invention will become more apparent to those skilled in the art after

Figure 2.A.1 (*Continued*)

5,710,682

considering the following detailed description in connection with the accompanying drawings, in which like reference numerals designate like parts throughout, wherein:

FIG. 1 is a top view of an integrated suspension assembly, pursuant to the invention;

FIG. 2 is a bottom view of the integrated suspension assembly, pursuant to the invention;

FIG. 3 is an exploded side view of the integrated suspension assembly, pursuant to the invention, with the load beam in its unloaded position;

FIG. 4 is a magnified view of a first alternative embodiment of the shorting bar and contact region, pursuant to the invention;

FIG. 5 is a magnified view of a second alternative embodiment of the shorting bar and contact region, pursuant to the invention;

FIG. 6 is a magnified view of a third alternative embodiment of the shorting bar and contact region, pursuant to the invention;

FIG. 7 is a magnified view of a fourth alternative embodiment of the shorting bar and contact region, pursuant to the invention;

FIG. 8 is a side view of the integrated suspension assembly, pursuant to the invention, with the load beam in its loaded position;

FIG. 9 is a side view of the integrated suspension assembly, pursuant to the invention, with the load beam in its unloaded position;

FIG. 10 is a top plan view showing various alternative embodiments of the suspension assembly pursuant to the invention; and

FIG. 11 is a simplified block diagram of a magnetic disk storage system embodying the present invention.

DETAILED DESCRIPTION OF THE PREFERRED EMBODIMENTS

Broadly, the present invention provides ESD protection for an MR recording head by shorting and consequently disabling the MR head. The invention includes an apparatus aspect, including one variation in which the mechanism for temporarily shorting the MR head is discarded prior operation of the MR head, and another embodiment in which the MR head is automatically activated in accordance with the position of the suspension assembly. The invention also includes a method aspect, including steps for protecting an MR recording head from ESD.

AUTOMATIC SHORTING MECHANISM

General Construction: Suspension Assembly

FIGS. 1–3 depict the hardware components and interconnections of one embodiment of the invention, from top, bottom, and side perspectives, respectively. More particularly, FIGS. 1–3 depict an integrated suspension assembly 199, which is one of many components in a magnetic data storage drive (not shown) such as a "hard drive." The assembly 199 defines a hole 124 that closely fits around an actuator bearing cartridge or another appropriate HDA component to mount the assembly 199 within the data storage drive.

The primary components of the assembly 199 include an MR head 128, a load beam 113, a flexure 116, and an actuator arm 102. When used herein, the term "MR head" denotes an integrated unit that includes an inductive write element and an MR read element. However, this does not exclude application of the invention to a read head that includes only an MR read element. Furthermore, the ESD protection features of the invention may be applied to other types of recording heads such as non-MR heads. The MR head 128 is mounted to the flexure 116. The flexure 116 extends largely coincident with the load beam 113. The load beam 113 includes an extension 104 for connecting it to the arm 102. The extension 104, for example, may be connected to the arm 102 by a number of welding points, e.g. 105. The load beam 113 includes a flexible hinge 117, which permits the distal tip of the load beam 113 to freely move in two directions as the load beam 113 bends about the hinge 107 with respect to the arm 102. The flexible hinge 117 preferably comprises a region of the load beam 113 that has been selectively narrowed to provide the load beam 113 with a desired level of flexibility relative to the arm 102. To further reduce the rigidity of the hinge 117, the load beam 113 may define an open area 107.

As shown in FIG. 3, hinge 117 permits the distal tip of the load beam 113 to move "downward" (direction 902, FIG. 9) and back "upward" again (direction 900, FIG. 9). Thus, as described in greater detail below, during operation of the suspension assembly 199 the hinge 117 permits the MR head 128 to closely track the surface of a magnetic recording disk above a thin air bearing, despite any ridges, valleys, or other imperfections in the disk surface.

In addition to the hinged load beam 113, the flexure 116 also helps the MR head 128 to closely track surfaces of magnetic recording disks. In particular, the flexure 116 preferably comprises a very thin layer of metal that generally extends coincident to the load beam 113. The flexure 116 may be about 1 mil thick, for example. Since the flexure 116 is only attached to the load beam 113 at its base, the distal tip of the flexure 116 can fluctuate with respect to the distal tip of the load beam 113. Hence, the flexure 116 helps the MR head 128 to closely track the recording disk, despite variations in the disk surface that might exceed the ability of the hinge 117 to allow sufficient movement of the load beam 113 in the downward 902 and upward 900 directions.

The assembly 199 further includes multiple MR lead wires 120 that electrically connect the MR head 128 to various circuits that assist the MR head 128 in reading and writing data from/to magnetic recording media. These circuits may include, for example, channel electronics, located apart from the assembly 199. The MR lead wires 120 preferably comprise copper wires, the assembly of which is referred to as a "wire harness" or "wire assembly". The MR lead wires 120 may be held together by a tubular sheath (as illustrated), configured in a wire "bundle", or arranged in another suitable manner.

The MR lead wires 120 are connected to various components of the MR head 128. Preferably, the wires 120 are electrically connected to relatively large conductive pads (not shown) on the MR head. Such pads typically are electrically connected to small MR head components, such as: (1) the MR head's inductive write coil, (2) the MR read stripe, and (3) shields, poles, and other elements with electrical connections external to the MR head. The wires 120 run from the MR head 128, along the flexure 116 and load beam 113, across a portion of the arm 102, and thereafter to off-assembly components via another interconnect such as a ribbon cable (not shown). The wires 120 may be affixed to the load beam 113 and the arm 102 by a series of anchors 108. The wires 120 ultimately extend to a terminal connecting side tab 138, where adhesive dots 136 anchor the wires 120 to the side tab 138 with predetermined spacing between the wires. The dots 136 may comprise

Figure 2.A.1 (Continued)

5,710,682

epoxy, glue, micro-thin ties, or another suitable adhesive device, for example. At the dots 136, the wires 120 are preferably spaced for alignment with an interconnect (not shown) such as a ribbon cable, flex cable, etc. Spacing of the wires may additionally be maintained at other points by adhesive dots 130, 132, or other suitable spacing devices. The assembly 199 may also include a reciprocal side tab 106, for applications in which the wires must exit the assembly 199 in the opposite direction.

Shorting Mechanism

In accordance with the automatic shorting embodiment of the invention, the assembly 199 also includes various features to automatically protect the MR head 128 from ESD damage by selectively shorting components of the MR head 128. In this embodiment, the MR lead wires 120 provide a contact region 114, at which electrical contact with the wires 120 is possible. Preferably, the contact region 114 comprises an area where each wire 120 has no insulation, and all wires 120 are spaced apart to permit individual electrical contact in the manner discussed below. Alternatively, the contact region 114 may comprise a region of planar conductive traces, of the type used in planar trace suspension assemblies. In applications where it is desirable to avoid shorting certain components of the head 128, the wires 120 coupled to these components may simply be provided with insulation at the contact region 114, or routed around the region 114.

A conductive shorting bar 112 is provided to electrically short the wires 120 of the contact region 114 at certain times. At these times the shorting bar 112 engages the MR lead wires 120 of the contact region 114 to electrically short the wires 120 together. This prevents any transient voltages from developing across components of the MR head 128 that are attached to the MR lead wires exposed at the contact region 114. These components may include, for example, the MR stripe, shields, poles, conductive coils, and other components for which ESD protection is desired. The shorting bar 112 preferably comprises an electrically conductive material, such as stainless steel, copper, or brass. From the standpoint of contamination, stainless steel is particularly appropriate for use with HDAs because it gives off few contaminant particles.

The shorting bar 112 is mounted to a portion of the assembly 199 that does not move with fluctuation of the flexure 116 and load beam 113. For example, the bar 112 may be mounted to the extension 104 or to the arm 102. As shown most clearly in FIG. 1, the bar 112 may comprise a "U" shape, with legs 112a–112b that are affixed to the extension 104 and interconnected by a contact member 112c. The contact member 112c and the contact region 114 meet and electrically connect when the load beam 113 bends sufficiently toward the contact member 112c about the hinge 117, i.e. when the MR head 128 moves a sufficient distance in the downward direction 902 (FIG. 9).

In addition to the example illustrated in FIGS. 1–3, the shorting bar 112 may be implemented in a number of other ways. One example is the "L" shaped contact bar 400, depicted in FIG. 4. This bar 400 includes a leg 400a and a contact member 400b. The leg 400a is preferably adhered to the extension 104 or another suitable stationary support by welding, glue, or another appropriate adhesive device. The contact member 400b is positioned to meet the contact region 114 when the load beam 113 bends sufficiently downward 902.

FIG. 5 illustrates another example, where a shorting bar 500 comprises a "U" shaped member. Although similar in shape to the bar 112, the bar 500 is mounted differently. The bar 500 includes legs 500a–500b, where the leg 500a is affixed to the extension 104 or another suitable stationary support by welding, glue, or another appropriate adhesive device. The leg 500b is positioned to meet the contact region 114 when the load beam 113 moves in its downward direction 902. One advantage of this embodiment is that the bar 500 can be mounted to the extension 104 at any desired angle about the leg 500a. Hence, the bar 500 may be mounted to establish greater or lesser travel distances for the flexure 116 to bring the contact region 114 against the leg 500b.

Another example of shorting bar appears in FIG. 6, where a shorting bar 600 comprises a straight member mounted to the extension 104 at an angle. The bar 600 includes a base region 600a having one or more points adhered to the extension 104 or another suitable stationary support by welding, glue, or another appropriate adhesive device. The bar 600 also includes a contact portion 600b positioned to meet the region 114 when the load beam 113 moves sufficiently in its downward direction 300.

FIG. 7 illustrates still another example of shorting bar contemplated by the invention. Specifically, the shorting bar 700 comprises a solid member, having an inner edge 700a adhered to the extension 104 or another suitable stationary support by welding, glue, or another appropriate adhesive device. The bar 700 also includes an outer edge 700b positioned to meet the region 114 when the load beam 113 moves in its downward direction 902. The solid contact bar 700 may also assume the shape of another solid polyhedron, a circle, a frame-shaped polyhedron with a cutout center, or another suitable configuration.

If desired, the shorting bars 112, 400, 500, 600, and 700 may be electrically grounded so that they ground the MR lead wires 120 upon contact in addition to shorting the wires 120 together. In the case of the bar 112 (FIG. 1), for example, when the load beam 113 moves sufficiently downward for the contact member 112c to meet the contact region 114, an electrically grounded shorting bar 112 would ground the wires 120 in addition to shorting them together.

Operation

In contrast to the various hardware components and interconnections that constitute the automatic shorting apparatus aspect of the present invention, a different aspect concerns a method for automatically protecting an MR head from ESD damage. The following description, which references FIGS. 8–9, exemplifies the operation of the integrated suspension assembly 199 with an automatic shorting mechanism as described above. For simplicity of illustration, FIGS. 8–9 omit the actuator bearing cartridge and other components of the HDA to which the assembly 199 is mounted.

FIG. 8 depicts the assembly 199 in its "loaded position". The assembly 199 is said to be "loaded" because the MR head 128 is supported, directly or indirectly, by a magnetic recording disk 800. The MR head is loaded, for example, when it "flies" above a thin air bearing on the surface of a rapidly spinning disk 800. Similarly, the MR head is loaded when it directly rests upon a disk 800 that is stationary, and thus has no air bearing. In the loaded position, not only is the MR head 128 supported, but the flexure 116 and the load beam 113 are also supported. Downward movement of the load beam 113 about the hinge 117 is therefore limited. In fact, this prevents the load beam 113 from drooping low enough to bring the contact region 114 into physical contact with the shorting bar 112. Therefore, the MR lead wires 120 are not shorted out and the MR head 128 is fully operational.

In contrast, the MR head 128 is not operational when the assembly 199 translates to its "unloaded position" (FIG. 9).

Figure 2.A.1 (*Continued*)

5,710,682

In FIG. 9, the assembly **199** is said to be "unloaded" because the MR head **128** is unsupported. This condition may occur in a number of situations, such as: (1) during manufacturing of the HDA, prior to installation of the assembly **199**, (2) after removal of the assembly **119** from the HDA during repair, pre/post-assembly quality testing, or maintenance, (3) during times when the disk **800** is removed from the storage drive, or (4) various other situations.

When the MR head **128** is unloaded due to removal of the assembly **199** or the disk **800** from the HDA, the sensitive components of the MR head **128** are automatically shorted and hence protected. Removal of support for the MR head **128** permits the load beam **113** to droop by bending about the hinge **117**. This drooping of the load beam **113** brings the MR lead wires **120** of the contact region **114** into contact with the shorting bar **112**, which does no droop since it is rigidly mounted to the arm **102**. When electrical connection is made between the contact region **114** and the bar **112**, the wires **120** are shorted together (and also grounded, if the bar **112** is grounded). All components of the MR head **128** coupled to the wires **120** are protected at this time.

TEMPORARY SHORTING MECHANISMS

As alternatives to FIGS. 1–9, the invention also encompasses various temporary shorting mechanisms to temporarily disable an MR head during manufacturing. With these mechanisms, the MR head is activated prior to operation by removing and discarding the temporary shorting mechanisms.

Conductive Finger Shorting

With conductive finger shorting (FIG. 10), the wires **120** include uninsulated contact points (not shown) in the span between the extension **104** and the side tab **138**. The uninsulated contact points touch a shorting finger **1000**, which extends from the arm **102** over the contact points. The wires **120** may be attached to the side tab **138** at predetermined spacings using the adhesive dots **136**, as discussed above.

The shorting finger **1000** comprises a metal structure made of stainless steel, or another material similar to the shorting bar **112** discussed above. The finger **1000** includes a base **1000**a and a contact member **1000**b. In one embodiment (as illustrated) the finger **1000** may comprise an extension of the material of the arm **102**. Alternatively, the finger **1000** may comprise an extension of the load beam **113**, or a piece separate from load beam **113** and arm **102** where the base is attached to the load beam **113** or arm **102** by welding, glue, or another suitable adhesive device. In either embodiment, the attachment between the base **1000**a and the load beam **113** or arm **102** is sufficiently weak to permit a user to cleanly break the finger **1000** therefrom.

The attachment of the finger **1000** to the actuator arm **102** (as illustrated) biases the contact member **1000**b toward the contact points of the wires **120**. This creates tension at the contact points of the wires **120**, as the wires are held in place at two ends by the adhesive dots **136** and spacer **132**. When the shorting finger **1000** touches the wires **120**, it shorts the wires **120** together. The wires **120** may additionally be grounded, if the shorting finger **1000** is electrically grounded.

After the integrated suspension assembly **199** is installed into the HDA, the MR head **128** is activated by performing the following steps. First, the MR lead wires **120** are connected to a corresponding number of conductors of an interconnect (not shown) leading to an appropriate electrical component such as a disk controller. This step is facilitated by the adhesive dots **136** holding the wires **120** in predetermined spacings, which optimally correspond to the spacing of the cable's conductors. Connection is made on a portion of the wires **120** prior to the adhesive dots **136**. Advantageously, the presence of the conductive finger **100** ensures that wires **120** are shorted while they are being connected to the cable.

Next, connection of the wires **120** to the side tab **138** is cut, and the connection between the side tab **138** and the load beam **113** is also broken. The side tab **138** is then removed from the load beam **113**. Next, the attachment between the base **1000**a and the actuator arm **102** is broken, and the shorting finger **1000** is removed from the actuator arm **102**. The above-described order of installation may be changed in a number of respects, however, to suit the application. For example, the shorting finger **1000** may be removed prior to the other steps, if desired. Alternatively, depending upon the particular application, the shorting finger **1000** may even be left in place, if space permits.

Side Tab Shorting

With side tab shorting (FIG. 10), the MR lead wires **120** are conductively adhered to the side tab **138** (as illustrated) or the side tab **106**, the side tabs **138**, **106** being made from an electrically conductive material. The following description explains side tab shorting using the side tab **138** as an example. The side tab **138** includes points **138**a–**138**b where it attaches to the extension **104** of the load beam **113**. The side tab **138** also defines a cutout region **138**c between the attachment points **138**a–**138**b, along the boundary between the side tab **138** and the load beam **113**. The cutout region **138**c ensures that the attachment between the side tab **138** and the load beam **113** is sufficiently weak to permit a user to cleanly break the side tab **138** off from the load beam **113**.

The MR lead wires **120** are provided with uninsulated regions (not shown), adhered to the side tab **138** by the adhesive dots **136**. In this embodiment, the dots **136** are made of a conductive material such as solder, conductive epoxy, or another appropriate adhesive device. This contact ensures that the components of the MR head **128** attached to the MR lead wires **120** are shorted together (and grounded, if the side tab **138** is electrically grounded). In this state, the MR head **128** is therefore disabled. This contact also ensures that the MR lead wires **120** are rigidly held with a predetermined inter-wire spacing, for convenient alignment and mating to an interconnect (not shown) such as a ribbon cable, flex cable, etc.

After the integrated suspension assembly **199** is installed into the HDA, the MR head **128** is activated by performing the following steps. First, the wires **120** are connected to a corresponding number of conductive elements of an interconnect leading to an appropriate electrical component such as a disk controller. This step is facilitated by the wires **120** being held in their predetermined spacing, which optimally corresponds to the spacing of the interconnect's conductors. Then, the attachment between the side tab **138** and the load beam **113** is broken, and the MR lead wires **120** are disconnected from the side tab **138**.

The above-described order of installation may be changed in a number of respects to suit the application. For example, the MR lead wires **120** may be disconnected from the side tab **138** before connecting them to the interconnect.

DRIVE COMPONENTS

In addition to the invention's automatic and temporary shorting mechanisms, another aspect of the invention con-

Figure 2.A.1 (*Continued*)

5,710,682

cerns a magnetic disk storage system embodying such shorting mechanisms. More particularly, FIG. 11 depicts an exemplary magnetic disk storage system **1106** embodying the invention. Ordinarily skilled artisans will recognize, however, that invention is also applicable to other magnetic recording systems than the specific embodiment **1100** illustrated in FIG. 11.

A magnetic disk storage comprises at least one rotatable magnetic disk **1102** is supported on a spindle **1104** and rotated by a disk drive motor **1106** with at least one slider **1108** positioned on the disk **1102**, each slider **1108** supporting one or more magnetic read/write heads. The magnetic recording media on each disk is in the form of an annular pattern of concentric data tracks (not shown) on the disk **1102**. As the disk **1102** rotates, the sliders **1108** are moved radially in and out over the disk surface **1110** so that the heads **1112** may access different portions of the disk where desired data is recorded. Each slider **1108** is attached to an actuator arm **1114** by means of a suspension **1116**. The suspension **1116** provides a slight spring force which biases the slider **1108** against the disk surface **1110**. Preferably, the actuator arm **1114**, suspension **1116**, and slider **1108** are embodied in an integrated suspension assembly constructed in accordance with the invention, such as ones of the various embodiments described in detail above. Each actuator arm **1114** is attached to an actuator means **1118**. The actuator means **1118** as shown in FIG. 1 may be a voice coil motor (VCM), for example. The VCM comprises a coil moveable within a fixed magnetic field, the direction and velocity of the coil movements being controlled by the motor current signals supplied by a controller. During operation of the disk storage system, the rotation of the disk **1102** generates an air bearing between the slider **1108** and the disk surface **1110** which exerts an upward force or lift on the slider. The air bearing thus counterbalances the slight spring force of the suspension **1116** and supports the slider **1108** off and slightly above the disk surface by a small, substantially constant spacing during operation.

The various components of the disk storage system are controlled in operation by control signals generated by control unit **1120**, such as access control signals and internal clock signals. Typically, the control unit **1120** comprises logic control circuits, storage means and a microprocessor, for example. The control unit **1120** generates control signals to control various system operations such as drive motor control signals on line **1122** and head position and seek control signals on line **1124**. The control signals on line **1124** provide the desired current profiles to optimally move and position a selected slider **1108** to the desired data track on the associated disk **1102**. Read and write signals are communicated to and from read/write heads **1112** by means of recording channel **1126**.

The above description of a typical magnetic disk storage system, and the accompanying illustration of FIG. 1 are for representation purposes only. It should be apparent that disk storage systems may contain a large number of disks and actuators, and each actuator may support a number of sliders.

OTHER EMBODIMENTS

While there have been shown what are presently considered to be preferred embodiments of the invention, it will be apparent to those skilled in the art that various changes and modifications can be made herein without departing from the scope of the invention as defined by the appended claims.

What is claimed is:

1. An apparatus for protecting an MR head from electrostatic discharge, comprising:
   an actuator arm for mounting to a head disk assembly;
   a load beam having a distal end and a proximal end, said proximal end being flexibly joined at a joint to the actuator arm to permit the load beam to bend with respect to the actuator arm about the joint;
   a flexible flexure having a distal end and a proximal end, said flexure being mounted to the load beam and extending therealong;
   an MR head mounted to the flexure's distal end, said MR head including MR components;
   a side tab having a projection and a base, said base being attached to the proximal end of the load beam and said projection being spaced apart a predetermined distance from the actuator arm, the side tab extending from the load beam and being substantially coplanar therewith;
   multiple MR lead wires, a number of which are coupled to said MR components, said MR lead wires running from the load beam's distal end along the load beam and the actuator arm to the side tab, said MR lead wires being adhered to a first surface of the side tab at predetermined spacings, and said MR lead wires being uninsulated at contact points positioned between the side tab and the actuator arm; and
   a conductive finger breakably connected to the actuator arm and positioned such that the conductive finger electrically shorts the MR lead wires at their contact points by pressing against the MR lead wires between the side tab and the actuator arm, the conductive finger extending from the actuator arm and being substantially coplanar therewith.

2. The apparatus of claim 1, the conductive finger being breakably connected to the actuator arm through a breakable connection to the load beam.

3. The apparatus of claim 1, the conductive finger being electrically grounded.

4. The apparatus of claim 1, the contact points being adhered to the side tab with glue.

5. The apparatus of claim 1, the side tab being integral with the load beam's proximal end.

6. The apparatus of claim 1, the predetermined spacings being selected to align the MR lead wires for attachment to an interconnect of a predetermined configuration.

7. A method for protecting an MR head from electrostatic discharge, said MR head being provided at an end of a hinged suspension assembly and said MR head having selected MR components coupled to multiple MR lead wires running along the suspension assembly, said method steps comprising:
   electrically shorting the MR lead wires by:
      running the MR lead wires to a side tab attached to an actuator arm of the suspension assembly, the side tab extending from the actuator arm and being substantially coplanar therewith;
      adhering the MR lead wires to the side tab at predetermined spacings;
      providing the MR lead wires with uninsulated contact points;
      providing a conductive finger breakably connected to the actuator arm and extending therefrom to electri-

Figure 2.A.1 (*Continued*)

5,710,682

11

cally short the MR lead wires at their contact points by pressing against the MR lead wires between the side tab and the actuator arm, the conductive finger extending from the actuator arm and being substantially coplanar therewith;

installing the suspension assembly into a head disk assembly of a disk storage drive;

electrically connecting the MR lead wires to a corresponding number of conductive elements of an interconnect leading to disk drive electronics; and

breaking the connection of the conductive finger to the suspension assembly and removing the conductive finger.

8. The method of claim 7, the conductive finger being electrically grounded and the step of electrically shorting the MR lead wires further comprising electrically grounding the MR lead wires.

9. The method of claim 7, the step of adhering the contact points using epoxy.

10. The method of claim 7, further comprising the steps of breaking the attachment between the side tab and the actuator arm and disconnecting the MR lead wires from the side tab.

11. A magnetic storage system, comprising:

a magnetic storage medium having a plurality of tracks for recording data;

an integrated suspension assembly, comprising:

an actuator arm for mounting to a head disk assembly;

a load beam having a distal end and a proximal end, said proximal end being flexibly joined at a joint to the actuator arm to permit the load beam to bend with respect to the actuator arm about the joint;

a flexible flexure having a distal end and a proximal end, said flexure being mounted to the load beam and extending therealong;

an MR head affixed to the flexure's distal end and including MR components, said MR head being maintained in a closely spaced position relative to

12

the magnetic storage medium during relative motion between the MR head and the magnetic storage medium;

a side tab having a projection and a base, said base being attached to the proximal end of the load beam and said projection being spaced apart a predetermined distance from the actuator arm, the side tab extending from the load beam and being substantially coplanar therewith;

multiple MR lead wires, a number of which are coupled to said MR components, said MR lead wires running from the load beam's distal end along the load beam and the actuator arm to the side tab, said MR lead wires being adhered to a first surface of the side tab at predetermined spacings, and said MR lead wires being uninsulated at contact points positioned between the side tab and the actuator arm; and

a conductive finger breakably connected to the actuator arm and positioned such that the conductive finger electrically shorts the MR lead wires at their contact points by pressing against the MR lead wires between the side tab and the actuator arm, the conductive finger extending from the actuator arm and being substantially coplanar therewith.

12. The system of claim 11, the conductive finger being breakably connected to the actuator arm through a breakable connection to the load beam.

13. The system of claim 11, the conductive finger being electrically grounded.

14. The system of claim 11, the contact points being adhered to the side tab with glue.

15. The system of claim 11, the side tab being integral with the load beam's proximal end.

16. The system of claim 11, the predetermined spacings being selected to align the MR lead wires for attachment to an interconnect of a predetermined configuration.

* * * * *

Figure 2.A.1 *(Continued)*

(12) United States Patent
Voldman

(10) Patent No.: **US 6,574,078 B1**
(45) Date of Patent: **Jun. 3, 2003**

(54) **METHOD AND APPARATUS FOR PROVIDING ELECTROSTATIC DISCHARGE PROTECTION OF A MAGNETIC HEAD USING A MECHANICAL SWITCH AND AN ELECTROSTATIC DISCHARGE DEVICE NETWORK**

(75) Inventor: **Steven H. Voldman**, South Burlington, VT (US)

(73) Assignee: **International Business Machines Corporation**, Armonk, NY (US)

(*) Notice: Subject to any disclaimer, the term of this patent is extended or adjusted under 35 U.S.C. 154(b) by 0 days.

(21) Appl. No.: **09/602,809**

(22) Filed: **Jun. 23, 2000**

(51) Int. Cl.[7] **G11B 5/39**; G11B 5/40
(52) U.S. Cl. **360/323**; 360/245.8; 360/128
(58) Field of Search 360/128, 323, 360/245.8

(56) **References Cited**

U.S. PATENT DOCUMENTS

5,680,274 A	10/1997	Palmer	360/245.9
5,710,682 A	1/1998	Arya et al.	360/245.8
6,163,437 A *	12/2000	Inage et al.	360/128
6,233,127 B1 *	5/2001	Shimazawa	360/319
6,259,573 B1 *	7/2001	Tsuwako et al.	360/244.1
6,275,361 B1 *	8/2001	Wallash et al.	360/323

* cited by examiner

Primary Examiner—Jefferson Evans
(74) *Attorney, Agent, or Firm*—Altera Law Group, LLC

(57) **ABSTRACT**

A magnetic head using a mechanical switch and an electrostatic discharge device network is disclosed. A mechanical switch is in series and parallel configuration with a silicon electrostatic discharge network. The electrostatic discharge network can be controlled in an on or off state for testing, evaluation or diagnostics of the armature or MR head.

18 Claims, 6 Drawing Sheets

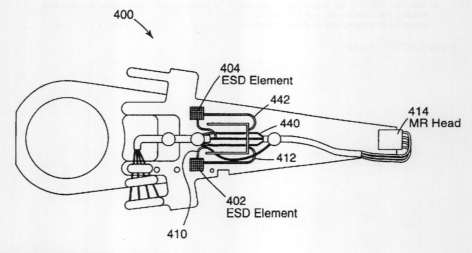

Figure 2.A.2 US Patent No. 5,521,115.

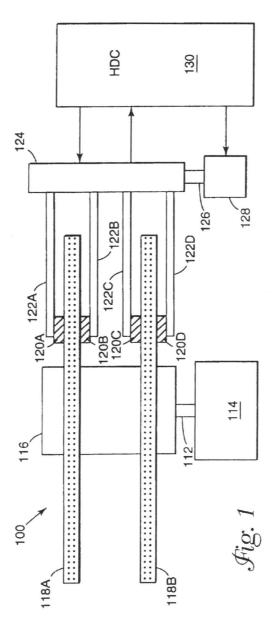

Figure 2.A.2 (*Continued*)

U.S. Patent Jun. 3, 2003 Sheet 2 of 6 US 6,574,078 B1

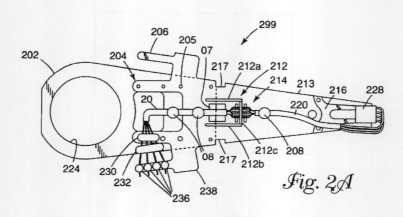

Fig. 2A

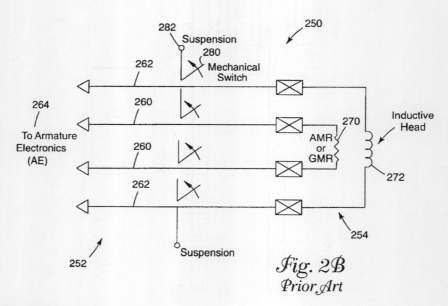

Fig. 2B
Prior Art

Figure 2.A.2 (*Continued*)

U.S. Patent　　Jun. 3, 2003　　Sheet 3 of 6　　US 6,574,078 B1

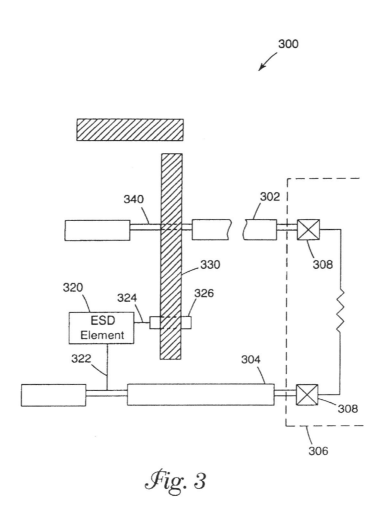

Fig. 3

Figure 2.A.2 *(Continued)*

U.S. Patent Jun. 3, 2003 Sheet 4 of 6 US 6,574,078 B1

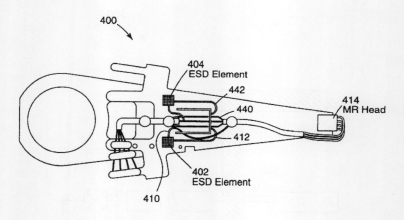

Fig. 4

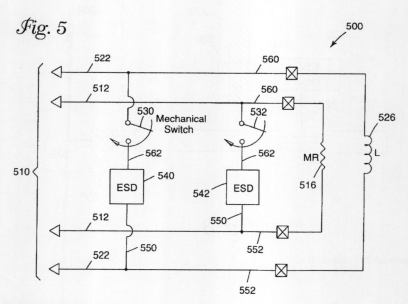

Fig. 5

Figure 2.A.2 (*Continued*)

U.S. Patent　　Jun. 3, 2003　　Sheet 5 of 6　　US 6,574,078 B1

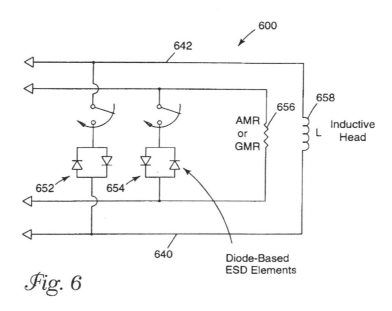

Fig. 6

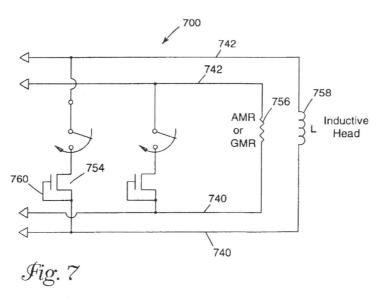

Fig. 7

Figure 2.A.2 (*Continued*)

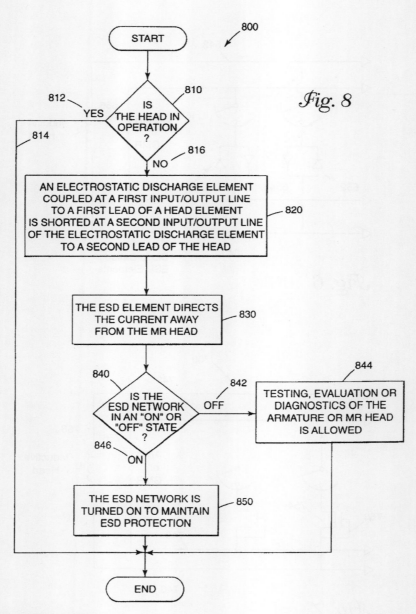

Figure 2.A.2 (*Continued*)

US 6,574,078 B1

METHOD AND APPARATUS FOR PROVIDING ELECTROSTATIC DISCHARGE PROTECTION OF A MAGNETIC HEAD USING A MECHANICAL SWITCH AND AN ELECTROSTATIC DISCHARGE DEVICE NETWORK

BACKGROUND OF THE INVENTION

1. Field of the Invention

This invention relates in general to a protection of magnetoresistive heads, and more particularly to a method and apparatus for providing electrostatic discharge protection of a magnetic head using a mechanical switch and an electrostatic discharge device network.

2. Description of Related Art

Most disc drives built today use conventional thin-film recording heads. Magnetic heads typically consist of a titanium carbide ceramic or silicon slider body and a transducer. Today, most disk drives employ MR heads. MR heads provide improved performance in a number of important respects.

MR heads operate on according to a phenomenon known as the magneto-resistive effect. Certain metals, when exposed to a magnetic field, change their resistance to the flow of electricity. This property is exploited in creating the read element of an MR head. Reading information for the media is accomplished by constantly passing a sense current through the read element of the head. When the head passes over a magnetic field on the media, the head changes its resistance, which is detected by the change in amperage of the sense current.

A major problem that is found during the manufacture of magnetic recording heads, particularly of the thin film type, is the spurious discharge of static electricity which has been undesirably generated. Static charges may be produced by the presence of certain materials, such as plastics, which are present in the surroundings at the place of manufacture of the magnetic heads. Further, static charges may be present during human handling and tooling of the magnetic recording heads. Compared to previous generation thin-film heads, MR heads are typically 200 times more sensitive to damage caused by electrostatic discharge (ESD).

When there is a static discharge, between a magnetic pole piece and an adjacent conductive layer, the pole piece may be damaged, particularly at a critical sensing portion, such as at the tip of the pole piece which is exposed and disposed adjacent to the transducing gap facing the record medium. In addition, the dielectric or insulating material that surrounds the magnetic head coil could break down from the discharge effect. As a result, the head assembly is subject to deterioration and degradation so that it is rendered virtually useless.

Approaches to alleviate this problem have involved the grounding of operators, table tops, or the use of ion producing fans and air hose nozzle application. Also, the materials used for storage containers and work trays must be carefully selected. However, the basic problem of spurious discharge at the critical pole tip area has not been completely solved by these approaches.

In addition, many approaches have been used for protecting magnetic heads from ESD destruction. For example, U.S. Pat. No. 5,710,682, issued Jan. 20, 1998, to Arya et al., entitled "ELECTROSTATIC DISCHARGE PROTECTION SYSTEM FOR MR HEADS", and incorporated herein by reference, discloses a shorting bar, which comprises an electrically conductive member attached to the actuator arm, for automatically connecting the MR wire leads at an exposed contact region of the MR head wire leads when absence of support of the MR head permits the load beam to bend sufficiently toward the shorting bar. However, the shorting of the leads by the shorting bar prevents testing and other diagnosis of the MR head.

It can be seen that there is a need for a method and apparatus for providing electrostatic discharge protection of a magnetic head without preventing electrostatic discharge testing or other diagnosis of the magnetic head.

It can also be seen that there is a need for a method and apparatus for providing electrostatic discharge protection of a magnetic head using a mechanical switch and an electrostatic discharge device network.

SUMMARY OF THE INVENTION

To overcome the limitations in the prior art described above, and to overcome other limitations that will become apparent upon reading and understanding the present specification, the present invention discloses a method and apparatus for providing electrostatic discharge protection of a magnetic head using a mechanical switch and an electrostatic discharge device network.

The present invention solves the above-described problems by providing a mechanical switch that is in series and parallel configuration with a silicon electrostatic discharge network. The electrostatic discharge network can be controlled in an on or off state for testing, evaluation or diagnostics of the armature or MR head.

A method in accordance with the principles of the present invention includes engaging an electrostatic discharge element between leads from a head element in response to absence of support to the head.

Other embodiments of a method in accordance with the principles of the invention may include alternative or optional additional aspects. One such aspect of the present invention is that the method further includes disengaging the electrostatic discharge element in response to support being applied to the head.

Another aspect of the present invention is that the electrostatic discharge element is turned on to maintain electrostatic discharge protection for the head.

Another aspect of the present invention is that the electrostatic discharge element is turned off to allow operation of the head.

Another aspect of the present invention is that the operation of the head includes testing, evaluation and diagnostics of the head.

Another aspect of the present invention is that the electrostatic discharge element is coupled to a first lead from a head element via a first input/output line, the engaging further including shorting a second lead from the head element to a second input/output line of the electrostatic discharge element.

Another aspect of the present invention is that the shorting connects the electrostatic discharge element being to the first and second leads from the head element in parallel with the head element.

Another aspect of the present invention is that the electrostatic discharge element includes a pair of back-to-back, parallel diodes for accommodating both positive and negative potentials across the element.

Another aspect of the present invention is that the electrostatic discharge element includes an N-type MOSFET

Figure 2.A.2 (*Continued*)

US 6,574,078 B1

transistor disposed between the first and second lead of the head element, a gate of the MOSFET transistor being tied to the first lead of the head element.

Another aspect of the present invention is that the head element includes an MR element.

Another aspect of the present invention is that the head element includes an inductive coil element.

Another embodiment of the present invention includes an apparatus for protecting an MR head from electrostatic discharge, wherein the apparatus includes a first lead from a head element, a second lead from a head element having a contact point, an electrostatic discharge element, coupled to the first lead from the head element via a first input/output line, the electrostatic discharge element having a second input/output line with a contact point and a conductive member positioned proximate the contact point of the second lead and the contact point of the second input/output line, the conductive member engaging the contact point of the second lead and the contact point of the second input/output line in response to absence of support to the head, the engagement of the conductive member with the contact point of the second lead and the contact point of the second input/output line connecting the electrostatic discharge element to the first and second leads from the head element in parallel with the head element.

Another embodiment of the present invention includes a disk drive system, wherein the disk drive includes a magnetic storage disk for storing data thereon, a MR head located proximate to the disk for reading and writing data to and from the disk, a disk movement device, coupled to the disk, for rotating the disk, an actuator arm, coupled to the MR head, for supporting the MR head and an actuator, coupled to the access arm, for moving actuator arm to position the MR head relative to the disk; wherein the actuator arm further includes a first lead from a head element, a second lead from a head element having a contact point, an electrostatic discharge element, coupled to the first lead from the head element via a first input/output line, the electrostatic discharge element having a second input/output line with a contact point, and a conductive member positioned proximate the contact point of the second lead and the contact point of the second input/output line, the conductive member engaging the contact point of the second lead and the contact point of the second input/output line in response to absence of support to the head, the engagement of the conductive member with the contact point of the second lead and the contact point of the second input/output line connecting the electrostatic discharge element to the first and second leads from the head element in parallel with the head element.

These and various other advantages and features of novelty which characterize the invention are pointed out with particularity in the claims annexed hereto and form a part hereof. However, for a better understanding of the invention, its advantages, and the objects obtained by its use, reference should be made to the drawings which form a further part hereof, and to accompanying descriptive matter, in which there are illustrated and described specific examples of an apparatus in accordance with the invention.

BRIEF DESCRIPTION OF THE DRAWINGS

Referring now to the drawings in which like reference numbers represent corresponding parts throughout:

FIG. 1 illustrates a hard disk drive (HDD) according to the present invention;

FIG. 2a depicts the hardware components and interconnections of a prior ESD protection system;

FIG. 2b is an electrical schematic of the ESD protection system described with reference to FIG. 2a above;

FIG. 3 illustrates the apparatus for providing electrostatic discharge protection of a magnetic head;

FIG. 4 illustrates the hardware components and interconnections of the ESD protection system according to the present invention;

FIG. 5 is an electrical schematic of the ESD protection system according to the present invention as described with reference to FIGS. 3–4 above;

FIG. 6 illustrates the ESD protection system according to the present invention wherein the ESD elements are diodes;

FIG. 7 illustrates the ESD protection system according to the present invention wherein the ESD elements are transistors; and

FIG. 8 illustrates a method for providing electrostatic discharge protection of a magnetic head using a mechanical switch and an electrostatic discharge device network according to the present invention.

DETAILED DESCRIPTION OF THE INVENTION

In the following description of the exemplary embodiment, reference is made to the accompanying drawings which form a part hereof, and in which is shown by way of illustration the specific embodiment in which the invention may be practiced. It is to be understood that other embodiments may be utilized as structural changes may be made without departing from the scope of the present invention.

The present invention provides a magnetic head using a mechanical switch and an electrostatic discharge device network. The present invention provides a mechanical switch that is in series and parallel configuration with a silicon electrostatic discharge network. The electrostatic discharge network can be controlled in an on or off state for testing, evaluation or diagnostics of the armature or MR head.

FIG. 1 illustrates a hard disk drive (HDD) 100 including disks 118A, 118B according to the present invention. The HDD 100 includes a disk 118 and a hard disk controller (hereinafter referred to as HDC) 130. The disk part has a motor 114 for rotating a shaft 112 at a high speed. A cylindrical support 116 is attached to the shaft 112 so that the their axes are in coincidence. One or more information recording disks 118A and 118B are mounted between support 116. Magnetic heads 120A, 120B, 120C and 120D are respectively provided to face the disk surface, and these magnetic heads are supported from an actuator 124 by access arms 122A, 122B, 122C, and 122D, respectively. The individual magnetic heads 120A to 120D receive the drive force transmitted from an actuator drive device 128 by a shaft 126 and rotates about the shaft 126 as the axis of rotation, and fly over the disk 118 to a predetermined position.

FIG. 2a depicts the hardware components and interconnections of a prior ESD protection system. FIG. 2a depicts an integrated suspension assembly 299, which is one of many components in a magnetic data storage drive. The assembly 299 defines a hole 224 that closely fits around an actuator bearing cartridge or another appropriate HDA component to mount the assembly 299 within the data storage drive.

The primary components of the assembly 299 include an MR head 228, a load beam 213, a flexure 216, and an actuator arm **202**. The MR head **228** Is mounted to the flexure **216**. The flexure **216** extends largely coincident with the load beam **213**. The load beam **213** includes an extension **204** for connecting it to the arm **202**. The load beam **213** includes a flexible hinge **217**, which permits the distal tip of the load beam **213** to freely move in two directions as the load beam **213** bends about the hinge **207** with respect to the arm **202**. The flexible hinge **217** preferably comprises a region of the load beam **213** that has been selectively narrowed to provide the load beam **213** with a desired level of flexibility relative to the arm **202**. To further reduce the rigidity of the hinge **217**, the load beam **213** may define an open area **207**.

Hinge **217** permits the distal tip of the load beam **213** to move "downward" and back "upward" again. Thus, as described in great detail below, during operation of the suspension assembly **299** the hinge **217** permits the MR head **228** to closely track the surface of a magnetic recording disk above a thin air bearing, despite any ridges, valleys, or other imperfections in the disk surface.

In addition to the hinged load beam **213**, the flexure **216** also helps the MR head **228** to closely track surfaces of magnetic recording disks. In particular, the flexure **216** preferably comprises a very thin layer of metal that generally extends coincident to the load beam **213**. Since the flexure **216** is only attached to the load beam **213** at its base, the distal tip of the flexure **216** can fluctuate with respect to the distal tip of the load beam **213**. Hence, the flexure **216** helps the MR head **228** to closely track the recording disk, despite variations in the disk surface that might exceed the ability of the hinge **217** to allow sufficient movement of the load beam **213** in the downward and upward directions.

The assembly **290** further includes multiple MR lead wires **220** that electrically connect the MR head **228** to various circuits that assist the MR head **228** in reading and writing data from/to magnetic recording media. These circuits may include, far example, channel electronics, locate apart from the assembly **299**. The MR lead wires **220** preferably comprise wires, the assembly of which is referred to as a "wire harness" or "wire assembly". The MR lead wires **220** may be held together by a tubular sheath (as illustrated), configured in a wire "bundle", or arranged in another suitable manner.

The MR lead wires **220** are connected to various components of the MR head **228**. The wires **220** run from the MR head **228**, along the flexure **216** and load beam **213**, across a portion of the arm **202**, and thereafter to off-assembly components via another interconnect such as a ribbon cable (not shown). The wires **220** may be affixed to the load beam **213** and the arm **202** by a series of anchors **208**. The wires **220** ultimately extend to a terminal connecting sidetab **238**, where adhesive dots **236** anchor the wires **220** to the side tab **238** with predetermined spacing between the wires.

A contact region **214** is provided so that electrical contact with the wires **220** is possible. A conductive member or shorting bar **212** is provided to electrically short the wires **220** of the contact region **214** at certain times. At these times the shorting bar **212** engages the MR lead wires **220** of the contact region **214** to electrically short the wires **220** together. This prevents any transient voltages from developing across components of the MR head **228** that arc attached to the MR lead wires exposed at the contact region **214**. The shorting bar **212** comprises an electrical conductive material, such as stainless steel, copper or brass. The shorting bar **212** is mounted to a portion of the assembly **299** that does not move with fluctuation of the flexure **216** and load beam **213**. For example, the bar **212** may be mounted to the extension **204** or to the arm **202**. As shown most clearly in FIG. 2a, the bar **212** may comprise a "U" shape, with legs **212a**–**212b** that are affixed to the extension **204** and interconnected by a contact member **212c**. The contact member **212c** and the contact region **214** meet and electrically connect when the load beam **213** bends sufficiently toward the contact member **212c** about the hinge **217**. i.e. when the MR head **228** moves a sufficient distance in the downward direction.

FIG. 2b is an electrical schematic **250** of the ESD protection system described with reference to FIG. 2a above. In FIG. 2b, a four lead **252** MR/inductive head **254** is shown. Two leads **260** extend to the armature electronics (AE) **264** from the MR head **270**. Note that the MR head **270** may be an anisotropic MR head or a giant MR head. In addition, two leads **262** extend from the inductive head **272** to the armature electronics. The shorting bar described with reference to FIG. 2a is shown in FIG. 2b as a series of switches **280**. The switches **280** are closed to short the four leads **252** to the suspension **282**. This is effectively the result of the shorting bar engaging the leads **252** from the head **254**.

FIG. **3** illustrates the apparatus **300** for providing electrostatic discharge protection of a magnetic head according to the present invention. In FIG. **3**, the leads **302**, **304** from a MR head **306** to the armature electronics (not shown) are illustrated as being attached to the pads **308** of the MR head **306**. The leads **302**, **304** may include an insulation covering the actual conductive material. An electrostatic discharge device **320** includes a first input/output line **322** that is coupled to a first lead **304**. The electrostatic discharge element **320** includes a second input/output line **324** that provides a shorting connection or contact point **326**. A mechanical switch, in the form of a conductive member **330**, aligns with contact points **326** of the electrostatic discharge element **320** and the contact points **340** of a second lead **302** from the MR head **306**. The mechanical switch **330** is engaged during the absence of support to the MR head **306** and disengaged during the presence of support to the MR head **306**. Thus, the MR head **306** (GMR or AMR) receives ESD protection from a mechanical mechanism **330** which is in a series and parallel configuration with a silicon ESD device network **320**. The mechanical device **330** consists of a paddle board or suspension assembly that acts as a mechanical switch, which is closed when the head is not in operation, i.e. supported.

The ESD network **320** is in series with the mechanical switch. **330** to direct the current away from the MR head. In a series configuration, the ESD network **320** can be in series with the mechanical switch **330**. The ESD network **320** can be controlled in an "on" or "off" state for testing, evaluation or diagnostics of the armature or MR head **308**. Furthermore, in a series configuration, the mechanical switch **330** can unload the ESD network **320** to avoid capacitance and functional loading on the MR head **306** so that GMR and AMR functional performance is not degraded. The mechanical device **330** is electrically isolated.

Accordingly, the mechanical switch **330** provides ESD protection to the MR head **306** while avoiding unwanted performance impact of the ESD network **320** on head performance. The present invention allows for ESD testing without direct shorting across the MR head **306**, whereas the prior systems shorts the leads, which prevents testing and other means of diagnosis.

FIG. **4** illustrates the hardware components and interconnections of the ESD protection system **400** according to the

Figure 2.A.2 (*Continued*)

present invention. In FIG. 4, the ESD elements **402, 404** are shown with a first input/output line **410** connected to a first wire **412** from the MR head **414**. The mechanical shorting bar, or conductive member **430**, is shown shorting together the second wires **440** from the MR head **414** to the second input/output line **442** of the ESD elements **402, 404**.

FIG. 5 is an electrical schematic **500** of the ESD protection system according to the present invention as described with reference to FIGS. 3–4 above. In FIG. 5, a four lead **510** MR/inductive head is shown. A first pair of leads **512** extend to the armature electronics (AE) (not shown) from the MR head **516**. Again, those skilled in the art will recognize that the present invention is not meant to be limited to a particular type of head, but that the MR head **516** may be any type of head including an anisotropic MR head or a giant MR head. In addition to the first pair of leads **512**, a second pair of leads **522** extend from the inductive head **526** to the armature electronics. The shorting bar, or conductive member, described with reference to FIGS. 3–4 is shown in FIG. 5 as a switches **530, 532**. The ESD elements **540, 542** are shown with first input/output lines **550** connected to first wires **552** from the MR **516** and inductive **526** head. The mechanical shorting bar shorts together the second wires **560** from the MR **516** and inductive **526** head to the second input/output lines **562** of the ESD elements **540, 542**. The ESD elements **540, 542** can be controlled in an "on" or "off" state for testing, evaluation or diagnostics of the armature or MR head. Furthermore, the mechanical switch **530, 532** can unload the ESD network **540, 542** to avoid capacitance and functional loading on the MR head **516** so that GMR and AMR functional performance is not degraded. The shorting bar **530, 532** is electrically isolated so the leads **510** of the MR **516** and inductive **526** heads are not shorted together or the leads **510** of the MR **516** and inductive **526** heads are not shorted to the suspension.

FIG. 6 illustrates the ESD protection system **600** according to the present invention wherein the ESD elements are diodes. Diodes pairs **652, 654** may be coupled in parallel to the MR **656** and inductive **658** element. The diodes pairs **652, 654** provide a low conduction path for the fast discharging of electrostatic charge buildup. The diodes pairs **652, 654** provide the electrical separation necessary to provide proper local isolation between the first **640** and last **642** turn of inductive element **658** and between the leads of the MR **656** element so that normal operation of the head is maintained. The diodes in the diode pairs **652, 654** are arranged in a back-to-back, parallel fashion to accommodate both positive and negative potentials across the MR **656** and inductive **658** element. When electrostatic discharge occurs across the MR **656** and inductive **658** element, the charge is dissipated across the diodes **652, 654** when the potential of the electrostatic charge rises above (in an absolute sense) the threshold voltage (typically 0.7 volts) on either of the back-to-back diodes.

FIG. 7 illustrates the ESD protection system **700** wherein the ESD elements are transistors. In FIG. 7, an N-type MOSFET **754** is disposed between the first **740** and last **742** lead of the MR **756** and inductive **758** element. The gate **760** of the MOSFET **754** is tied to the first lead **740** of the MR **756** and inductive **758** element.

FIG. 8 illustrates a method **800** for providing electrostatic discharge protection of a magnetic head using a mechanical switch and an electrostatic discharge device network according to the present invention. First, is the head in operation **810**, i.e., is the MR head supported. If yes **812**, the shorting and engagement of the ESD element are not performed **814**. If no **816**, an electrostatic discharge element coupled at a first input/output line to a first lead of a head element engages a contact point at a second input/output line of the electrostatic discharge element and a contact point at a second lead of the head to short the two leads together **820**. The ESD element is then positioned in parallel with the head element. The ESD element directs the current away from the MR head **830**. The ESD network can be controlled in an "on" or "off" state **840**. The ESD network may be turned off **842** to allow testing, evaluation or diagnostics of the armature or MR head **844**. The ESD network may be turned on **846** to maintain ESD protection **850**.

A magnetic head using a mechanical switch and an electrostatic discharge device network is disclosed. A mechanical switch is in series and parallel configuration with a silicon electrostatic discharge network. The electrostatic discharge network can be controlled in an on or off state for testing, evaluation or diagnostics of the armature or MR head. Accordingly, the mechanical switch provides ESD protection to the MR head while avoiding unwanted performance impact of the ESD network on head performance. The present invention allows for ESD testing without direct shorting across the MR head, whereas the prior systems shorts the leads together, which prevents testing and other means of diagnosis.

The foregoing description of the exemplary embodiment of the invention has been presented for the purposes of illustration and description. It is not intended to be exhaustive or to limit the invention to the precise form disclosed. Many modifications and variations are possible in light of the above teaching. It is intended that the scope of the invention be limited not with this detailed description, but rather by the claims appended hereto.

What is claimed is:

1. An apparatus for protecting an MR head from electrostatic discharge, comprising:

a first lead from a head element;

a second lead from a head element having a contact point;

an electrostatic discharge element, coupled to the first lead from the head element via a first input/output line, the electrostatic discharge element having a second input/output line with a contact point; and

a conductive member positioned proximate the contact point of the second lead and the contact point of the second input/output line, the conductive member engaging the contact point of the second lead and the contact point of the second input/output line in response to absence of a load being applied to the head, the engagement of the conductive member with the contact point of the second lead and the contact point of the second input/output line connecting the electrostatic discharge element to the first and second leads from the head element in parallel with the head element.

2. The apparatus of claim **1** wherein the conductive member disengages the contact point of the second lead and the contact point of the second input/output line in response to a load being applied to the head.

3. The apparatus of claim **1** wherein the electrostatic discharge element is turned on to maintain electrostatic discharge protection for the head.

4. The apparatus of claim **1** wherein the electrostatic discharge element is turned off to allow operation of the head.

5. The apparatus of claim **4** wherein the operation of the head includes testing, evaluation and diagnostics of the head.

Figure 2.A.2 (*Continued*)

US 6,574,078 B1

6. The apparatus of claim 1 wherein the electrostatic discharge element comprises a pair of back-to-back, parallel diodes for accommodating both positive and negative potentials across the element.

7. The apparatus of claim 1 wherein the electrostatic discharge element comprises an N-type MOSFET transistor disposed between the first and second lead of the head element, a gate of the MOSFET transistor being tied to the first lead of the head element.

8. The apparatus of claim 1 wherein the head element comprises an MR element.

9. The apparatus of claim 1 wherein the head element comprises an inductive coil element.

10. A disk drive system, comprising:
a magnetic storage disk for storing data thereon;
a MR head located proximate to the disk for reading and writing data to and from the disk;
a disk movement device, coupled to the disk, for rotating the disk;
an actuator arm, coupled to the MR head, for supporting the MR head; and
an actuator, coupled to the access arm, for moving actuator arm to position the MR head relative to the disk;
wherein the actuator arm further comprises:
a first lead from a head element;
a second lead from a head element having a contact point;
an electrostatic discharge element, coupled to the first lead from the head element via a first input/output line, the electrostatic discharge element having a second input/output line with a contact point; and
a conductive member positioned proximate the contact point of the second lead and the contact point of the second input/output line, the conductive member engaging the contact point of the second lead and the contact point of the second input/output line in response to absence of a load being applied to the head, the engagement of the conductive member with the contact point of the second lead and the contact point of the second input/output line connecting the electrostatic discharge element to the first and second leads from the head element in parallel with the head element.

11. The disk drive system of claim 10 wherein the conductive member disengages the contact point of the second lead and the contact point of the second input/output line in response to a load being applied to the head.

12. The disk drive system of claim 10 wherein the electrostatic discharge element is turned on to maintain electrostatic discharge protection for the head.

13. The disk drive system of claim 10 wherein the electrostatic discharge element is turned off to allow operation of the head.

14. The disk drive system of claim 13 wherein the operation of the head includes testing, evaluation and diagnostics of the head.

15. The disk drive system of claim 10 wherein the electrostatic discharge element comprises a pair of back-to-back, parallel diodes for accommodating both positive and negative potentials across the element.

16. The disk drive system of claim 10 wherein the electrostatic discharge element comprises an N-type MOSFET transistor disposed between the first and second lead of the head element, a gate of the MOSFET transistor being tied to the first lead of the head element.

17. The disk drive system of claim 10 wherein the head element comprises an MR element.

18. The disk drive system of claim 10 wherein the head element comprises an inductive coil element.

* * * * *

Figure 2.A.2 (*Continued*)

第 3 章
专利和专利语言

3.1 引言

本章将讨论专利搜索引擎和专利语言。为了理解这个过程,了解专利语言很重要。本章讨论了美国专利商标局[1-5]、专利搜索引擎[6,7]、专利审查[8]以及专利草案[9-25]。

3.2 专利搜索引擎

专利搜索引擎对于学习如何进行专利检索非常重要[1-7,26](如图 3-1 所示)。使用专利搜索引擎,发明人可以回顾其他存在的专利,以了解如何编写自己的专利。专利搜索引擎可以让您了解专利的结构、编写和处理方式。您可以将搜索引擎作为学习工具,以便更好地定义您的发明和工具。人们也可以将搜索引擎用于许多其他用途。

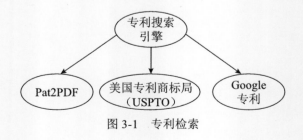

图 3-1 专利检索

从发明到专利——科学家和工程师指南.第一版.史蒂文·H.沃尔德曼 ©2018 John Wiley & Sons 公司,2018 年由 John Wiley & Sons 公司出版。

3.2.1 美国专利商标局（USPTO）

美国专利商标局（USPTO）网站[1]是一个很好的起点。USPTO 网站具有展示有关专利语言的信息以及进行专利检索的能力。现在有一个搜索引擎能够检索已发布的现有专利和尚未发布的专利申请。可以执行使用 AND 和 OR 功能的布尔搜索。可以将关键词放置在图形单元界面（GUI）中进行搜索。

3.2.2 Pat2PDF

从历史上看，专利代理人必须支付专利技术的使用费并提供搜索服务。今天，有免费的专利搜索引擎可供使用。例如，被称为"Pat2PDF"[6]的专利搜索引擎，可以将它搜索到的专利打印为 PDF 文件。该网站的网址是 https://www.pat2pdf.org[6]。如果您拥有发布的专利号，您可以在该网站搜索特定的专利。

3.2.3 Google 专利

Google 发起了一个简单的搜索引擎。Google 专利搜索引擎的网址是 https://www.google.com/patents [7]。在谷歌搜索引擎界面，可以输入"谷歌专利"，然后出现一个url。谷歌浏览器在谷歌标签和输入栏中也有"专利"一词。在输入栏中，可以输入专利号或关键字，从而启动搜索引擎。

3.3 专利语言

在本节中，将介绍专利语言，以便发明人能够与专利代理人进行沟通，并开始了解如何读懂专利局文件。

3.3.1 说明书

说明书也称为交底书，是对一项发明的书面描述（如图 3-2 所示）。撰写专利说明书是为了满足专利授权的书面要求以及界定权利要求的保护范围[9-25]。

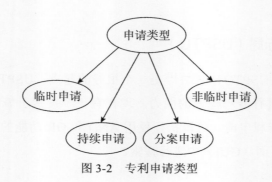

图 3-2　专利申请类型

专利说明书的内容如下所述。

发明名称：本发明的名称旨在用几个词来描述本发明的本质。

相关申请的交叉引用：在美国，要求专利申请人提交标题为"相关申请的交叉引用"的部分。在这部分，申请人列出了他们要求优先权的所有临时专利申请，如果申请是继续申请，则列出母案申请号。

关于联邦资助研究的声明：在美国，如果发明是根据政府合同进行的，或者是联邦拨款资助研究，则还要求申请人提供"关于联邦资助研究的声明"。

背景技术：本发明的背景技术选定一个技术领域进行讨论，以强调本发明的不同之处，并指出本发明所提供的必要改进。

发明内容：发明内容意在讨论本发明（即权利要求书）而不是整个公开内容。通常，该内容将讨论本发明的优点或其如何解决在本发明背景中提出的问题。

附图说明：如果附图包含在申请中，则需要对每张附图进行简要说明。

具体实施方式：本发明的具体实施方式是充分和准确地描述本发明。实施方式提供了对本发明的解释以及如何实现它。其次，给出了如何实施本发明的具体实施案例。

摘要：摘要是整个说明书的简要概述。

3.3.2　权利要求

"权利要求"是指专利申请人陈述发明的内容。这是专利申请中最重要的部分之一。权利要求放在专利申请[16-18]的权利要求部分。权利要求部分放在摘要前的专利申请结尾附近。权利要求分为两种类型：独立权利要求和从属权利要求。

3.3.2.1 独立权利要求

"独立权利要求"是专利中最广泛的权利要求,并不依赖于任何其他权利要求[1,16-18]。独立权利要求不涉及对其他权利要求的依赖。在独立权利要求的前序部分,独立权利要求以"A"开头。

3.3.2.2 从属权利要求

"从属权利要求"一词与独立权利要求[1]相关。从属权利要求将在从属权利要求的前序部分体现对独立权利要求(或其他从属权利要求)的依赖。此外,从属权利要求以前序部分中的"The"开头。从属权利要求添加了更多的特征,比独立权利要求更具体。在专利申请中,从属权利要求将体现独立权利要求,并被归类在一起。

3.3.3 发明人

"发明人"是一个对发明构想有贡献的个人[1]。美国专利法规定,专利的申请人必须是发明人。发明人通过成为至少一项权利要求的贡献者而对本发明的构想做出贡献。

3.3.4 共同发明人

当一项发明有多个发明人时,贡献者被称为"共同发明人"。在法律上,"共同发明人"是指在至少一项权利要求中,对发明构想做出贡献的个人[1]。如果"共同发明人"为这项发明工作,但没有提出权利要求,那么从技术上讲,他不是一个发明人。这对于在团队环境中工作的工程师来说可能看起来很奇怪。如果团队成员或经理没有直接参与权利要求,则该成员就不是"共同发明人"。

3.3.5 临时申请

根据美国专利法,临时申请是向美国专利商标局提交的法律文件,该申请确立了早期申请日期,但未进入专利公开阶段,除非申请人在一年内提交正式的非临时性专利申请。不存在所谓的"临时专利"[1](见图3-2)。

一项临时申请包括说明书,即发明的实施方式和附图(为了理解想要获得专利权的主题,必要时提供附图),但不包括正式的权利要求、发明人的宣誓或声明或任何信息披露声明(IDS)。此外,由于没有根据现有技术对申请的可专利性进行审查,所以提交临时专利申请的 USPTO 费用明显低于提交标准非临时专利申请的费用。一个临时申请可以为一个或多个持续专利申请确定一个早期有效的申请日期，在持续申请和在前的临时申请具有一个或多个相同发明人的情况下，持续申请可以要求在前的临时申请的优先权日期[1]。

这个词条在过去和现在其他国家的专利法中具有不同的含义。

3.3.6 非临时申请

非临时性专利申请确定申请日期并启动审查程序。非临时性专利申请必须包括说明书、权利要求和附图。此外，申报过程中必须包括声明和申报费用[1]。

3.3.7 分案申请

分案专利申请（有时称为分案申请或简称为分案），包含来自先前提交的申请（所谓的母案申请）的内容。虽然分案申请迟于母案申请，但它可以保留其母案的申请日期，并且通常会申请相同的优先权[1]。

分案申请通常用于母案申请可能缺乏发明单一性的情况。也就是说，母案申请描述了多项发明，并且申请人被要求将母案分成一个或多个分案申请，每份申请只要求一项发明。《保护工业产权巴黎公约》"（简称《巴黎公约》）第 4G 条规定了在发明缺乏单一性的情况下提交分案申请的权利。在美国，分案申请被认为是一种持续专利申请[1]。

3.3.8 持续专利申请

在美国专利法中，持续专利申请是一种专利申请类型，其可以要求在前提交的专利申请的优先权。

持续专利申请可以是以下三种类型：继续申请、分案申请或部分继续申请（CIP）。虽然继续申请和部分继续申请通常仅在美国可用，但其他国家也可以有分案专利申请；《巴黎公约》[1]第 4G 条规定了这种可用性。

3.4 专利语言——状态和操作

在本节中，讨论了有关专利状态和操作的专利语言。重点是审查意见通知书和与意见通知书相关的其他操作。

3.4.1 审查意见通知书

"审查意见通知书"是一封来自专利审查员的确定专利申请法律地位的信函。专利审查员利用"美国专利审查指南（Manual of Patent Examining Procedure，MPEP）"文件来定义美国专利商标局专利审查员，用来判定您的想法是否具有专利性的规则和程序[8]。专利审查员在 MPEP 文件中明确规定了专利审查的可专利性。专利审查员有几种类型的审查意见通知书[1]：

- 修改通知书
- 优先权通知书
- 非最终通知书
- 最终通知书
- 中止审查询问通知书

3.4.1.1 第一次审查意见通知书

"第一次审查意见通知书"是来自专利审查员的一封信，其规定了专利申请的法律状态[1]。这是专利局第一次就专利申请的授权或驳回情况做出回应。

3.4.1.2 最终审查意见通知书

"最终审查意见通知书"是来自专利审查员的一封信，其规定了专利申请的法律状态。这是专利局最后一次就专利申请的授权或驳回情况做出回应。

3.4.2 美国专利审查指南

美国专利审查指南（MPEP）是美国专利商标局专利审查员用来决定您的想法是否具有专利性的规则和程序。专利审查员在 MPEP 文件中陈述了具体的规定，为专利审查的可专利性提供参考[8]。

现有技术的含义如下所述。

"现有技术"一词与公开领域中的内容相关，这些内容在专利审查过程中用于确定专利申请的可专利性。现有技术可以包括已授权的专利、专利申请、期刊文章、会议论文、书籍或文件，专利审查员可以使用这些文件对专利进行审查[1]。

3.4.3 授权

"授权"一词与在专利过程中被评估的权利要求有关,这些权利要求确实满足可专利性的要求。权利要求的授权可能与现有技术缺乏共性有关[1]。在专利审查员的审查过程中,如果确定专利申请中的权利要求具有的内容不属于公开领域,也不属于先前的专利内容,则该权利要求可以被授权。依赖于独立权利要求的从属权利要求也可以因为从属于授权权利要求而被授权。

3.4.4 驳回

"驳回"一词与在专利过程中被评估的权利要求有关,这些权利要求不满足可专利性的要求[1]。权利要求的驳回可能与现有技术存在共性。在专利审查员的审查过程中,如果确定专利申请中的权利要求具有的内容属于公开领域,或者属于先前的专利内容,则该权利要求被驳回。依赖于独立权利要求的从属权利要求也可以因为从属于被驳回的权利要求而被驳回。

驳回通知书的含义如下所述。

"驳回通知书"一词是指在第二次审查意见通知书或第 N 次审查意见通知书后,专利审查员对非临时性专利申请准备结束审查时发出的驳回通知。

申请人对驳回通知书的答复可通过向专利局上诉与干预处理委员会提出申诉予以解决。此外,还可以通过修改权利要求来解决,以符合审查意见通知书提出的建议。

3.4.5 撤回

"撤回"一词与从专利申请流程中删除的未被评估的权利要求相关[1]。在专利申请中,一次只能处理一项发明。专利申请可以包括结构、装置或方法。这些都是独立专利。专利代理人必须选择他/她想要评估的案件,并且必须"撤回"未被评估的权利要求。然后专利代理人要重新提交材料作为一个单独的案件。

3.4.6 授权通知书

授权通知书(NOA)是向专利申请人发出的通知,申请人有权根据法律获得专利[1]。根据相应的地址向申请人发送授权通知书。授权通知书还应指明办登费用和所有的公布费用的总和,其中办登费用和所有的公布费用必须在邮寄 NOA 之日起三个月内支付,以避免申请被放弃;这三个月的期限是不可延长的。

3.4.7 专利未决

"专利未决"(有时缩写为"pat. pend."或者"pat. pending")是指提交产品或方法的专利申请后,在专利授权或申请被放弃之前[1],用于产品或方法的法律标示或表示。该标记用于通知公众、企业或可能复制本发明的侵权人,一旦专利被授权,他们可能要承担损害赔偿(包括追溯赔偿)、扣押和禁令。

3.5 专利草案

本节讨论专利说明书的专利草案和专利草案结构。

3.5.1 专利草案——结构

专利草案具有发明人必须遵循的特定结构。另外,专利说明书的每个部分都有规则。专利草案的结构如图 3-3 所示 [1, 16-18, 22]。

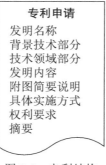

图 3-3 专利结构

3.5.2 专利草案——发明名称

专利草案的名称应该简短具体,并且长度不得超过 500 个字。名称中的单词通常是大写[1]。

3.5.3 专利草案——背景技术部分

背景技术部分是针对专利草案展开的。本节首先介绍发明的技术领域，其次是现有技术，包括已授权的专利、专利申请、出版物和参考文献。

背景技术让读者准备了解专利公开的领域，并讨论了现有技术的优点和缺点[1]。请注意，在背景技术部分中，对发明本身不作讨论。本节讨论的任何内容将被视为现有技术。

3.5.4 专利草案——技术领域部分

技术领域部分通过与本发明领域相关的陈述对说明书背景技术部分进行展开，然后突出显示该领域内更具体的领域。它通常是单个句子[1]。

3.5.5 专利草案——发明内容

在背景技术部分之后，是发明内容部分。在本节中，它包含以下内容[1]：
- 发明的优点
- 发明的对象

发明内容部分讨论了如何解决前一节中讨论的问题，还讨论了现有技术没有解决的问题。

3.5.6 专利草案——附图简要说明

附图简要说明是说明书的一部分，它提供了一个简单的附图列表，其中包含附图的说明，这将在以后的小节中讨论[1]。附图包括现有技术的附图以及将要公开的本发明的附图。附图列表将用于本发明的具体实施方式，以讨论附图。

3.5.7 专利草案——具体实施方式

具体实施方式部分是描述本发明的专利草案的关键部分[1, 22]。该部分提供了本发明的全部细节。本节讨论与现有技术的差异、优点和改进之处。

本节讨论专利草案中各个附图的所有内容。每个附图至少有一个单独的段落。段落通常从指定一个特定的附图开始。此外，图中的所有元素都在讨论中被标识。

重要的是，在权利要求部分中讨论的所有内容都应包含在本发明的具体实施方式中。因此，许多专利草案将"反映"本节的权利要求，以证实它们包含在本节中。

3.5.8 专利草案——权利要求

权利要求部分是专利草案最重要的部分之一。权利要求部分是发明人确定主题的地方，该主题被认为是发明[16-18, 22]。权利要求定义了本发明的范围，在已知和未知之间建立了界限。

权利要求有从属权利要求和独立权利要求之分。每个从属权利要求必须确定与之相关的独立权利要求。通常，从属权利要求与相应的独立权利要求归为一类。所有从属权利要求均应引用独立权利要求。发明的非临时申请必须至少包含一项权利要求。

在撰写权利要求时，要在一份列表中列举每项权利要求。每项权利要求应该是一个句子（以句号结尾）。权利要求中的每个元素都用划分符号（使用分号）分隔。最后一个元素前面有一个分号和"and"。

在方法权利要求中，每个步骤都应该用划分符号来区分。每一步的第一个词应该是一个"主动词"。类似于其他权利要求，方法权利要求的每个步骤以分号结尾，并且以句号结束句子。

每项权利要求的"前序"也非常重要。前序确定权利要求是结构权利要求，装置权利要求还是方法权利要求。

3.5.9 专利草案——摘要

摘要在专利草案中是必需的。一般情况下，它被放置在专利草案的末尾，在权利要求之后。摘要的目的是让专利局和公众确定发明的主题和性质。摘要通常是对专利的简要总结，并指出你的发明中有什么新东西。许多专利代理人都有将独立权利要求置于摘要中的做法。因此，第一个"独立要求"可以放在摘要中[1]。

对于 USPTO 来说，摘要仅限于一个段落，最多不超过 150 个字。在多数情况下，专利现有技术的检索会使用摘要中的关键词。

3.6 总结

第 3 章强调了专利草案本身与专利过程相关的专利语言。本章首先讨论了专利搜索引擎，包括专利网站和 Google 专利。为了与专利代理人、专利审查员和专利局沟通，专利语言是很重要的。本章首先对专利搜索引擎进行了讨论。其次对专利语言进行了讨论，以使人们能够理解专利过程的语言。最后本章还讨论了专利草案的结构。

在下一章中，我们将开始讨论专利，包括不同类型的专利和专利本身的细节。该章将讨论发明专利、外观专利和植物专利。对于每种专利类型，其结构、内容和要求都不相同。讨论内容包括说明书、附图和权利要求部分。

问题

1. 什么样的人是发明家？如何才能成为发明家？
2. 什么是搜索引擎？
3. 你能用专利局的网站作为搜索引擎吗？
4. 什么是说明书？
5. 什么是权利要求？
6. 权利要求的类型有哪些？权利要求是如何构成的？
7. 什么是授权？
8. 什么是驳回？
9. 什么是异议？
10. 什么是撤回？
11. 什么是授权通知书？
12. 有多少权利要求被接受才能授予专利？
13. 什么是专利摘要？
14. 美国专利申请有哪些部分？
15. 描述美国专利的各个部分。

案例研究

案例研究 A

针对 MOSFET 设备撰写一组独立权利要求和从属权利要求。MOSFET 具有源极、漏极、栅极电介质、栅极叠层、间隔物、井和衬底。附图中的每种形状都要包含这些元素。

画一张附图，然后撰写权利要求。间隔物可以添加到从属权利要求中。

撰写 MOSFET 设备的发明专利，包括本发明的标题、背景技术、技术领域、具体实施方式、权利要求和摘要。

案例研究 B

针对双极面结型晶体管（BJT）设备撰写一组独立权利要求和从属权利要求。BJT 具有发射极、基极和集电极以及衬底。附图中的每种形状都要包含这些元素。在一组从属权利要求中加入这些部分的连接关系。

撰写 BJT 设备的发明专利，包括本发明的标题、背景技术、技术领域、具体实施方式、权利要求和摘要。

案例研究 C

为包括阳极和阴极的 p-n 二极管的发明专利撰写权利要求。在索赔中包括井区。写出独立权利要求和从属权利要求。

撰写 p-n 二极管的发明专利，包括本发明的标题、背景技术、技术领域、具体实施方式、权利要求和摘要。

参考文献

1. U.S. Patent Office (USPTO). https://www.uspto.gov (accessed 21 December 2017).

2. European Patent Office (EPO). https://www.epo.org (accessed 21 December 2017).

3. World Intellectual Property Organization (WIPO). https://www.wipo.int (accessed 21 December 2017).

4. Organisation Africaine de la Propriete Intellectuelle (OAPI). https://www.oapi.int (accessed 21 December 2017).

5. African Regional Intellectual Property Organization (ARIPO). https://www.airpo.org (accessed 21 December 2017).

6. Pat2PDF Search Engine. https://www.pat2pdf.org (accessed 21 December 2017).

7. Google Search Engine. https://www.google.com/patents (accessed 21 December 2017).

8. USPTO (1883). Manual of patent examination procedures (MPEP). https:www.uspto.gov/web/offices/pac/mpep (accessed 21 December 2017).

9. U.S. Department of Commerce (2000). *Patents and How to Get One: A Practical Handbook*. US

Department of Commerce, Courier Corporation.

10. Stim, R. (2016). *Patent, Copyright, and Trademark: An Intellectual Property Desk Reference*. Nolo. ISBN: 978-1-4133-2221-7.

11. Slusky, R. (2013). *Invention Analysis and Claiming: A Patent Lawyer's Guide*, 2e. American Bar Association. ISBN: 13 978-1614385615.

12. Rosenberg, M. (2016). *Essentials of Patent Claim Drafting*, LexisNexis IP Law and Strategy Series. Matthew Bender.

13. Adams, D.O. (2015). *Patents Demystified: An Insider's Guide to Protecting Ideas and Invention*. American Bar Association. ISBN: 13 978-163425679.

14. Pressman, D. and Tuytschaevers, T. (2016). *Patent It Yourself: Your Step-by-Step Guide to filing at the U.S. Patent office*. Nolo.

15. Charmsson, H.J.A. and Buchaca, J. (2008). *Patents, Copyrights, and Trademarks for Dummies*. Wiley.

16. Voldman, S. (2014). Short Course, *Innovating, Inventing, and Patenting*, Dr. Steven H. Voldman LLC, Ministry of Science Technology and Innovation (MOSTI), Putrajaya, Malaysia, May 2014.

17. Voldman, S. (2015). Short Course, *Innovating, Inventing, and Patenting*, Dr. Steven H. Voldman LLC, FITIS, Sri Lanka, February 2015.

18. Voldman, S. (2016). Short Course, *Writing and Generating Patents*, Dr. Steven H. Voldman LLC, FITIS, Sri Lanka, February 2016.

19. DeMatteis, B., Gibb, A., and Neustal, M. (2006). *The Patent Writer: How to Write Successful Patent Applications*. Garden City Park, NY: Patents for Commerce, Square One Publishers.

20. Stim, R. and Pressman, D. (2015). *Patent Pending in 24 Hours*, 7e. Nolo. ISBN: 978-1-4133-2201-9.

21. Lo, J. and Pressman, D. (2015). *How to Make Patent Drawings*, 7e. Nolo. ISBN: 978-1-4133-2156-2.

22. Amernick, B.A. (1991). *Patent Law for the Nonlawyer: A Guide for the Engineer, Technologist, and Manager*, 2e. Von Nostrand Reinhold. ISBN: 13 978-0442001773.

23. Durham, A.L. (2013). *Patent Law Essentials: A Concise Guide*, 4e. Oxford: Praeger, ABC-CLIO, LLC. ISBN: 13 978-1440828782.

24. Mueller, J.M. (2016). *Patent Law*, 5e. Wolter Kluwer Publications.

25. Sutton, E. *Software Patents: A Practical Perspective*, 2016. CreateSpace Independent Publishing Platform.

26. WIPO *Paris Convention for the Protection of Industrial Property*. World Intellectual Property Organization (WIPO). https://www.wipo.int (accessed 21 December 2017).

第 4 章 专　利

4.1　引言

本章主要研究专利和专利结构[1-26]。对不同类型的专利进行了讨论,重点介绍了它们之间的不同之处。

4.2　专利类型

世界各地的专利局提供不同的专利申请。在美国专利商标局(USPTO)中,提供了三种类型的专利。它们是发明专利、外观专利和植物专利[1, 9, 10](如图4-1所示)。这些专利将在下面的章节中进行讨论,重点介绍它们之间的不同之处。

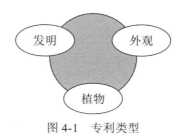

图4-1　专利类型

4.2.1 发明专利

发明专利是一项保护有用的步骤、机器、制造物品和物质组成的专利[1,10,21-23]。发明专利可以是结构、装置或方法。在科学和工程领域，大多数专利申请都是发明专利。在电气工程中，发明专利可以是半导体器件、电路和系统（如图4-2所示）。

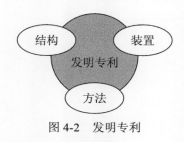

图4-2　发明专利

4.2.2 外观专利

外观专利是保护外观的专利。外观专利是一种保护新颖独特外观的知识产权[1]。外观专利可以帮助发明人在专利保护期内通过阻止竞争对手复制外观设计来获得更高的利润。外观可以是手机、汽车车身、容器或瓶子的任何部分。外观专利包含外观的所有不同视角。外观专利由"D××××"代表其专利号。"D"表示外观专利。在美国，外观专利是由USPTO授予发明者或发明者的继承人的[1]。

4.2.3 植物专利

植物专利是给定的新植物的专利。植物专利由"PP×××"代表其专利号。"PP"表示植物专利。植物专利是一种知识产权，可以保护新颖独特的植物的关键特征不被他人复制、再生产、出售或使用[1]。植物专利可以通过阻止竞争者使用植物来帮助发明人在专利保护期内获得更高的利润。在美国，植物专利是由USPTO授予发明人或发明人的继承人的[1]。

除了一些特殊情况，植物专利申请与发明专利申请的要求相同。《联邦政府管理条例》第37条第1.163（a）节要求，说明书必须尽可能完整地包含有关植物的植物学描述，以及区别于已知相关植物的特征[1]。植物专利申请的组成部分类似于发明专利申请。

4.3 发明专利的专利结构

在本节中,讨论了发明专利的专利结构。发明专利的专利结构详见图 4-3 [1, 10, 21-23]。

图 4-3 发明专利结构

在这些部分中,每个部分都有相应的指南说明和规则。

4.3.1 发明专利——名称

专利草案的名称应简短明确。名称长度必须少于 500 个字符。名称中的单词通常是大写的 [1]。

4.3.2 发明专利——背景技术部分

背景技术部分是针对专利草案展开的。本部分首先介绍本发明的技术领域,其次是现有技术,包括已授权的专利、专利申请、出版物和参考文献 [1]。

背景技术让读者准备了解专利公开的领域,并讨论了现有技术的优点和缺点。请注意,在本背景技术部分中,对本发明本身不作讨论。本节讨论的所有内容都将被视为现有技术。

4.3.3 发明专利——技术领域

技术领域部分通过对本发明领域相关的陈述来对说明书背景技术部分进行展开,然后突出显示该领域内更具体的领域。它通常是单个句子 [1, 10]。

4.3.4 发明专利——发明内容

在背景技术部分之后,是发明内容部分。在本部分中,它包含以下内容:

- 发明的优点
- 发明的对象

发明内容部分讨论了如何解决前一部分中讨论的问题,并讨论了现有技术没有解决的问题。

4.3.5 发明专利——附图简要说明

在附图简要说明中,提供了附图列表。在背景技术部分中的现有技术中的附图以及与具体实施方式相关的附图都包含在该列表中[1]。

每个附图的格式以"FIG.X"开头,其中 X 是附图编号。每个列出的附图用分号结尾,而最后一个附图以句子结尾处的句号结束。

4.3.6 发明专利——具体实施方式

具体实施方式部分是描述本发明的专利草案的关键部分。该部分提供了本发明的全部细节。本部分讨论与现有技术的差异、优点和改进之处。

本部分讨论专利草案中各个附图的所有内容。每个附图至少有一个单独的段落。段落通常从指定一个特定的附图开始。此外,图中的所有元素在讨论中都能被标识。

重要的是,在权利要求部分中讨论的所有内容都应包含在本发明的具体实施方式中。因此,许多专利草案将"反映"本节的权利要求,以证实它们包含在本节中。

4.3.7 发明专利——附图

在发明专利中,交底书中的现有技术和本发明都包括在附图中。现有技术附图以及与本发明相关的附图都包含在专利的说明书中。现有技术的附图用于提供公开领域中已知的现有附图和发明新公开内容之间的比较。附图用于提供比较并突出差异[1]。

4.3.7.1 附图——现有技术

现有技术的附图通常在专利申请的背景技术部分中讨论。因此,这些附图通常放在附图列表中的前面。现有技术的附图通常在图号下标记为"现有技术"。这些附图在图本身下方被指定为"图××"。

4.3.7.2 附图——发明

专利的实施案例包含在专利申请的具体实施方式讨论的附图中。因此,这些附图通常放在附图列表列出的现有技术附图之后。这些附图在图本身下方被指定为"图××"。

所有附图中的物理形状需要能够被识别出来。形状用数字线或"波浪线"标识,并将数字连接到物理形状。

4.3.8 发明专利——权利要求

专利草案最重要的部分之一是权利要求部分。权利要求部分是发明人确定主题的地方,该主题被认为是发明。权利要求定义了本发明的范围,在已知和未知之间建立了界限[10, 21-23]。

权利要求有从属权利要求和独立权利要求之分。每个从属权利要求必须确定与之相关的独立权利要求。通常,从属权利要求与相应的独立权利要求归为一类。所有从属权利要求均应引用独立权利要求。发明的非临时申请必须至少包含一项权利要求。

在撰写权利要求时,在一份列表中列举每项权利要求。每项权利要求应该是一个句子(以句号结尾)。权利要求中的每个元素都用划分符号(使用分号)分隔。最后一个元素前面有一个分号和"and"[1]。

在方法权利要求中,每个步骤都应该用划分符号来区分。每一步的第一个词应该是一个"主动词"。 类似于其他权利要求,方法权利要求的每个步骤以分号结尾,并且以句号结束句子。

每项权利要求的"前序"非常重要。前序确定权利要求是结构权利要求、装置权利要求还是方法权利要求。

4.3.9 发明专利——摘要

在发明专利中,专利申请要求有一份摘要。专利申请的摘要必须少于 150 字[1]。

4.4 外观专利

为知识产权申请专利的一种方法是申请外观专利。外观专利可防止他人抄袭外观并伪造产品。本节将讨论外观专利。

4.4.1 外观专利——专利名称设计

外观专利的专利号与发明专利不同。外观专利可以通过其第一个字母"D"来识别。

接下来是外观专利发行编号。例如，苹果电脑外观专利被指定为"D762 208"[26]。

4.4.2 外观专利——名称

外观专利的名称简短且具有描述性。例如，外观专利 D762 208 是"具有图形用户界面的便携式显示设备"[26]。

4.4.3 外观专利——发明人

发明人和共同发明人都会列在专利上。一旦确定了声明书中的宣誓书，发明人的名字就会被列在专利上。在外观专利 D762 208 中，它被列为"Akana, et al."[26]。表 4-1 是本外观专利的发明人列表。名字都是以姓开头的，有时只用中间名的首字母。它还列出了在提交专利申请时发明者的居住地[26]。

表 4-1 发明人列表

Inventors	Akana; Jody (San Francisco, CA), Andre; Bartley K. (Palo Alto, CA), Coster; Daniel J. (San Francisco, CA), Cranfill; Elizabeth Caroline (San Francisco, CA), De Iuliis; Daniele (San Francisco, CA), Hankey; M. Evans (San Francisco, CA), Howarth; Richard P. (San Francisco, CA), Inose; Mikio (Cupertino, CA), Ive; Jonathan P. (San Francisco, CA), Jobs; Steven P. (Palo Alto, CA), Kerr; Duncan Robert (San Francisco, CA), Lemay; Stephen O. (San Francisco, CA), Nishibori; Shin (Kailua, HI), Rohrbach; Matthew Dean (San Francisco, CA), Russell‐Clarke; Peter (San Francisco, CA), Stringer; Christopher J. (Woodside, CA), Whang; Eugene Antony (San Francisco, CA), Zorkendorfer; Rico (San Francisco, CA)

有些公司按字母顺序列出发明人的名字。这有两个目的。首先，它可以防止就谁应该是第一个发明者这个问题引起的冲突。其次，鉴于专利进入专利诉讼，法院不知道谁是主要发明人。

4.4.4 外观专利——申请人和受让人

在专利申请中，指定了申请人和受让人。在本节中，表 4-2 列出了申请人名称、城市、州和国家。请注意，此部分还包含专利申请号以及申请日期。

表 4-2 专利申请信息

申请人	名称	城市	州	国家	类型
	苹果	库比蒂诺	加州	美国	
申请号	D/491,128				
申请日期	2014 年 5 月 16 日				

4.4.5 外观专利——参考文献

对于美国专利申请，美国专利文献列表能提供参考文献信息。列表中，列出了专利号、授权日期和发明人名字。在上述专利（如 D762 208）中，同一发明人以及其他发明人的外观专利有许多页。

4.4.6 外观专利——外国专利文献

对于美国专利申请，美国专利商标局还列出了外国专利文献。列表中，列出了专利号、授权日期和来源国。在上述专利（如 D762 208）中，列举了许多来自不同国家的外观专利。

4.4.7 外观专利——其他参考文献

对于美国专利申请，还提供了"其他参考文献"的列表。列表中，它可以是公开领域中与参考书籍、参考文献、新闻稿、文章和网站相关的任何内容。

4.4.8 外观专利——简要说明

对于外观专利，简要说明与发明专利的说明书不同。简要说明主要是对将在附图中展示的各种视图进行的描述。专利 D762 208 的一个示例说明（如图 4-4 所示）[26] 如下。

对于外观专利，它的重点是描述外观的图形。

```
说明书
    图 1 是一种在我们新设计中展示的带有
图形用户界面的便携式显示设备的正视图；
    图 2 是后视图；
    图 3 是顶视图；
    图 4 是左侧视图；
    图 5 是右侧视图；
    图 6 是底视图；
    附图中的虚线表示未要求保护的附图，
其展示了带有图形用户界面的便携式显示设
备的一部分，其并不是外观设计要求的组成
部分。图 1 中的阴影线表示透明度，并不是
表面装饰。
```

图 4-4　外观设计说明书

4.4.9 外观专利——附图

外观专利的重点是可视化描述外观的图形。外观应包含顾及完整描述外观的所有视图，以允许将外观转移到公开领域[1]。这包括后视图、顶视图、底视图、左侧视图和右侧视图。对于圆形或对称图形，可以用顶视图、底视图和侧视图充分描述外观专利。

4.4.10 外观专利——权利要求

外观专利的权利要求部分与发明专利的权利要求明显不同。由于该专利主要是附图，因此没有类似于发明专利或植物专利所用风格的权利要求。例如，在权利要求部分中，外观专利 D762 208 的文本可以如下所述[26]（如图 4-5 所示）。

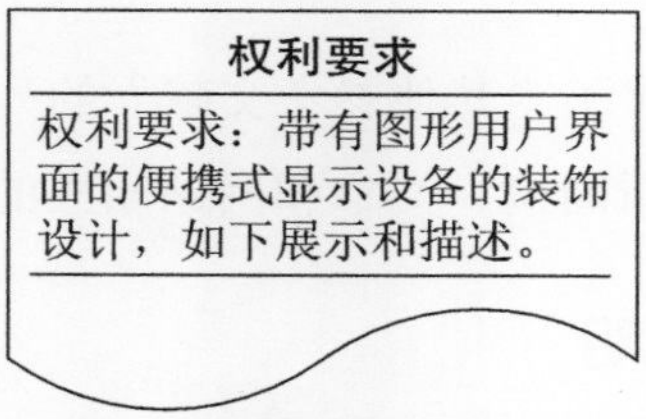

> **权利要求**
> 权利要求：带有图形用户界面的便携式显示设备的装饰设计，如下展示和描述。

图 4-5 外观设计权利要求

外观专利的权利要求部分指出它是一种外观，以及其外观结构或功能是什么。这又与描述发明的发明专利有很大的不同。

4.5 植物专利

可申请专利的植物可以是天然的、培育的或体细胞培养的（由植物的非生殖细胞产生）。它可以被发明或发现，但如果该发现是在栽培区域进行的，植物专利将仅授予发现植物专利。该植物必须是无性繁殖的，并且繁殖必须在遗传上与原始植物相同，并通过诸如根插条、鳞茎、分裂或嫁接和芽接等方法进行，以确定植物的稳定性[1]。

这种植物可以是藻类或大真菌，但细菌不具备资格。块茎，如土豆和耶路撒冷洋蓟，没有资格申请植物专利，因为不是由于生长条件或土壤肥力而独特的植物[1]。

与其他发明一样，植物必须是非显而易见的才有资格获得专利。另一种专利——发明专利，适用于某些植物、种子和植物繁殖过程。

发明人在销售或公开植物的一年内可以申请植物专利。美国专利商标局会授予植物专利，前提是发明人提供完整、完善的植物学描述，阐释植物的独特性，并包括展示植物独特特征的附图。申请人还必须遵守专利申请的其他详细要求，并支付相关费用。

一个植物专利可以有两个发明人：一个发现植物的发明人，一个无性繁殖的发明人。如果发明是一个团队的努力，那么团队中的每个成员都可以作为共同发明人[1]。

虽然植物专利保护发明人的知识产权是从申请专利的日期起 20 年，但是专利申请在最早的专利申请日期之后的 18 个月就公开了，这意味着竞争对手将能更早地了解发明的细节[1]。植物专利具有与发明专利相似的结构（如图 4-6 所示）。

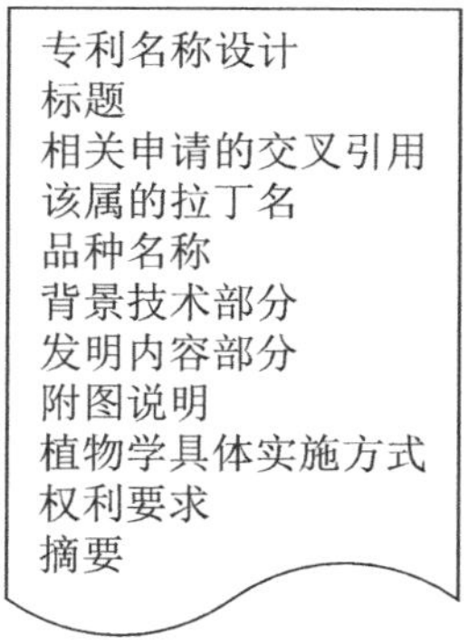

```
专利名称设计
标题
相关申请的交叉引用
该属的拉丁名
品种名称
背景技术部分
发明内容部分
附图说明
植物学具体实施方式
权利要求
摘要
```

图 4-6　植物专利结构

除了申请植物专利外，发明人可能还需要申请发明专利或外观专利才能完全保护植物。例如，如果新的植物品种具有独特的外观，那么发明人既要申请植物专利，又要申请外观专利。

4.5.1　植物专利——专利名称设计

对于植物专利，植物名称的启用必须能标示这一新植物。标题的名称可以是植物的名称。

第一个植物专利，植物专利 1，于 1930 年 8 月 6 日提交，于 1931 年 8 月 18 日授权。请注意，这明显晚于结构专利（例如，在 1791 年授权）。发明人伯森伯格[27]发明的这项专利名称为"Climbing or Trailing Rose"。然而，在文中，发明人将玫瑰命名为"新黎明"。

4.5.2 植物专利——标题

本发明的标题可以包括一个导言部分，说明申请人姓名、公民身份和居住地介绍。

4.5.3 植物专利——相关申请的交叉引用

在植物专利中，相关的申请应该交叉引用[1]。包括以下内容：
- 一种发明申请，其中要求保护的植物是分案申请的主题
- 当亲本申请未被允许进入亲缘品种时，对同一植物的继续申请（同时待审的申请，新提交的申请）
- 与未被授权的原始申请没有同时待审的申请
- 由同一育种计划开发的亲缘或类似植物的待审申请

4.5.4 植物专利——该属的拉丁名

在植物专利中，重要的是指出该属的拉丁名称以及所要保护的植物的种类。

4.5.5 植物专利——品种名称

在植物专利中，还应在专利申请中确定品种名称。

4.5.6 植物专利——背景技术部分

在本发明的背景技术部分中，公开了发明领域和现有技术的描述。

发明领域旨在确定本发明的植物学和市场类别，并反映该植物将如何使用[1,10,21-23]。本部分应按科属和种类说明植物的学名，并应说明植物的市场类别[1]。

在该背景技术部分中，讨论了要求保护的植物的亲本或所涉及的已知植物。亲本植物在背景部分中鉴定，并描述其最重要或最显著的特征。可以将要求保护的植物与亲本植物进行比较。宣布一种新的植物时，区分亲本和新品种的特性是很重要的。如果不知道亲本植物，可能的亲本或亲本植物会被认为是新品种。

本节详细说明植物是如何获得的。还应详细说明植物无性繁殖的方式和地点。该背景部分还必须包括声明所要求保护植物的克隆或繁殖体的所有特征与原始植物相同，这是为了确定要求保护的植物是稳定的。

4.5.7 植物专利——发明内容部分

在发明内容部分，披露了植物的主要特征。特征可以表示为[1]：

- 一系列新颖的特征；
- 对特征的叙述性描述；
- 植物在其他植物学类别和市场类植物中独一无二的植物特性。

对于早期的专利，格式有很大不同。

这是来自美国植物专利1[27]的说明书。

> "My invention relates to improvements in roses of the type known as climbing or trailing roses in which the central or main stalks acquire considerable length and when given moderate support "climb" and branch out in various directions.
> In roses it is very desirable to have a long period of blooming. This has been acquire in non-climbing roses of the type ordinarily called monthly roses or everblooming roses.
> My invention now gives the true everblooming character to climbing roses.
> The following description and accompanying illustrations apply to my improvements upon the well known variety Dr. Van Fleet, with which my new plant is identical as respects color and form of flower, general climbing qualities, foliage and hardiness, but from which it differs radically in flowering habits—but the same everblooming habits may be attained by breeding this new quality into other varieties of climbing roses."[1]

因此，在早期专利中，这些要求明显低于后来的专利。

4.5.8 植物专利——附图说明

在附图说明部分中，应该提供单独的简要描述来描述各个视图或附图的内容。一项植物专利的附图必须以足够的比例展示出该植物最显著的特征，以便在减少多达50%比例时仍然可识别。该比例应足够大，以区分与其亲本植物不同的植物特性[1]。

附图应该是照片形式的，而且必须是彩色的，因为着色是一个显著特征。在发明专利中，通常不允许使用照片，并且只在必要时才使用颜色。在植物专利中，照片和着色更为常见。在早期的专利中，包括植物的附图，而不是照片[1]。

在植物专利中，区别于母体植物的特征很重要。如果叶子、树皮、花或果实的特征是有区别的，植物的这些部分应该在一个或多个附图中清楚地描绘出来。除非审查员明确要求，否则图中的数字无须编号。在植物专利中，必须放置附图以满足与发明申请中的附图相同的要求。

4.5.9 植物专利——植物学具体实施方式

本说明书中植物学具体实施方式提供了完整的植物学描述。该说明书如图4-7所示[1]。

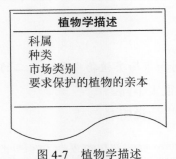

图4-7 植物学描述

该植物的特征如图4-8所示[1]。

图4-8 植物特征

不能明确书面描述或不能清楚显示的植物特征必须在申请的这部分给予实质性的着重关注。这些包括但不限于[1]以下内容：

- 香味
- 味道
- 抗病性
- 生产力
- 早熟性
- 活力

即使特征被很好地描述，也必须对植物学特征进行实质性描述。本节中的描述应该是自然植物学的，并且应该按照植物艺术方面进行描述。

本节的细节应足以防止其他人通过简单地描述植物的更多细节并指称原始专利没有进一步描述特征而试图为同一植物申请专利。

4.5.10 植物专利——权利要求

植物专利仅限于一项权利要求[1]。这与发明专利非常不同。权利要求应按照所示和所述的植物的正式格式进行撰写，因为权利要求是按照法律规定向植物整体提出的。权利要求还可以参考植物的一种或多种不寻常的特征，但是不可以请求保护植物的部分或产品。权利要求必须是单句形式。

在早期专利中，发明人习惯于在文件上提供证词和签名。在第一个植物专利 USPP1 中，权利要求内容如下[27]（如图 4-9 所示）。

一种蔷薇于此所示，其特征在于四季开花的习性。我在此签名为证。

享利·F. 博森尹伯格

权利要求：
一种蔷薇于此所示，其特征在于其四季开花的习性。

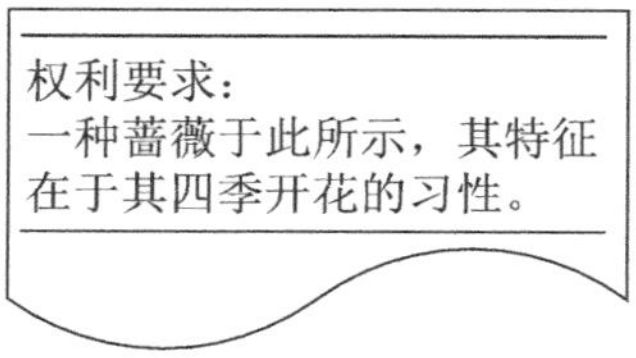

图 4-9　第一个植物专利 USPP1 的权利要求实例

对于植物专利，名为"Lois"的锗植物，USPP10000，一般特征和权利要求如下所示[28]。

GENERAL CHARACTERISTICS

Lois adds a new color pattern to the Regal geranium group. The interesting blotching and veining gives Lois' flowers a multi - color presentation. Lois exhibits ease and quickness of rooting, good cutting production, a controlled plant habit, rich medium green foliage and longer post - harvest production, all of which are unique to Lois among Regal geraniums. Given all these positive properties, this variety should please both the grower and the gardener.
I claim:
1. A new and distinct variety of Geranium plant, substantially as shown and described.

4.5.11 植物专利——摘要

摘要是植物的简要描述，以及所要求保护的植物的最显著或最新颖和最重要的特征。植物的描述是对植物的最显著特征的简要回顾或呈现。

对于植物专利，一种名为"Lois"的锗植物，USPP10000，摘要如下[28]（见图4-10）。

摘要

该品种特征在于，其独特的多色彩的呈现形式。它的顽强生长特性以及持续开花可以提供出色的插枝，也可以在潮湿天气中快速恢复。花期相对于其他常规品种更长一些。

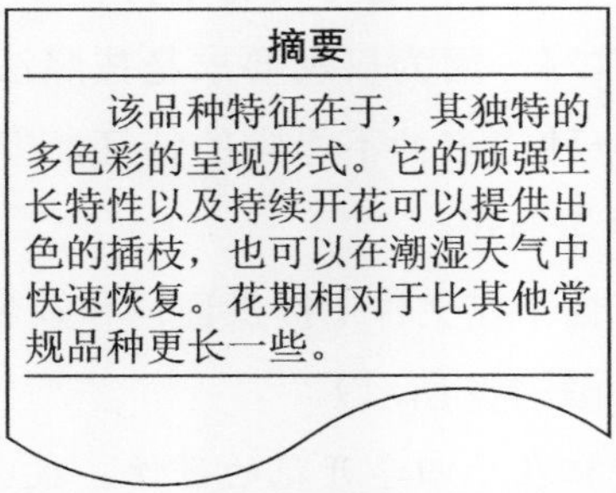

图 4-10　植物专利摘要的实例

4.6　专利示例

在该部分中，给出了结构、装置和方法的示例。美国专利 No.7 173 310 展示了用于未来技术的 FinFET 器件的结构和方法。美国专利 No.6 549 061 描述了一种用于半导体器件中的静电放电保护装置。美国专利 No.6 762 918 是无故障熔丝网络的示例（见图 4.A.1-4.A.3）。

4.7　总结

第 4 章阐述了发明专利、外观专利和植物专利三种不同类型专利的内容和区别，这三种专利对于每种情况，其结构、内容和要求都是不同的。讨论内容包括说明书、附图和权利要求部分的内容。

在第 5 章，重点将放在专利附图上，将给出结构、装置和方法附图的示例。并展示外观专利附图和植物专利附图，讨论专利附图的规则和要求，以及与拒绝接受附图相关的一些问题。

问题

1. 什么是发明专利？
2. 什么是外观专利？
3. 什么是植物专利？
4. 发明专利与外观专利有什么区别？它们有什么不同？
5. 发明专利的权利要求有什么要求？
6. 外观专利是否必须有权利要求？请描述。
7. 植物专利中的附图有什么规则？
8. 一项发明可以有多种类型的专利吗？
9. 发明专利与植物专利有什么区别？
10. 外观专利的附图有什么要求？
11. 发明专利的摘要有什么要求？植物专利有什么要求？外观专利有什么要求？
12. 您是否需要专利申请的发明模型？
13. 发明专利、外观专利和植物专利的附图有什么区别？
14. 画一张发明专利附图。
15. 画一张方法专利附图。
16. 画一张植物专利附图。
17. 画一张外观专利附图。
18. 您在哪里可以找到在美国提交专利的规则？
19. 您在哪里可以找到在欧洲提交专利的规则？

案例研究

案例研究 A

提供了一个跑车的发明专利。显示跑车的基本部件。

案例研究 B

提供了一个跑车的外观专利。从不同的角度展示外观专利（如前部、后部、侧面）。

参考文献

1. U.S. Patent Office (USPTO). https://www.uspto.gov (accessed 20 December 2017).
2. European Patent Office (EPO). https://www.epo.org (accessed 20 December 2017).
3. Japan Patent Office. https://www.jpo.go.jp (accessed 20 December 2017).
4. Malaysian Patent Office (MyIPO). https://www.myipo.gov/my (accessed 20 December 2017).
5. World Intellectual Property Organization (WIPO). https://www.wipo.int (accessed 20 December 2017).
6. Organisation Africaine de la Propriete Intellectuelle (OAPI). https://www.oapi.int (accessed 20 December 2017).
7. African Regional Intellectual Property Organization (ARIPO). https://www.airpo.org (accessed 20 December 2017).
8. U.S. Department of Commerce (2000). *Patents and How to Get One: A Practical Handbook*. U.S. Department of Commerce Courier Corporation.
9. Pressman, D. and Tuytschaevers, T. (2016). *Patent It Yourself: Your Step-by-Step Guide to filing at the U.S. Patent office*. Nolo.
10. Charmsson, H.J.A. and Buchaca, J. (2008). *Patents, Copyrights, and Trademarks for Dummies*. Wiley.
11. Adams, D.O. (2015). *Patents Demystified: An Insider's Guide to Protecting Ideas and Invention*. American Bar Association. ISBN: 13 978-163425679.
12. Stim, R. and Pressman, D. (2015). *Patent Pending in 24 Hours*, 7e. Nolo. ISBN: 978-1-4133-2201-9.
13. Amernick, B.A. (1991). *Patent Law for the Nonlawyer: A Guide for the Engineer, Technologist, and Manager*, 2e. Von Nostrand Reinhold. ISBN: 13 978-0442001773.
14. DeMatteis, B., Gibb, A., and Neustal, M. (2006). *The Patent Writer: How to Write Successful Patent Applications*. Garden City Park, NY: Patents for Commerce, Square One Publishers.
15. Sutton, E. *Software Patents: A Practical Perspective*, 2016. CreateSpace Independent Publishing Platform.
16. Mueller, J.M. (2016). *Patent Law*, 5e. Wolter Kluwer Publications.
17. Stim, R. (2016). *Patent, Copyright, and Trademark: An Intellectual Property Desk Reference*. Nolo. ISBN: 978-1-4133-2221-7.
18. Slusky, R. (2013). *Invention Analysis and Claiming: A Patent Lawyer's Guide*, 2e. ABA Book Publishing. ISBN: 13 978-1614385615.
19. Rosenberg, M. (2016). *Essentials of Patent Claim Drafting*, LexisNexis IP Law and Strategy Series. Matthew Bender.

20. Lo, J. and Pressman, D. (2015). *How to Make Patent Drawings*, 7e. Nolo. ISBN: 978-1-4133-2156-2.

21. Voldman, S. (2014). Short Course, *Innovating, Inventing, and Patenting*, Dr. Steven H. Voldman LLC, Ministry of Science Technology and Innovation (MOSTI), Putrajaya, Malaysia, May 2014.

22. Voldman, S. (2015). Short Course, *Innovating, Inventing, and Patenting*, Dr. Steven H. Voldman LLC, FITIS, Sri Lanka, February 2015.

23. Voldman, S. (2016). Short Course, *Writing and Generating Patents*, Dr. Steven H. Voldman LLC, FITIS, Sri Lanka, February 2016.

24. Durham, A.L. (2013). *Patent Law Essentials: A Concise Guide*, 4e. Oxford: Praeger, ABC-CLIO, LLC. ISBN: 13 978-1440828782.

25. Jackson Knight, H. (2013). *Patent Strategy for Researchers and Research Managers*, 3rde. Chichester, England: Wiley.

26. Akana et al. (2016). Portable display device with graphical user interface. US Patent D762, 208, 26 July, 2016.

27. H.F Bosenberg (1931). Climbing or trailing rose. USPP1P, 18 August 1931.

28. D. Lemon (1997). Germanium plant named 'Lois'. USPP10000P, 12 August 1997.

附录 4.A

US007173310B2

(12) United States Patent
Voldman et al.

(10) Patent No.: US 7,173,310 B2
(45) Date of Patent: Feb. 6, 2007

(54) **LATERAL LUBISTOR STRUCTURE AND METHOD**

(75) Inventors: **Steven H. Voldman**, South Burlington, VT (US); **Jack A. Mandelman**, Flat Rock, NC (US)

(73) Assignee: **International Business Machines Corporation**, Armonk, NY (US)

(*) Notice: Subject to any disclaimer, the term of this patent is extended or adjusted under 35 U.S.C. 154(b) by 0 days.

(21) Appl. No.: **10/908,961**

(22) Filed: **Jun. 2, 2005**

(65) **Prior Publication Data**
US 2006/0273372 A1 Dec. 7, 2006

Related U.S. Application Data

(63) Continuation of application No. PCT/US02/38546, filed on Dec. 3, 2002.

(51) Int. Cl.
H01L 27/01 (2006.01)
H01L 27/12 (2006.01)
H01L 29/00 (2006.01)
H01L 31/0392 (2006.01)

(52) U.S. Cl. 257/350; 257/347; 257/401; 257/449; 257/536

(58) Field of Classification Search 257/347, 257/350, 401, 449, 536
See application file for complete search history.

(56) **References Cited**

U.S. PATENT DOCUMENTS

4,037,140 A	7/1977	Eaton, Jr.	
4,282,556 A	8/1981	Ipri	
5,287,377 A *	2/1994	Fukuzawa et al.	372/45.01
5,610,790 A	3/1997	Staab et al.	
5,616,944 A	4/1997	Mizutani et al.	
6,404,269 B1 *	6/2002	Voldman	327/534
6,413,802 B1	7/2002	Hu et al.	
6,894,324 B2 *	5/2005	Ker et al.	257/199
6,967,363 B1 *	11/2005	Buller	257/288
7,064,413 B2 *	6/2006	Fried et al.	257/536
2002/0109253 A1	8/2002	Sanyal	
2004/0105203 A1 *	6/2004	Ker et al.	361/56
2004/0159910 A1 *	8/2004	Fried et al.	257/536
2004/0188705 A1 *	9/2004	Yeo et al.	257/170
2005/0035410 A1 *	2/2005	Yeo et al.	257/355
2005/0035415 A1 *	2/2005	Yeo et al.	257/401

* cited by examiner

Primary Examiner—Kenneth Parker
Assistant Examiner—Jesse A. Fenty
(74) *Attorney, Agent, or Firm*—Joseph P. Abate

(57) **ABSTRACT**

An ESD LUBISTOR structure based on FINFET technology employs a vertical fin (a thin vertical member containing the source, drain and body of the device) in alternatives with and without a gate. The gate may be connected to the external electrode being protected to make a self-activating device or may be connected to a reference voltage. The device may be used in digital or analog circuits.

14 Claims, 10 Drawing Sheets

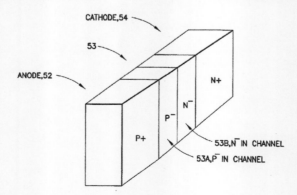

Figure 4.A.1

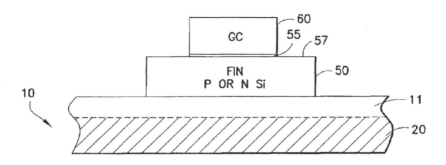

FIG.1A

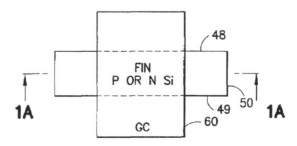

FIG.1B

Figure 4.A.1 (*Continued*)

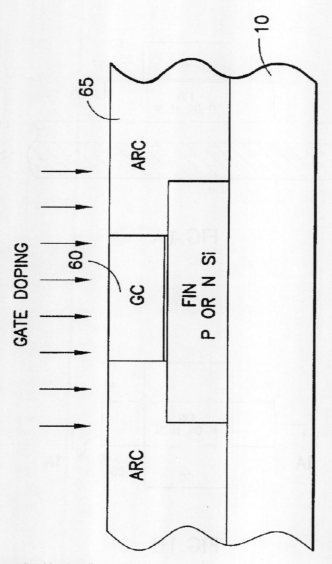

Figure 4.A.1 (Continued)

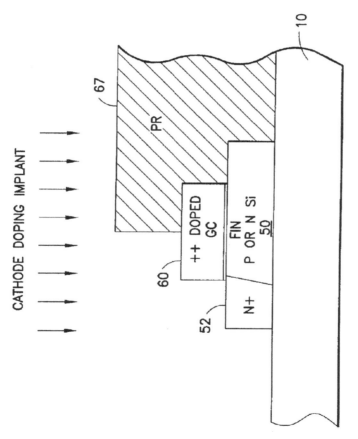

Figure 4.A.1 (Continued)

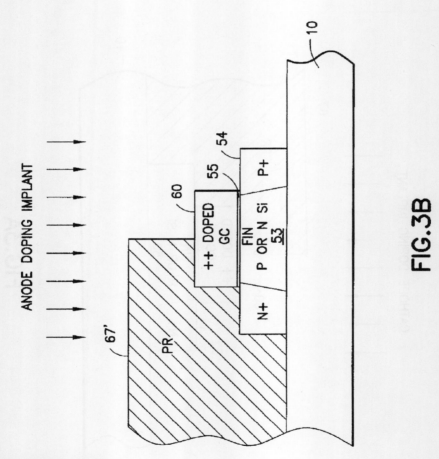

Figure 4.A.1 (*Continued*)

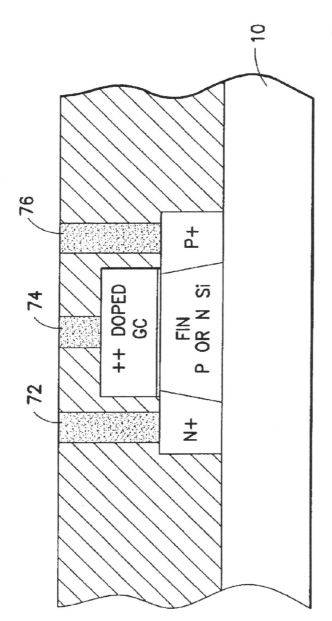

Figure 4.A.1 (Continued)

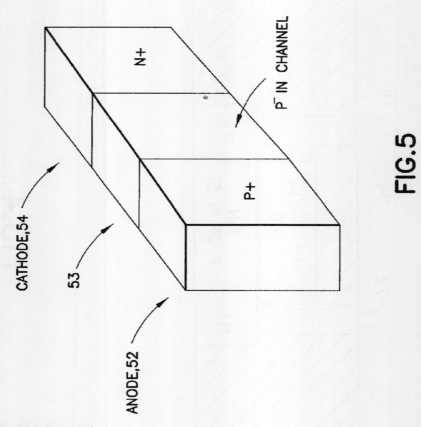

Figure 4.A.1 (Continued)

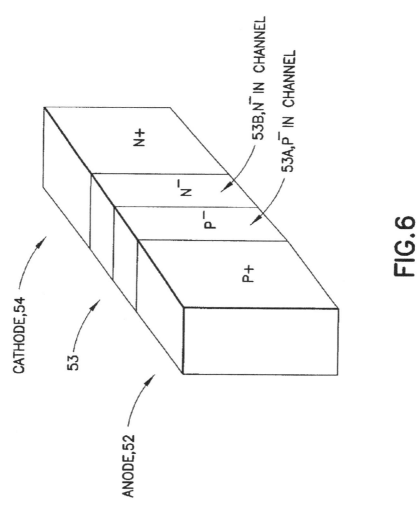

Figure 4.A.1 (Continued)

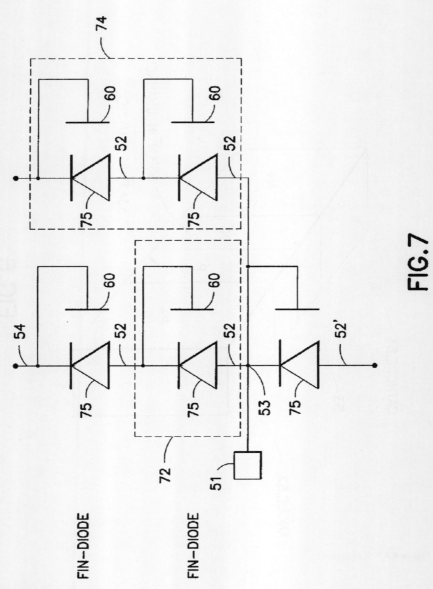

Figure 4.A.1 (Continued)

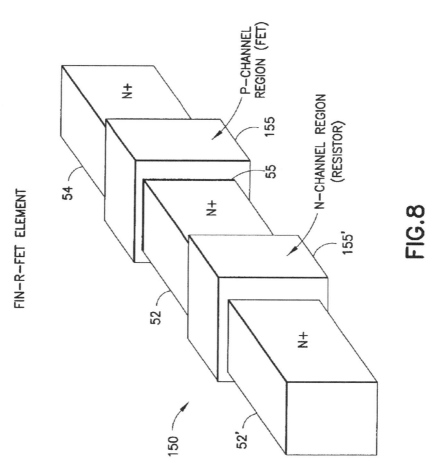

Figure 4.A.1 (Continued)

112 | 从发明到专利——科学家和工程师指南

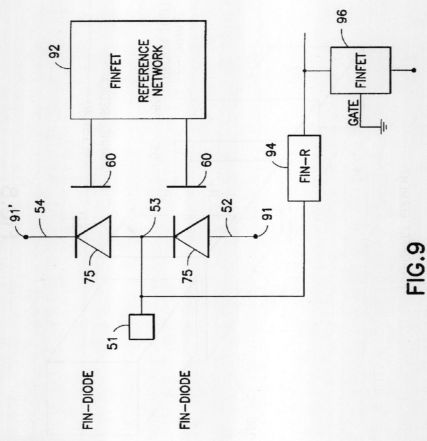

Figure 4.A.1 (Continued)

US 7,173,310 B2

LATERAL LUBISTOR STRUCTURE AND METHOD

This application is a continuation of PCT application #PCT/US2002/038546 filed 3 Dec. 2002.

TECHNICAL FIELD

The present invention generally relates to the field of integrated circuit fabrication, in particular of fabricating devices for electrostatic discharge protection (ESD) in integrated circuit technologies that use FINFETs.

BACKGROUND ART

The FINFET is a promising integrated circuit technology that employs a thin (10 nm–100 nm) vertical member as the source, drain and body of a field effect transistor (FET) and has a gate that is next to two vertical sides and the top of the channel. With such a thin body, there is very strong gate coupling, so that fully depleted operation is readily achieved. These structures will require overvoltage protection from electrical overstress (EOS), such as electrostatic discharge (ESD), as well as other voltage or current related stress events that are present in the semiconductor manufacturing, shipping and test processes. EOS events include over-current stress, latchup, and high current that occurs during testing and stressing. ESD events such as those occurring in the course of the human body model (HBM), machine model (MM), charged device model (CDM), transient latchup (TLU), cable discharge model, cassette model (CM) as well as other events can lead to electrical failure of FINFET structures.

Hence, it is evident that EOS and ESD protection of FINFET structures are necessary to provide adequate ESD protection for these structures.

U.S. Pat. No. 6,015,993 illustrates construction techniques for lateral ESD devices having a gated diode, where the channel is formed in bulk silicon or in the device layer of an SOI wafer. This structure is not compatible with FINFET structures and FINFET processing.

DISCLOSURE OF THE INVENTION

The invention relates to structures that provide EOS and ESD protection to FINFET technology.

According to an aspect of the invention, an ESD LUBISTOR structure based on FINFET technology employs a vertical fin (a thin vertical member containing the source, drain and body of the device) in alternative embodiments with and without a gate. The gate may be connected to the external electrode being protected to make a self-activating device or may be connected to a reference voltage. The device may be used in digital or analog circuits.

Accordingly, a structure is provided in an integrated circuit based on a substrate, which includes an elongated vertical member including a semiconductor, projecting from the substrate and having a top and two opposite elongated sides. A first electrode is formed in a first end of the vertical member and a second electrode, of opposite polarity, is formed in a second, opposite, end of the vertical member. The first and second electrodes are doped with an electrode concentration greater than a dopant concentration in a central portion of the vertical member, between the first and second electrodes.

BRIEF DESCRIPTION OF THE DRAWING

FIGS. **1A** and **1B** show plan and cross sections of a device according to the invention in an early stage.

FIGS. **2–4** show cross sections of the same device at further stages.

FIGS. **5** and **6** show examples of alternative embodiments.

FIG. **7** shows a schematic representation of the device in an ESD application.

FIG. **8** shows a view of a Fin-resistor integrated with a FINFET.

FIG. **9** shows another ESD application.

BEST MODE FOR CARRYING OUT THE INVENTION

An ESD LUBISTOR structure based on FINFET technology employs a vertical fin (a thin vertical member containing the source, drain and body of the device) in alternatives with and without a gate. The gate may be connected to the external electrode being protected to make a self-activating device or may be connected to a reference voltage. The device may be used in digital or analog circuits.

Among the benefits which may possibly arise from one or more preferred embodiments of the invention are the following:

provision of ESD-robust structures which are compatible with FINFET semiconductor processing and structures;

use of ESD-robust FINFET structures and supporting structures;

provision of a fin having diode terminals separated by a body, controlled by a gate or ungated with a doping structure such as $p^+/p^-/n^+$, $p^+/n^-/n^+$, or $p^+/p^-/n^-/n^+$;

provision of a lateral gated diode formed on a layer of insulator and having a $p^+/p^-/n^+$ structure or $p^+/n^-/n^+$, or $p^+/p^-/n^-/n^+$ with a body contact to the lightly doped body;

provision of a FINFET structure having a body contact that allows for a dynamic threshold FINFET device for use as an ESD protection element; and

provision of a FINFET resistor element (gated or ungated) to provide electrical and thermal stability of a FINFET device for ESD protection.

Referring now to the drawings, and more particularly to FIG. **1**, a process sequence according to the invention involves preliminary steps of forming the fin or vertical member (for the FIN-Diode structure) that are conventional in FINFET technology. Typically, a hard mask of appropriate width (less than 10 nm) is formed, e.g. by forming a nitride sidewall on a dummy oxide mesa that has been formed on a (single crystal or epitaxial film) silicon layer. The silicon film may be single crystal silicon (including an epitaxial layer). Polysilicon, selective silicon, strained silicon on a silicon germanium film, or other films may also be used. The silicon is etched in a directional dry etch that leaves a thin vertical member, illustratively 10 nm thick, 1 um wide and 0.1 um long that will provide the electrodes and the body of the device.

Referring to FIGS. **1A** and **1B**, the top view in FIG. **1B** shows a gate **60** above the fin **50**, the gate extending in front of and behind the plane of the cross section of FIG. **1A**. In FIG. **1A**, substrate **10** has fin **50** disposed on it, separated by gate dielectric **55**, illustratively 1 nm of oxide, from gate **60**. In this example, fin **50** rests directly on the silicon substrate, but some versions of the invention may have a dielectric layer between the substrate and the fin, e.g. the buried

Figure 4.A.1 (*Continued*)

insulator in a silicon on insulator (SOI) wafer. In this example, the substrate is an SOI substrate and buried oxide 20 is shown as below device layer 10. In some versions, the fin may be formed from the device layer and rest on the buried oxide. Illustratively, fin 50 is initially doped p⁻, as is layer 10. Gate 60 is polysilicon (poly), doped later by implantation.

The gate implantation step is shown in FIG. 2, with temporary layer 65, illustratively an anti-reflection coating that has been deposited as a step in forming other devices in the circuit and has been planarized, e.g. by chemical-mechanical polishing to the level of gate 60. Gate 60 is implanted with a heavy dose of ions, either p or n. Preferably, gate 60 receives an N^{++} dose of about $10^{21}/cm^2$, e.g. two order of magnitude greater than an N^+ dose of $5 \times 10^{19}/cm^2$. With this degree of difference, any further doping that the gate receives will not significantly affect its work function.

In FIGS. 3A and 3B, a non-critical aperture has been opened to expose the cathode 52, which is implanted N^+ (with a dose at least one order of magnitude less than the gate implant). Optionally, an aperture can be opened in the ARC or any other convenient mask such as a layer of photoresist 67 can be put down and patterned. FIG. 3B shows the same process, with anode 54 being implanted. Again, the dose (P^+) is one tenth that of the gate.

The fin has been implanted at an earlier stage. It can be implanted when it is put down if it is polysilicon and a single polarity is required. Optionally, the fin can be formed before the well implants and apertures can be opened in the photoresist for the well implants so that the fin receives P and/or N implants simultaneously with the wells.

Referring now to FIG. 4, there is shown the FIN-Diode device after deposition of the final interlayer dielectric, formation of apertures for contacts 72, 74 and 76 and deposition of the contact material. Since these contacts are at a low level, it is appropriate to use tungsten (W), if that is being used for other contacts at this level. If poly is being used at this level, then poly contacts are adequate. For electrical interconnects, standard interconnects (Al or Cu) and inter-level dielectrics (ILD) processes can be used. Aluminum interconnect structures can consist of an adhesive refractory metal (eg. Ti), a refractory metal (eg. Ti, TiNi, Co), and an aluminum structure for adhesion, diffusion barriers and to provide good electrical conductivity. Copper interconnect structures can consist of an adhesive film (e.g. TaN), refractory metal (e.g. Ta) and a copper interconnect. Typically, for Cu interconnect structures, the structures are formed using a single damascene or dual damascene process. For ESD and resistor ballasting in these structures, refractory metals can be used because of their high melting temperatures.

The advantage of the gate in this FIN-Diode structure shown in FIG. 4 is that the current in the gated $p+/n-/n+$ structure can be modulated by electrical control of the gate structure. Hence, the leakage, bias, and electrical stress can be modulated by connection of the gate structure to an anode or cathode node, a ground plane or power supply, a voltage or current reference circuit, or an electrical network. A disadvantage of the gate is that the gate insulator may be damaged. The circuit designer will make a choice based on trading off advantages and drawbacks.

A set of several FIN-Diode structures can be placed in parallel to provide a lower total series resistance, and higher total current carrying capability, and a higher power-to-failure of the ESD structure. For example, the anode and cathode connections can all be such as to allow electrical connections of the parallel FINFET diode structures. These parallel structures may or may not use the same gate electrode. There can also be personalization and customization of the number of parallel elements based on the ESD requirement or performance objectives. Additionally, there can be resistor ballasting, and different gate biases can be established to allow improved current uniformity, or providing a means to turn-on or turn-off the elements. The advantage of the parallel elements compared to a prior art device is: 1) three-dimensional capability, 2) improved current ballasting control, and 3) improved current uniformity control. In two-dimensional single finger Lubistor structures, current uniformity is not inherent in the design, causing a weakening of the ESD robustness per unit micron of cross sectional area. In these structures, the thermal heating of each FIN-Diode structure is isolated from adjacent regions. This prevents thermal coupling between adjacent regions from providing a uniform thermal profile and ESD robustness uniformity in each FIN-Diode parallel element.

Additionally, these FIN-Diode structures can be designed as $p+/p-/n+$ elements or $p+/n-/n+$ elements. The difference in the location of the metallurgical junction makes one implementation superior to the other for different purposes. This has been shown experimentally by the inventor and is a function of the doping concentration and application. The choice will be affected by the capacitance-resistance tradeoff and by the possibility of using an implant for the FIN-Diode that is originally intended for some other application. When a lower resistance is better suited for the purpose at hand, and the available implant has a relatively low dose, a $p+/n-/n+$ structure is preferred because of the higher mobility of electrons. Conversely, when the dose of the available implant is relatively high, a $p+/p-/n+$ structure would be preferred.

Halo implants can be established in these devices to allow improved lateral conduction, better junction capacitance, and improved breakdown characteristics. In this case, a halo is preferably provided for only one doping polarity to prevent a parasitic diode formed by the wrong halo implant in the wrong polarity.

FIG. 5 illustrates an alternative version of the FIN-Diode structure in which the channel is doped P^- and there is no separate gate. Its advantage is that the gate is not exposed to the ESD voltage stress. Electrical overstress in the gate dielectric can be eliminated by not allowing the gate structure to be present.

CDM failure mechanisms can occur due to the electrical connection of the gate structure for FINFET ESD protection networks. Whereas the prior embodiment containing a gate allowed for electrical control, that embodiment also required more electrical connections and/or design area for electrical circuitry. In the case of this embodiment, less electrical connections are necessary, allowing for a denser circuit.

In the embodiment of FIG. 5, a plurality of parallel FIN-Diode structures can be placed closely to allow for a high ESD robustness per unit area. Additionally, resistor ballasting and current uniformity control can be addressed by varying the effective resistance in the individual FIN-Diode structures. With the physical isolation of the adjacent FIN-Diode structures, the thermal coupling between adjacent elements can be reduced. Optimization of the spacing between adjacent elements can be insured by proper spacing and non-uniform adjacent spacing conditions to provide the optimum thermal result. This provides a thermal methodology allowing for optimization of the elements. This thermal methodology can not be utilized in the two-dimensional

Figure 4.A.1 (*Continued*)

Lubistor element but is a natural methodology in construction of parallel FIN-Diode structures.

Similarly, FIG. 6 illustrates a version in which the body is divided into two doped areas; a first P⁻ and a second N⁻ body region. This FIN-Diode structure allows for the optimization and placement of the metallurgical junction independent of the gate structure. This implant can be the p-well and/or n-well implant, or provided by halo type implantation (e.g. Angled, twist or straight), or other known implantation or diffusion process steps. The gradual profile introduced by the p+/p− transition, and the n+/n− transition provide a less abrupt junction and can lead to an improved ESD robustness.

The version of the devices that have a gate may be divided into several categories:

1) An N⁺/P FIN-Gated diode with the body contacting the substrate 10. In this case, there is a path to the substrate.
2) An N⁺/P FIN-Gated diode with a floating body on SOI.
3) An N⁺/P FIN-Gated diode on SOI with the gate contacting the (P⁺) body permits dynamic control of the anode potential.
4) An N⁺/P FIN-Gated diode on SOI with the gate contacting the N⁺ cathode.

To provide ESD protection to a FINFET device it is also advantageous to provide resistor elements integrated and/or unintegrated into FINFET devices.

Referring to FIG. 8, a FINFET device can be formed by similar techniques used in previous embodiments and having a source and drain implant of the same polarity separated by a body of opposite polarity. The body is covered by a gate insulator 55 and gate 155. This structure can be formed by a symmetric or asymmetric implant to provide ESD advantage. Additionally, to provide ESD robust FINFET structures, a resistor can be combined in the same structure. Illustratively, a second gate 155' can be placed in series with the drain structure where the second gate structure provides blockage of the heavily doped source/drain implant so that the lightly doped fin provides resistance. The gate structure serves two purposes: first, it provides a resistive region in the source or drain region; second it provides a means to block the salicide film placed on the source or drain region from shorting the resistor. This forms a _ballasting resistor_ inherently integrated with the FINFET. We will refer to this structure as the FIN-R-FET structure.

Additionally, this second gate structure 155' can be removed from the FIN-R-FET, as was done in the FIN-Diode structure. The removal of the second gate structure after salicidation allows for the prevention of electrical overstress or ESD issues with the resistor element.

This 150 element used in the FIN-R-FET device can also be constructed as a stand-alone resistor element. This is achieved by placing a n-channel FINFET into a n-well or n-body region. This resistor, or FIN-R device, can be used to provide ESD robustness for FINFETs, FIN-Diodes or used in circuit applications. As previously discussed, the gate can be removed to avoid electrical overstress in the physical element.

Additionally, to improve the ESD robustness of the FINFET device, the salicide can be removed from the source, drain and gate regions. Since the gate length of the device is relatively small compared to planar devices, the salicide can be removed in the gate regions.

Referring now to FIG. 7, there is shown a schematic of a typical arrangement for protecting a circuit from ESD on terminal 51. The dotted lines denoted with numerals 72 and 74 indicate options discussed below. Two FIN-LUBISTORs according to the invention are connected between the protected node 53 and the voltage terminals at 54 and 52'. In this case, gates 60 are connected to terminal 54, so that an ESD event dynamically reduces the resistance of one of the diodes. Alternatively, the terminals 60 could be connected to power supplies. To avoid electrical overstress to the gate structures, electrical circuits comprising of FINFET devices can be used to electrically isolate the gate structures from the electrical overstress. Electrical circuits with FINFET-based inverters, or FINFET-based reference control networks to provide electrical isolation from the power supplies, prevent overstress and establish a potential to avoid leakage.

To utilize as an ESD network for human body model (HBM), machine model (MM), and other ESD events, a plurality of lateral FIN-Diode structures must be used in parallel to minimize series resistance and to be able to discharge a large current through the structure without failure occurring in the FIN-Diode element or the circuitry. Hence a plurality of parallel FIN-Diode elements is placed connected between the input pin and the power supplies.

For voltage tolerance, an ESD network is shown in FIG. 7 constructed of FIN-Diode elements in a series configuration, now including the FIN-Diode 75 within dotted line 72. FIN-Diode structures can be constructed where the first FIN-Diode element anode is connected to a first pad, and the cathode is connected to a second FIN-Diode anode. This can continue in a string or series configuration. For each stage in series, a plurality of parallel FIN-Diode elements can be placed for each _stage_ of the string of FIN-Diodes. These strings can be placed between input pad and a power-supply, between two common power supply pads (e.g. VDD1 and VDD1), any two dissimilar power supply pads (e.g. VCC and VDD), any ground rails (e.g. VSS1 and VSS2), and any dissimilar ground rails (e.g. VSS and VEE). These FIN-Diode series elements can be configured as a single series string or back-to-back configuration to allow for bidirectional current flow between the two pads. For input pads to power supplies, typically, only a single FIN-Diode string will exist to provide uni-directional current flow.

For HBM and charged device model (CDM) events, an ESD circuit consisting of a FIN-Diode element, a FIN-resistor (FIN-R) element, and a FINFET may be used to improve ESD results. FIG. 9 is an example of a circuit to provide ESD protection utilizing a FIN-Diode element 75, a FIN-resistor 94 and a FINFET 96 with its gate connected to ground. Illustratively, the gate voltage of the FIN-Diodes is provided by a reference network, not by the ESD voltage itself. This permits better control of the current capacity of the diodes 75.

Additionally, ESD protection can be provided utilizing a resistor ballasted FIN-R-FET element. This circuit can be implemented in two fashions. First, utilizing a FIN-R resistor in series with a FINFET. To provide ESD protection a plurality of parallel FIN-R resistors are placed in series with a plurality of FINFET devices. Another implementation can use a plurality of parallel FIN-R-FET structures for ESD protection. These aforementioned structures can be placed in a cascode configuration for higher snapback voltage or voltage tolerance. For ESD protection, as in the case of FIN-Diode elements, a series of stages of FINFETs with FIN-R resistor elements can be connected where each stage includes a parallel set of elements.

Devices constructed according to the invention are not restricted to ESD uses and may also be employed in a conventional role in circuits—digital, analog, and radio frequency (RF) circuits. The invention is not restricted to silicon wafers and other wafers, such as SiGe alloy or GaAs may be used. These structures can be placed on a strained

Figure 4.A.1 (*Continued*)

silicon film, utilizing SiGe deposited or grown films. These structures are suitable for silicon on insulator (SOI), RF SOI, and ultra-thin SOI (UTSOI).

While the invention has been described in terms of a single preferred embodiment, those skilled in the art will recognize that the invention can be practiced in various versions within the spirit and scope of the following claims.

INDUSTRIAL APPLICABILITY

The invention has applicability to integrated circuit electronic devices and their fabrication.

The invention claimed is:

1. A structure in an integrated circuit, comprising:
   an elongated vertical fin member having a thickness in a range of 10 nm to 100 nm and comprising a semiconductor, said elongated vertical fin member projecting from a bulk semiconductor substrate and having a top and two opposite elongated sides, in which
   a first electrode is formed in a first end of said elongated vertical fin member and
   a second electrode, of opposite polarity to said first electrode, is formed in a second end of said elongated vertical fin member opposite said first end,
   said first and said second electrodes being doped with an electrode concentration greater than a dopant concentration in a central portion between said first and second electrodes.

2. A structure according to claim 1, in which one of said electrodes is doped p^+ and the other of said electrodes is doped n^+.

3. A structure according to claim 1, in which one of said electrodes is doped p^+, said central portion is doped p^- and the other of said electrodes is doped n^+.

4. A structure according to claim 2, in which a first sub-portion of said central portion adjacent to a first electrode is doped with the same polarity as said first electrode and with a lower concentration and a second sub-portion of said central portion adjacent to said second electrode is doped with the same polarity as said second electrode and with a lower concentration.

5. A structure according to claim 2, in which said dopants are arranged in a sequence $p^+/p^-/n^-/n^+$.

6. A structure according to claim 1, further comprising:
   a gate disposed over a central portion of said top and in proximity to central portions of said two sides, said gate being separated from said vertical member by a dielectric gate layer.

7. An electrostatic discharge (ESD) protection circuit being attached to an external terminal of an integrated circuit and including a structure according to claim 1.

8. An ESD protection circuit according to claim 7, further comprising two devices, in which said external terminal is connected to an anode of a first device and to a cathode of the other device.

9. An ESD protection circuit according to claim 7, in which said substrate is an SOI substrate having a layer of buried insulator and said vertical fin member is disposed directly on said buried insulator layer.

10. An ESD protection circuit according to claim 7, in which said substrate is a bulk substrate and said vertical member is disposed directly on said bulk substrate.

11. An ESD protection circuit according to claim 7 comprising a plurality of FIN-Diode structures connected in parallel between an external terminal and a voltage terminal.

12. An ESD protection circuit according to claim 7 comprising at least one FIN-Diode structure in a series configuration between two external terminals.

13. An ESD protection circuit according to claim 7 comprising at least one FIN-Diode structure, and at least one FIN-R resistor element.

14. An ESD protection circuit according to claim 7 comprising at least one FIN-Diode structure, at least one FIN-R resistor element, and at least one FINFET element.

* * * * *

Figure 4.A.1 (*Continued*)

(12) United States Patent
Voldman et al.

(10) Patent No.: **US 6,549,061 B2**
(45) Date of Patent: **Apr. 15, 2003**

(54) **ELECTROSTATIC DISCHARGE POWER CLAMP CIRCUIT**

(75) Inventors: **Steven Howard Voldman**, South Burlington, VT (US); **Alan Bernard Botula**, Essex Junction, VT (US); **David TinSun Hui**, Poughkeepsie, NY (US)

(73) Assignee: **International Business Machines Corporation**, Armonk, NY (US)

(*) Notice: Subject to any disclaimer, the term of this patent is extended or adjusted under 35 U.S.C. 154(b) by 0 days.

(21) Appl. No.: **10/026,308**

(22) Filed: **Dec. 20, 2001**

(65) **Prior Publication Data**

US 2002/0186068 A1 Dec. 12, 2002

Related U.S. Application Data

(63) Continuation-in-part of application No. 09/681,667, filed on May 18, 2001, now Pat. No. 6,429,489.

(51) Int. Cl.[7] H03K 17/615
(52) U.S. Cl. **327/483**; 327/575; 327/324
(58) Field of Search 327/478, 482–492, 327/575, 309, 321, 324, 327, 310

(56) **References Cited**

U.S. PATENT DOCUMENTS

3,535,532 A * 10/1970 Merryman 327/483
3,629,623 A * 12/1971 Sakurai et al. 327/575
3,671,833 A * 6/1972 Rakes 327/483
4,764,688 A * 8/1988 Matsumura 327/483
4,769,560 A * 9/1988 Tani et al. 327/575
5,684,427 A * 11/1997 Stoddard et al. 327/483

* cited by examiner

Primary Examiner—Toan Tran
(74) *Attorney, Agent, or Firm*—Richard A. Henkler

(57) **ABSTRACT**

An ESD clamping circuit arranged in a darlington configuration and constructed from SiGe or similar type material. The ESD clamping circuit includes additional level shifting circuitry in series with either the trigger or clamping device or both, thus allowing non-native voltages that exceed the BVCEO of the trigger and/or clamp devices.

9 Claims, 8 Drawing Sheets

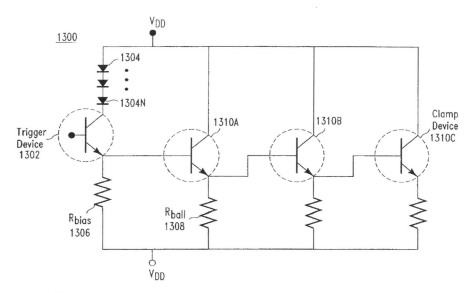

Figure 4.A.2

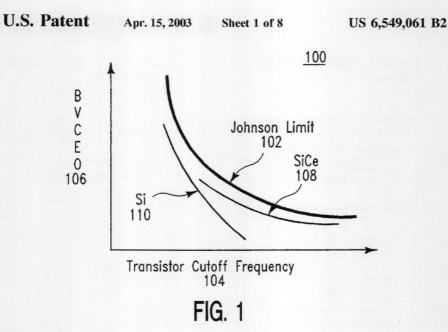

FIG. 1

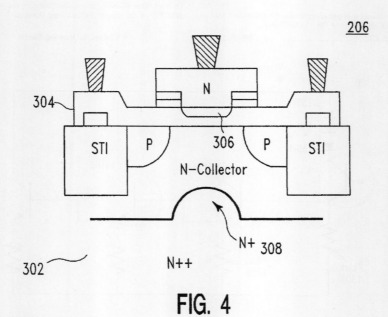

FIG. 4

Figure 4.A.2 (Continued)

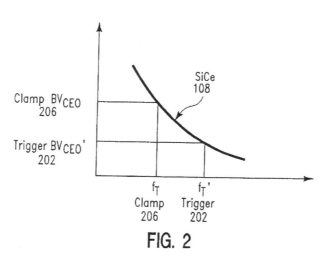

FIG. 2

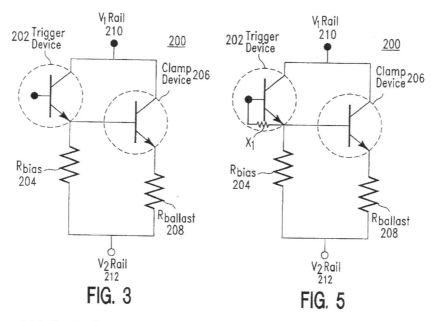

FIG. 3   FIG. 5

Figure 4.A.2 (*Continued*)

120 | 从发明到专利——科学家和工程师指南

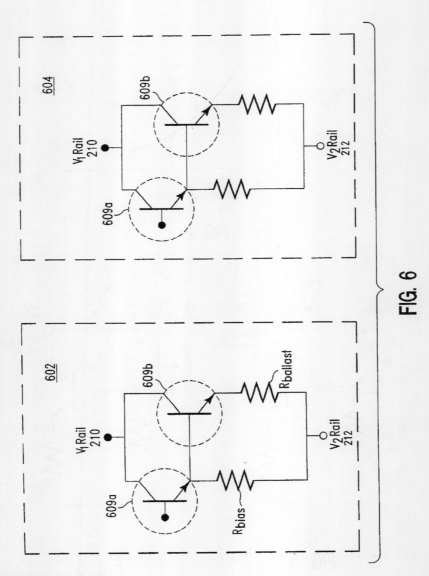

Figure 4.A.2 (*Continued*)

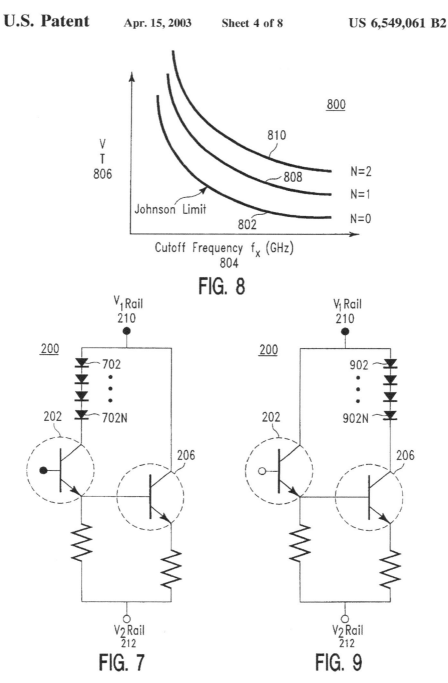

Figure 4.A.2 (*Continued*)

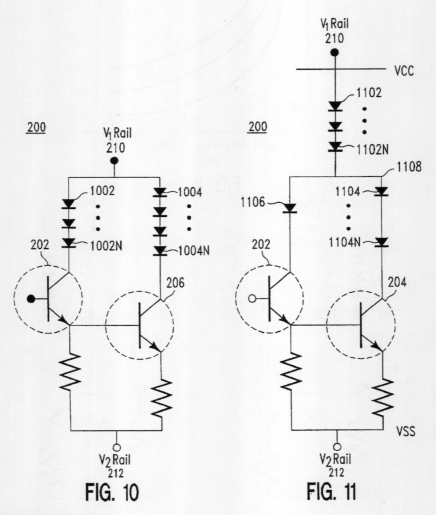

Figure 4.A.2 (Continued)

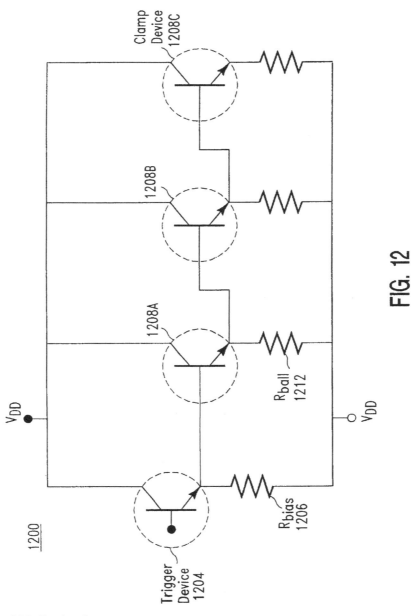

Figure 4.A.2 (Continued)

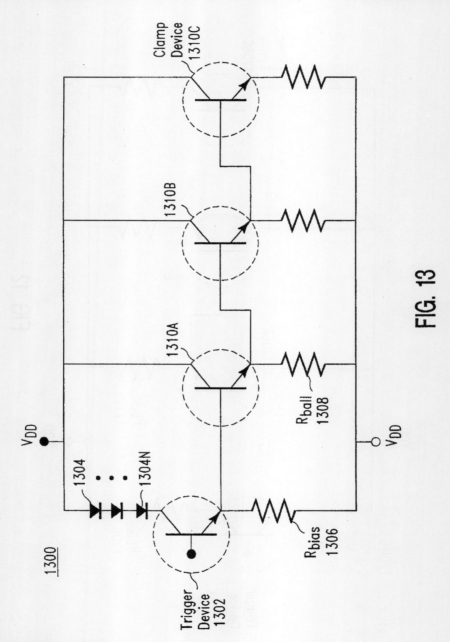

Figure 4.A.2 (*Continued*)

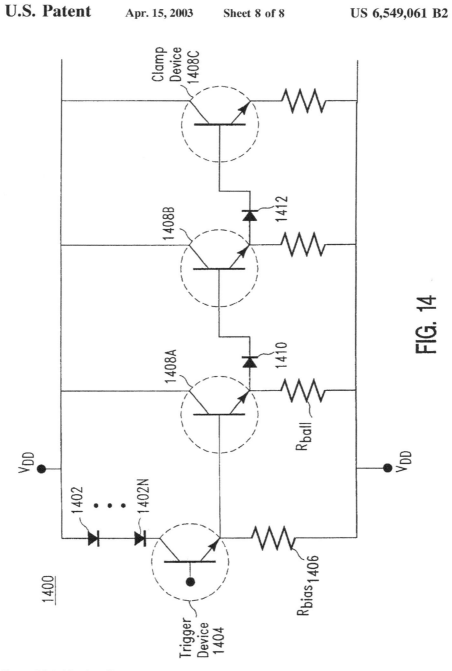

Figure 4.A.2 (Continued)

US 6,549,061 B2

ELECTROSTATIC DISCHARGE POWER CLAMP CIRCUIT

This is a Continuation-in-part of application Ser. No. 09/681,667, filed May 18, 2001, now U.S. Pat. No. 6,429,489.

BACKGROUND

TECHNICAL FIELD OF THE PRESENT INVENTION

The present invention generally relates to electrostatic discharge circuits, and more specifically, to electrostatic discharge power clamp circuits.

BACKGROUND OF THE PRESENT INVENTION

Electrostatic Discharge (ESD) events, which can occur both during and after manufacturing of the Integrated Circuit (IC), can cause substantial damage to the IC. ESD events have become particularly troublesome for CMOS and BiCMOS chips because of their low power requirements and extreme sensitivity.

A significant factor contributing to the ESD sensitivity is that the transistors of the circuits are formed from small regions of N-type materials, P-type materials, and thin gate oxides. When a transistor is exposed to an ESD event, the charge applied may cause an extremely high current flow to occur within the device which can, in turn cause permanent damage to the junctions, neighboring gate oxides, interconnects and/or other physical structures.

Because of this potential damage, on chip ESD protection circuits for CMOS and BiCMOS chips are essential. In general, such protection circuits require a high failure threshold, a small layout size and a low Resistive/Capacitive (RC) delay so as to allow high speed applications.

An ESD event within an IC can be caused by a static discharge occurring at one of the power lines or rails. In an effort to guard the circuit against damage from the static discharge, circuits referred to as ESD clamps are used. An effective ESD clamp will maintain the voltage at the power line to a value which is known to be safe for the operating circuits, and not interfere with their operation under normal conditions.

An ESD clamp circuit is typically constructed between a positive power supply (e.g. VDD) and a ground plane, or a ground plane and a negative power supply (VSS). The main purpose of the ESD clamp is to reduce the impedance between the rails VDD and VSS so as to reduce the impedance between the input pad and the VSS rail (i.e. discharge of current between the input to VSS), and to protect the power rails themselves from ESD events.

The never ending demand by the consumer for increased speed in Radio Frequency (RF) devices has resulted in some unique challenges for providing ESD protection in these high speed applications. More specifically, the physical size (e.g. Breakdown voltage) and loading effects of the ESD devices must now be considered in such high speed applications (e.g. 1–200 Giga Hertz range). The capacitive loading of the ESD device itself becomes a major concern for chips running at high frequencies, since the capacitive loading has an adverse effect on performance. For example, the capacitive loading effect of a typical ESD clamp at a frequency of 1 Hz is 0.5 pF, 10 GHz–0.1 pF, and at 100 GHz–0.05pF, 200 Hz–0.01 pF).

In bipolar transistors, there is an inverse relationship between the breakdown voltage and the current gain cutoff frequency known as the Johnson Limit. In each technology generation, the cutoff frequency increases leading to lower collector to emitter breakdown voltages BVCEO. At the same time, mixed voltage applications exist where chips of non-native power supply voltages need to be applied above the BVCEO of the transistor. The term "non-native" as used herein refers to any power supply that is greater than that for which the transistor is constructed.

It would, therefore, be a distinct advantage to have an ESD clamp that could provide substantial benefits in high speed devices while limiting any performance degradation from capacitive loading. It would be further advantageous to have an ESD clamp that provides the ability to raise the clamp and/or trigger condition above the native power supply voltages. The present invention provides such an ESD clamp.

SUMMARY OF THE PRESENT INVENTION

In one aspect, the present invention is an ESD device that is useful in high speed radio frequency applications where size and loading effects are a concern. The ESD device is preferably constructed on a SiGe, SiGeC or equivalent type material that nearly approximates the Johnson Limit curve, and constructed in a Darlington type configuration. In the preferred embodiment of the present invention, the trigger device has a collector-to-emitter breakdown voltage (BVCEO) that is lower than that of the clamping device, and a frequency cutoff that is higher than that of the clamping device.

In yet another aspect, the present invention is an ESD device preferably constructed on a SiGe, SiGeC or equivalent type material that nearly approximates the Johnson Limit curve, constructed in a Darlington type configuration, and that allows a trigger condition below the BVCEO of the clamp element and above the BVCEO of the trigger element.

In yet a further aspect, the present invention is an ESD device preferably constructed on a SiGe, SiGeC, or equivalent type material that nearly approximates the Johnson Limit curve, constructed in a Darlington type configuration, and where the trigger/clamp rail is level shifted relative to the power supply rail conditions, avoiding the Johnson limit constraint on the trigger and clamp elements.

BRIEF DESCRIPTION OF THE DRAWINGS

The present invention will be better understood and its advantages will become more apparent to those skilled in the art by reference to the following drawings, in conjunction with the accompanying specification, in which:

FIG. 1 is a pictorial diagram illustrating in greater detail the Johnson Limit curve;

FIG. 2 is a pictorial diagram illustrating the frequency cutoff and BVCEO characteristics of a trigger and a clamp device constructed in accordance with the teachings of the present invention;

FIG. 3 is a schematic diagram illustrating a Darlington ESD clamp constructed in accordance with the teachings of the present invention;

FIG. 4 is a cross sectional view diagram of an implementation of the trigger device of the clamp of FIG. 3 as a SiGe Heterojunction Bipolar Transistor (HBT) according to the teachings of the present invention;

FIG. 5 is a circuit diagram of the Darlington ESD clamp of FIG. 3 as modified according to the teachings of the present invention;

Figure 4.A.2 (*Continued*)

FIG. **6** is a schematic diagram of a dual stage darlington ESD clamp according to the teachings of the present invention;

FIG. **7** is a schematic diagram of the darlington ESD clamp of FIG. **3** modified according to the teachings of an alternative embodiment of the present invention;

FIG. **8** is a pictorial diagram illustrating the initial Johnson limit curve in relation to the trigger and clamp devices of FIG. **3** as modified to include new pnp elements according to the teachings of the present invention;

FIG. **9** is a schematic diagram of the darlington ESD clamp of FIG. **3** modified according to the teachings of an alternative embodiment of the present invention;

FIG. **10** is a schematic diagram of the darlington ESD clamp of FIG. **3** modified according to the teachings of an alternative embodiment of the present invention;

FIG. **11** is a schematic diagram of the darlington ESD clamp **200** of FIG. **3** modified according to the teachings of an alternative embodiment of the present invention;

FIG. **12** is a schematic diagram illustrating an alternative embodiment of an ESD clamp structure where the clamp network size is increased according to the teachings of the present invention;

FIG. **13** is a schematic diagram illustrating an alternative embodiment of an ESD clamp structure where the clamp network size is increased according to the teachings of the present invention; and

FIG. **14** is a schematic diagram illustrating an alternative embodiment of an ESD clamp structure where the clamp network size is increased according to the teachings of the present invention.

DETAILED DESCRIPTION OF THE PREFERRED EMBODIMENT OF THE PRESENT INVENTION

In the following description, numerous specific details are set forth, however, it will be obvious to those of ordinary skill in the art that the present invention can be practiced with different details. In other instances, well-known circuits have been shown in block diagram form in order not to obscure the present invention in unnecessary detail.

The present invention capitalizes upon the recognition that the structural and physical characteristics of Silicon Germanium (SiGe) material and other equivalent materials (e.g. Silicon Germanium Carbon "SiGeC") are ideal for use in an ESD clamp for high speed applications. More specifically, the present invention recognizes that the scaling of the SiGe heterojunction bipolar transistor is driven by both structural changes and the physical limitations of the transistor itself and such recognition's can be used where size and loading effects are important.

An equation $(P_m X_c)^{1/2} f_T = E_m V s / 2\pi$ known as the Johnson Limit describes a fundamental relationship between the frequency response of the transistor and the maximum power applied across the transistor element. P_m represents the maximum power, X_c represents the reactance ($X_c = 1/2\pi f_T C_{bc}$), f_T represents the unity current gain cutoff frequency, E_m represents the maximum electric field, and Vs represents the electron saturation velocity. The equation can be manipulated so that is it is expressed in terms of maximum voltage $V_m f_T = E_m V s / 2\pi$ to illustrate the inverse relationship between the transistor speed and the allowed breakdown voltage.

FIG. **1** is a pictorial diagram **100** illustrating in greater detail the Johnson Limit curve **102** and approximations of how transistors constructed of Silicon and SiGe would compare. In this diagram **100**, the x-axis represents f_T, and the y-axis represents the Breakdown Voltage of the transistor from the collector-to-emitter (BVCEO). The curve **102** demonstrates that the BVCEO of the transistor decreases with the increase in the unity current gain cutoff frequency (f_T). Approximations of how a transistor constructed of Si (Silicon) **110** and SiGe **108** have been transposed on the diagram **100**.

FIG. **2** is a pictorial diagram illustrating the frequency cutoff and BVCEO characteristics of a trigger and a clamp device constructed in accordance with the teachings of the present invention. As previously discussed and illustrated in FIG. **1**, a transistor constructed from SiGe material closely approximates the Johnson Limit curve. The present invention recognizes and capitalizes upon this recognition. More specifically, the present invention uses a first ESD device having a low BVCEO and a high f_T to trigger (Trigger **202**) a second ESD device having a f_T that is lower than that of the first ESD device, and a BVCEO that is higher than that of the first ESD device (Clamp **206**).

Because of the Johnson Limit relationship, the frequency response of the "trigger" transistor **202** is at a higher frequency for a device with a lower breakdown voltage.

In the preferred embodiment, the first and second devices are arranged in a common-collector configuration as explained in greater detail in connection with FIG. **3**.

FIG. **3** is a schematic diagram illustrating a Darlington ESD clamp **200** constructed in accordance with the teachings of the present invention. Specifically, the Darlington ESD clamp **200** is constructed between two power rails V1 **210** and V2 **212**. Power rails V1 and V2 can be, for example, Power and Ground or Ground and negative Power, respectively. The Darlington ESD clamp **200** includes a trigger device (npn SiGe transistor) **202**, a clamp device (npn SiGe transistor) **206**, bias resistor **204**, and ballast resistor **208**.

It should be noted that although a single stage Darlington ESD clamp **200** has been illustrated for ease of explanation purposes, the present invention is equally applicable to multiple staged Darlington pairs/stages as well (e.g. A plurality of trigger and/or clamp elements).

Trigger device **202** has a BVCEO that is lower than that of the clamp device **206** and a cutoff frequency that is higher than that of the clamp device **206**. During an ESD event, the trigger device **202**, upon reaching its BVCEO, will provide base current into the clamp device **206**.

Conversely, clamp device **206** has a BVCEO that is higher than that of the trigger device **202**, a cutoff frequency that is lower than that of the trigger device **202**, and discharges the current from the ESD event from Power rail V1 to V2.

In series with the clamp device **206** is ballast resistor **208**. Ballast resistor **208** is used in a conventional fashion for providing emitter stability, voltage limitations, thermal stability, and ESD stability.

In series with trigger device **202** is bias resistor **204**. Bias resistor **204** is used for keeping the base of the clamp device **206** to a low potential in order to limit the amount of current that flows through the trigger device **202** during an ESD event.

In the preferred embodiment of the present invention, the trigger device **202** is a SiGe heterojunction bipolar transistor (HBT) formed in the configuration explained in connection with FIG. **4** below.

FIG. **4** is a cross sectional view diagram of an implementation of the trigger device **202** of FIG. **3** as a SiGe HBT

Figure 4.A.2 (*Continued*)

according to the teachings of the present invention. The SiGe HBT **202** is formed on a n++ subcollector **302**. The SiGe epitaxial film is placed on the silicon surface forming the extrinsic base **304** over the STI isolation and the intrinsic base region **306** over the single crystal silicon region. An n+ pedestal implant **308** is formed through the emitter window.

The pedestal implant **308** is typically formed to reduce the Kirk effect. The Kirk effect is due to the high current density which forces the space charge region of the base-collector junction to get pushed into the collector region reducing the frequency response of the transistor. To prevent this, an extra "pedestal implant" is placed so as to maintain a high f_T device, which in turn causes a low BVCEO breakdown voltage. Obviously, more pedestal implants can be added to lower the BVCEO until a desired level is obtained.

In the preferred embodiment of the present invention, a first pedestal implant is placed in both the clamp **206** and trigger **202** devices. A second pedestal implant is used through the emitter window to form the high frequency trigger device **202**.

Alternative embodiments of both the clamp **206** and trigger **202** devices can be created by adding additional pedestal and/or CMOS N-well implants. For example, table below illustrates how such implants could be used to create three distinct transistors each of which have a differing f_T.

Device f_T	Implants
Low	First pedestal only
Medium	First pedestal and N-Well
High	First and second pedestals

From the above table it can be seen that a low f_T device is created by implanting a single pedestal, a medium (with respect to the low and high devices) f_T device by implanting a single pedestal and a N-Well implant from CMOS technology into the collector region, and a high f_T device by implanting two pedestals.

As previously stated, the present invention is equally applicable to multi-staged Darlington ESD clamps. FIG. 6 and its accompanying description further illustrate such applicability.

FIG. 6 is a schematic diagram of a dual stage darlington ESD clamp according to the teachings of the present invention. First ESD clamp **602** is constructed from a high f_T trigger device **602a** coupled to a medium f_T clamp device **602b**. The second ESD clamp **604** is constructed from a medium f_T trigger device **604a**, having its base coupled to the emitter of the clamp device **602b**, and its emitter coupled to the base of a low f_T clamp device **604b**.

FIG. 5 is a circuit diagram of the Darlington ESD clamp **200** of FIG. 3 as modified according to the teachings of the present invention. The Darlington ESD Clamp **200** of FIG. 3 has been modified to include a resistor R1 coupled between the base and emitter of the trigger device **202**. Resistor R1 provides changes the turn on characteristics of the trigger device **202** so that during Direct Current (DC) operation it is not activated by spurious signals.

From the above description it should be apparent to one skilled in the art that the present invention is applicable to high frequency devices where a trigger device triggers an ESD clamp, regardless of the particular configuration. The construction of the trigger and clamp devices can vary depending on the particular application, provided the trigger device has a BVCEO that is lower than the BVCEO of the clamping device, and a frequency cutoff that is higher than that of the clamping device. For example, if a high BVCEO is desired then this can be accomplished by having no pedestal or CMOS N-well structures in the transistor. If a medium BVCEO is desired then this can be accomplished by creating pedestal and/or CMOS N-well structures in the transistor. If a low BVCEO is desired, then this can be accomplished by implanting one or more pedestals and/or multiple CMOS N-well structures into the transistor.

Alternative Embodiments

In another embodiment, the trigger device **202** of either FIGS. 3 or 5 can be constructed with Silicon Germanium in the base region forming a SiGe heterojunction bipolar transistor (HBT). The clamp device **206** can be a silicon bipolar junction transistor (BJT). In this embodiment, the trigger device **202** will have a lower breakdown voltage and higher frequency threshold than the clamp device **206**.

In yet another embodiment, both the trigger **202** and clamp **206** devices are SiGe HBT devices. The trigger device **202** has been constructed with the pedestal implant as discussed in connection with FIG. 4.

In a further embodiment, both the trigger **202** and clamp **206** devices are SiGe HBT devices. The trigger device **202** has been constructed with a plurality of pedestal implants as discussed in connection with FIG. 4, or alternatively a plurality of CMOS n-well implants could be added to the collector region to provide a high frequency threshold device.

In yet another embodiment, the trigger device **202** can be a SiGeC HBT where the base region contains both Germanium and Carbon to provide a high frequency threshold and a low BVCEO. The clamp device **206** is a SiGe HBT.

Mixed Voltage Applications

In yet another embodiment, the present invention has the ability to address mixed voltage applications where the power supply conditions exceed the native breakdown voltages of the transistors. More specifically, the present invention is preferably applicable to ESD power clamps for BiCMOS applications and Bipolar applications that have voltage differentials between the power rails which are larger than the native power voltage conditions of the transistors themselves.

In these type of designs, the trigger condition for the trigger and clamp voltages needs to be modified to address these higher voltage conditions.

The present invention introduces a new relationship between the trigger voltage and the clamp voltage by placing level shifting elements between the collectors of the trigger and clamp devices and the power supply. The use of these level shifting elements creates a new condition for the trigger device **202**.

FIG. 7 is a schematic diagram of the ESD clamp **200** of FIG. 3 modified according to the teachings of an alternative embodiment of the present invention. The ESD clamp **200** of FIG. 3 has been modified by introducing a plurality of level shifting elements **702-N** in series with the collector of the Trigger device **202** and v1 rail **210**. The term level shifting elements as used herein refers to diodes, varactors, pin diode elements, Schottky diodes, pnp elements in a common collector configuration or MOSFETs. These level shifting elements can be pnp transistor elements or diode type elements, and can be constructed from SiGe, SiGeC, or CMOS devices.

Figure 4.A.2 (*Continued*)

US 6,549,061 B2

The addition of the level shifting elements in series with the trigger device **202** establishes a new trigger condition. More specifically, the addition of the level shifting elements **702-N** elevates (shifts) the BVCEO vs. f_T curve, forming a new "trigger condition" relationship as explained in greater detail in connection with FIG. **8**.

FIG. **8** is a pictorial diagram **800** illustrating the trigger condition of an ESD device constructed in accordance with the teachings of the present invention. In this diagram **800**, the x-axis represents the unity current cutoff frequency f_T, and the y-axis represents the "trigger condition". The addition of the level shifting elements **702-N** in series with the trigger device **202** develops a new variable trigger implementation.

In the case where pnp elements are placed in series with the trigger device **202**, the trigger condition is level shifted to a higher breakdown condition. Adding a string of pnp elements creates a new trigger condition defined as $V_T = E_m v_s / 2\Pi f_T + NV_f - (kT/q)(N-1)N/2 \ln(\beta+1)$ where N is the number of pnp elements, V_f is the forward diode voltage, and β is the pnp current gain and f_T is the unity current gain cutoff frequency of the trigger device. This trigger condition provides a set of design contours of trigger values where the number of level shifting elements and the cutoff frequency are the trigger parameters.

In the case where non-pnp elements are placed in series with the trigger device **202**, the above noted equation changes to $V_T = E_m v_s / 2\Pi f_T + NV_f$.

Curve **802** represents the situation where no level shifting elements are in series with the trigger device **202**. In this situation, the trigger condition equals the breakdown voltage, and the trigger condition is substantially equal to the Johnson limit curve. Curves **808**, and **810** illustrate the affects of adding one or more respectively pnp level shifting elements in series with the trigger device **202**.

As can be seen, the additional pnp level shifting elements (N=1,2) shift the curve in an upward direction on the diagram **800**. This allows the trigger device **202** to accommodate power supply voltages that previously exceeded its BVCEO. In this manner, the triggering of the circuit **200** allows for decoupling of the frequency response of the device (**202**) from the trigger condition of the circuit embodiment **200**.

FIG. **9** is a schematic diagram of the ESD clamp **200** of FIG. **3** modified according to the teachings of an alternative embodiment of the present invention. The ESD clamp **200** of FIG. **3** has been modified to include a plurality of pnp elements **902-N** in series with the collector of the clamp device **206**. In this configuration, a new breakdown voltage is established for the clamp device **206** that avoids over voltage when placed between power supplies for mixed voltage applications, or power supply differential voltages that exceed the breakdown voltage of the clamp device **206**, or voltages above the native power supply voltage of the semiconductor chip.

In the case where pnp elements are placed in series with the clamp device **206**, the clamp breakdown condition is shifted to a higher breakdown condition. Adding a string of pnp elements creates a new trigger condition as defined by the equation $V_c = E_m v_s / 2\Pi f_T + MV_f - (kT/q)(M-1)M/2 \ln(\beta+1)$ where M is the number of pnp elements, V_f is the forward diode voltage, and β is the pnp current gain and f_T is the unity current gain cutoff frequency of the clamp device. This breakdown condition provides a set of design contours of breakdown values where the number of elements and the cutoff frequency are the clamp element breakdown parameters.

In the case where non-pnp elements are placed in series with the clamp device **206**, the above noted equation changes to $V_c = E_m v_s / 2\Pi f_T + MV_f$.

FIG. **10** is a schematic of the ESD clamp **200** of FIG. **3** modified according to the teachings of an alternative embodiment of the present invention. The ESD clamp **200** of FIG. **3** has been modified to include a plurality of pnp elements **1002-N** and **1004-M** in series with both the trigger **202** and clamp **204** devices, respectively. In this configuration, both the trigger and clamp breakdown conditions allow for elevation of the differential voltage placed across the trigger device **202** and the clamp device **204**.

In the case where pnp elements are placed in series with the trigger device **202**, the trigger condition is shifted to a higher breakdown condition. Adding a string of pnp elements creates a new trigger condition as defined by the following equation $V_T = E_m v_s / 2\Pi f_T + NV_f - (kT/q)(M-1)M/2 \ln(\beta+1)$ where N is the number of pnp elements, V_f is the forward diode voltage, and β is the pnp current gain and f_T is the unity current gain cutoff frequency of the trigger device. This trigger condition provides a set of design contours of trigger values where the number of elements and the cutoff frequency are the trigger parameters.

In the case where non-pnp elements are placed in series with the trigger device **202**, the above equation changes to $V_T = E_m v_s / 2\Pi f_T + NV_f$.

In the case where pnp elements are placed in series with the clamp device **204**, the clamp breakdown condition is shifted to a higher breakdown condition. Adding a string of pnp elements creates a new trigger condition as defined by the following equation $V_c = E_m v_s / 2\Pi f_T + MV_f - (kT/q)(M-1)M/2 \ln(\beta+1)$ where M is the number of pnp elements, V_f is the forward diode voltage, and β is the pnp current gain, and f_T is the unity current gain cutoff frequency of the clamp device. This breakdown condition provides a set of design contours of breakdown values where the number of elements and the cutoff frequency are the clamp element breakdown parameters.

In the case where non-pnp elements are placed in series with the clamp device **204**, the above noted equation changes to $V_c = E_m v_s / 2\Pi f_T + MV_f$.

FIG. **11** is a schematic diagram of the ESD clamp **200** of FIG. **3** modified according to the teachings of an alternative embodiment of the present invention. The ESD clamp **200** of FIG. **3** has been modified to include a plurality of pnp elements **1102-P** placed in series between the v1 rail **210** (e.g. power supply) and the trigger/clamp rail **1108**. The addition of pnp elements **1102-P** avoids the Johnson limit condition for both the trigger **202** and the clamp **204** devices. Specifically, the pnp elements delay the turning on of the ESD clamp **200**, thus allowing a non-native power supply condition to be placed on the power rail VCC without the initiation of the ESD clamp **200**.

In this case, the trigger condition is defined by the following equation $V_T = E_m v_s / 2\Pi f_T + NV_f - (kT/q)(N-1)N/2 \ln(\beta+1) + PV_f - (kT/q)(M1)P/2 \ln(\beta+1)$ where N is the number of pnp elements, V_f is the forward diode voltage, and β is the pnp current gain and f_T is the unity current gain cutoff frequency of the trigger device, and P is the number of pnp elements of the string between rail **210** and rail **1108** whose V_f is the forward diode voltage, and β is the pnp current gain.

In this case, the clamp condition is defined by the following equation $V_c = E_m v_s / 2\Pi f_T + MV_f - (kT/q)(M-1)M/2 \ln(\beta+1) + PV_f - (kT/q)(P-1)P/2 \ln(\beta+1)$ where M is the number of pnp elements, V_f is the forward diode voltage, and β is the pnp current gain and f_T is the unity current gain cutoff

Figure 4.A.2 (*Continued*)

frequency of the clamp device, and P is the number of pnp elements of the string between rail **210** and rail **1108** whose V_f is the forward diode voltage, and β is the pnp current gain.

FIG. **12** is a schematic diagram illustrating an alternative embodiment of an ESD clamp structure **1200** where the clamp network size is increased according to the teachings of the present invention. As the clamp device size increases, the trigger device **1204** cannot supply the current to all base devices equally. The clamp device **1208** can be designed so as to have multiple clamp devices **1208A–B** as illustrated in FIG. **12**. More specifically, the emitter of clamp device **1208A** is connected to the base of the second clamp device **1208B**, and the emitter of the second clamp device **1208B** is connected to the base of the third clamp device **1208C**.

In this fashion, as the trigger device **1204** supplies current to the base of the first clamp device **1208A**, current flows from the emitter through the ballast resistor **1212**. As the current flows through the ballast resistor **1212**, the base **1208A** rises leading to a V_{be} to be established in the second clamp device **1208B**. This leads to the turn-on of the second clamp device **1208B**, and allowing the ESD current to flow through the second clamp element **1208B**. This process continues with third clamp device **1208C**.

FIG. **13** is a schematic diagram illustrating an alternative embodiment of an ESD clamp structure **1300** where the clamp network size is increased according to the teachings of the present invention. In this embodiment, a series of pnp elements **1304-N** are placed in series with the trigger device **1302** in order to delay the turn-on of the ESD clamp circuit **1300**.

FIG. **14** is a schematic diagram illustrating an alternative embodiment of an ESD clamp structure **1400** where the clamp network size is increased according to the teachings of the present invention. In this embodiment, a series of pnp elements are placed in series with the trigger device **1302** in order to delay the turn-on of the ESD clamp circuit **1400**. An additional pnp element **1410** and **1412** is placed between each successive stage of the clamp circuit **1400**. As the BVCEO is reached in the device, current will flow through the device into the base of the first clamp element. The addition of rectifying elements in series with the base of the successive clamp devices **1408A–B** will prevent current flowing between successive stages of the clamp network.

It is thus believed that the operation and construction of the present invention will be apparent from the foregoing description. While the method and system shown and described has been characterized as being preferred, it will be readily apparent that various changes and/or modifications could be made wherein without departing from the spirit and scope of the present invention as defined in the following claims.

What is claimed is:

1. An electrostatic discharge device comprising:
   a trigger transistor;
   a plurality of first level shifting elements coupled in series with the collector of the trigger transistor;
   a first power rail coupled to one of the level shifting elements;
   a clamp transistor having its base coupled to the emitter of the trigger transistor and its collector coupled to the first power rail, the clamp transistor having a frequency cutoff and a breakdown voltage that are both larger than that of the trigger transistor; and
   a second power rail coupled to the emitters of both the trigger and clamp transistors.

2. The electrostatic discharge device of claim **1** wherein the trigger and clamp transistors are constructed from silicon germanium material.

3. The electrostatic discharge device of claim **1** further comprising:
   a plurality of second level shifting elements coupled in series between the collector the clamp transistor and the first power rail.

4. The electrostatic discharge device of claim **3** wherein the trigger and clamp transistors are constructed from silicon germanium material.

5. An electrostatic discharge device comprising:
   a trigger transistor;
   a first power rail coupled to the collector of the trigger transistor;
   a plurality of first level shifting elements coupled in series with one another and the first power rail;
   a clamp transistor having its base coupled to the emitter of the trigger transistor and its collector coupled to one of the first level shifting elements, the clamp transistor having a frequency cutoff and a breakdown voltage that are both larger than that of the trigger transistor; and
   a second power rail coupled to the emitters of both the trigger and clamp transistors.

6. The electrostatic discharge device of claim **5** wherein the trigger and clamp transistors are constructed from silicon germanium material.

7. The electrostatic discharge device of claim **5** further comprising:
   a plurality of second level shifting elements coupled in series between the collector the trigger transistor and the first power rail.

8. The electrostatic discharge device of claim **7** wherein the trigger and clamp transistors are constructed from silicon germanium material.

9. An electrostatic discharge device comprising:
   a trigger transistor;
   a first power rail coupled to the collector of the trigger transistor;
   a first clamp transistor having a frequency cutoff and breakdown voltage that are both larger than that of the trigger transistor, the collector being coupled to the first power rail and the base being coupled to the emitter of the trigger transistor;
   a second clamp transistor having a frequency cutoff and a breakdown voltage that are both larger than that of the trigger transistor, the collector coupled to the first power rail and the base being coupled to the emitter of the first clamp transistor;
   a second power rail coupled to the emitters of the trigger, first and second clamp transistors.

* * * * *

Figure 4.A.2 (*Continued*)

第 4 章 专 利 131

US 006762918B2

(12) **United States Patent**
Voldman

(10) Patent No.: **US 6,762,918 B2**
(45) Date of Patent: **Jul. 13, 2004**

(54) **FAULT FREE FUSE NETWORK**

(75) Inventor: **Steven Howard Voldman**, South Burlington, VT (US)

(73) Assignee: **International Business Machines Corporation**, Armonk, NY (US)

(*) Notice: Subject to any disclaimer, the term of this patent is extended or adjusted under 35 U.S.C. 154(b) by 0 days.

(21) Appl. No.: **10/063,858**

(22) Filed: **May 20, 2002**

(65) **Prior Publication Data**
US 2003/0213978 A1 Nov. 20, 2003

(51) Int. Cl.^{7} **H02H 3/20**; H02H 9/04
(52) U.S. Cl. **361/91.1**; 257/355
(58) Field of Search 361/91, 90, 91.1, 361/57, 91.61, 56; 327/525, 252; 257/758, 355, 356, 357, 358, 359, 360, 361, 200, 365

(56) **References Cited**
U.S. PATENT DOCUMENTS

5,159,518 A	*	10/1992	Roy 361/56
5,452,171 A		9/1995	Metz et al. 361/56
5,610,790 A		3/1997	Staab et al. 361/56
5,880,917 A		3/1999	Casper et al. 361/56
5,910,874 A		6/1999	Iniewski et al. 361/56
5,956,219 A		9/1999	Maloney 361/56
5,963,409 A		10/1999	Chang 361/56
6,074,899 A		6/2000	Voldman 438/155
6,281,702 B1		8/2001	Hui 326/30
6,319,333 B1	*	11/2001	Noble 148/33.2
2003/0016074 A1	*	1/2003	Yung 327/525

FOREIGN PATENT DOCUMENTS

JP 2000216673 8/2000

* cited by examiner

Primary Examiner—Long Pham
Assistant Examiner—Dana Farahani
(74) *Attorney, Agent, or Firm*—Richard A. Henkler

(57) **ABSTRACT**

A fuse state circuit for reading the state of a fuse that is enhanced to reduce the circuits susceptibility to ESD, EOS or CDM events.

11 Claims, 4 Drawing Sheets

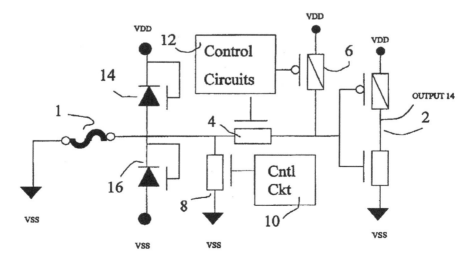

Figure 4.A.3

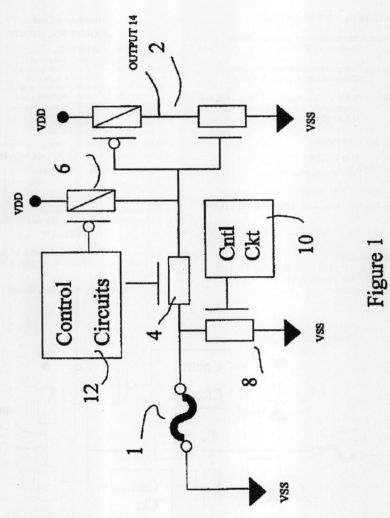

Figure 4.A.3 *(Continued)*

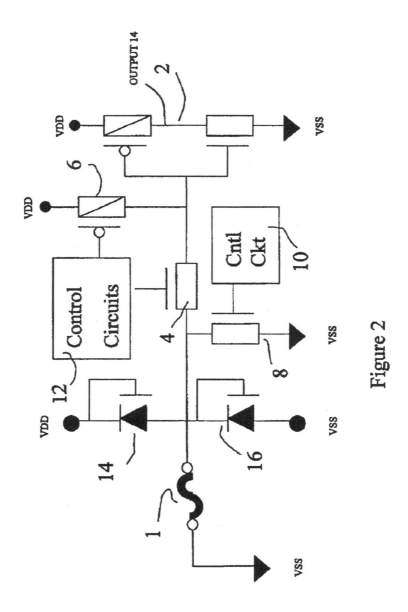

Figure 4.A.3 (Continued)

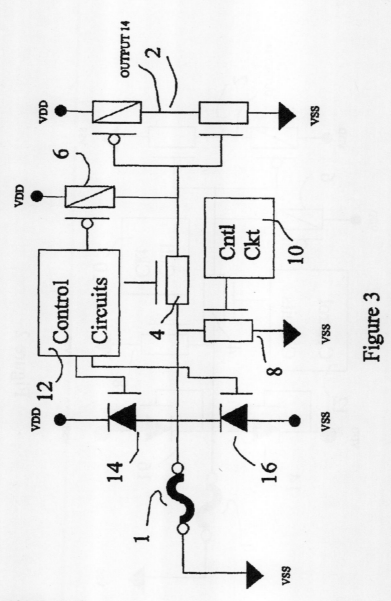

Figure 4.A.3 (Continued)

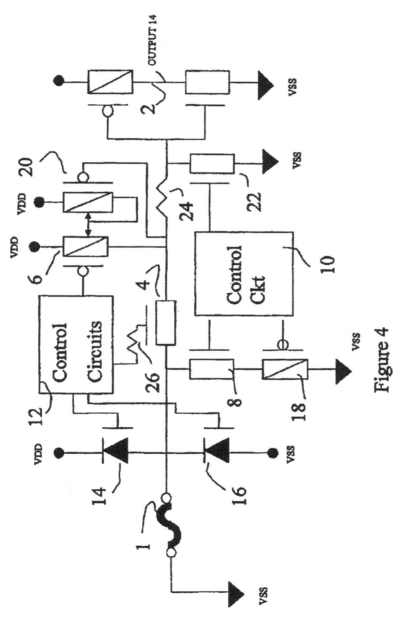

Figure 4.A.3 (*Continued*)

US 6,762,918 B2

1
FAULT FREE FUSE NETWORK

BACKGROUND OF THE INVENTION

1. Field of the Invention

The present invention generally relates to fuse networks, and, more particularly, to providing fault tolerance for fuse networks in integrated circuits.

2. Description of the Related Art

As integrated circuits have become more complicated offering greater functionality, fuse networks have become an integral part of the overall design process. Fuse networks are currently being used for enabling or disabling circuits/circuit blocks, and thereby, providing differing functionality (e.g. customized) without the need to have multiple masks or other redundant circuit processes. In addition, fuse networks are also being used to correct defects in circuits that result from various manufacturing deffiencies (e.g. photolithographis, etch, masking, process).

Obviously, in order to make the fuse(s) useful, some type of circuitry must be used to determine the state of the fuse (e.g. open/close). Fuses are typically blown (opened) via electrical means (e.g. electrical fuse blow) or optical means (e.g., Laser fuse blow). Unfortunately, the techniques used for blowing the fuse can induce enough energy to lead to Electrical OverStress (EOS) or ElectroStatic Discharge (ESD) failure of the circuitry used to read the state of the fuse ("Fuse state circuitry"). For example, electrical fuse blow can lead to currents which cause failure of the fuse element and the fuse state circuitry at the same time. In further example, laser fuse blow can lead to conversion of optical to thermal energy where the thermal energy can lead to an electrical current, forming a pulsed electrical event propagating into the fuse state circuitry.

It would, therefore, be a distinct advantage to have fuse state circuitry that was protected from both EOS and ESD events. The present invention provides such state circuitry.

BRIEF SUMMARY OF THE INVENTION

The present invention is applicable to fuse networks, and more specifically, to the circuitry that reads the state of the fuse(s) in such networks. The present invention provides ESD, EOS, and CDM protection to the circuitry that reads the state of such fuse(s).

BRIEF DESCRIPTION OF THE SEVERAL VIEWS OF THE DRAWINGS

The foregoing and other objects, aspects and advantages will be better understood from the following detailed description of a preferred embodiment of the invention with reference to the drawings, in which:

FIG. 1 is a schematic diagram representative of an environment in which a fuse network can be used;

FIG. 2 is a schematic diagram illustrating an example of how the fuse network of FIG. 1 can be modified to include ESD and EOS protection according to the teachings of the present invention;

FIG. 3 is a schematic diagram illustrating the fuse network of FIG. 1 modified to avoid ESD, EOS, and CDM failures according to the teachings of the present invention; and

FIG. 4 is a schematic diagram illustrating the fuse network of FIG. 1 modified to avoid the introduction of ESD, EOS, and CDM failures according to the teachings of the present invention.

2
DETAILED DESCRIPTION OF THE INVENTION

Detailed Description of a Preferred Embodiment of the Invention

Referring now to the drawings, and more particularly to FIG. 1, there is shown a schematic diagram representative of an environment in which a fuse network can be used. The fuse network includes, a fuse **1**, a receiver circuit **2**, half pass NFET **4**, NFET **8**, PFET **6**, and control circuitry **10** and **12**. In this example, fuse structure **1** is connected to ground potential on one side and to half pass NFET **4** and NFET **8** on the other side. NFET **8** and NFET **4** are coupled to PFET **6** and receiver **2**.

The receiver circuit **2** can be a simple inverter (PFET/NFET in the configuration as shown). If the fuse element **1** is intact, then the output side of the fuse will read "0" pulling the input to the NFET **4** to a low state. Provided the gate of the half pass NFET **4** is enabled on, the fuse **1** pulls the circuitry low forming a high state at the output **14** (i.e. state=fuse intact). If, however, the fuse **1** is blown, then the fuse pulls the circuitry high forming a low state at the output **14** (i.e. state=fuse blown).

Additional control circuit elements **12**, **10**, **8**, and **6** are included to determine and verify the state of the fuse **1**. Circuit element **8** is an NFET device tied to ground whose gate is connected to control circuity **10** for determining the fuse state. For example, when the control circuitry **10** is enabled to place a voltage on the gate of the NFET **8**, the input of the NFET pass transistor **4** can be pulled low by turning NFET **8** in an on state. PFET pull-up element **6** and control circuitry **12** can be used to pull the input node of the receiver **2** to a high state by enabling PFET pull-up element **6**.

The fuse network illustrated in FIG. 1 is subject to many different potential failure mechanisms from EOS or ESD, possibly leading to false readings of the state of fuse **1**. For example, it has been determined that pull-down NFET **8** can have a MOSFET second breakdown event between its drain and the source in bulk CMOS and SOI applications. The ESD failure of NFET **8** forces the node low, resulting in having the receiver indicating that the fuse is still intact, when in fact, it could be open.

Erroneous states can also result from the ESD failure of pull-up PFET **6**. For example, if VDD is grounded, and an ESD event occurs, the p-diffusion diode of the PFET **6** can forward bias. This leads to the current flow across the half pass NFET **4** leading to failure of both the PFET **6** and the NFET half pass NFET **4**.

Reference now being made to FIG. 2, a schematic diagram is shown illustrating an example of how the fuse network of FIG. 1 can be modified to include ESD and EOS protection according to the teachings of the present invention. Gated diode structures **14** and **16** (also referred to as Lubistors) have been added to provide ESD and OSD protection. Both gated diode structures **14** and **16** include an anode, cathode and gate structure. The gate structure includes a polysilicon film and a thin dielectric MOSFET gate structure. In this implementation, the polysilicon film is doped n+ and p+ on the cathode and anode side, respectively. The gate structure(s) are connected to their respective cathode. As an alternative embodiment, the gate structures can be removed using additional masks.

These elements can provide ESD protection improvements from human body model (HBM), machine model (MM) and other pulse waveforms entering from the fuse electrode output node. In this implementation, it can be constructed in a CMOS technology (e.g., single well CMOS, twin well CMOS, triple well CMOS, RF CMOS), a silicon-on-insulator (SOI) technology, or a BiCMOS technology.

In the case of the SOI technology, the bodies of the NFET and PFET transistor can be floating or connected with body contacts. In SOI technology, it has been shown by the inventor that different failure mechanisms occur which are dissimilar to the events in bulk CMOS. For example, it has been shown that charged device model (CDM) failures occur in the half pass NFET **4** from the gate to the input node of the half pass NFET **4** when the node is shorted (e.g. Intact fuse **1** connected to ground). In this case, the CDM event can destroy the half pass NFET **4** and also lead to failure of the fuse element.

In SOI technology, it has been shown by the inventor that ESD failure from a CDM event can occur in the gated diode element **14** and gated diode element **16**. When the ground rail (VSS) is charged negative, and the input node is grounded, the gated diode element **16** can have an CDM failure from the gate to the cathode. When the VDD is charged positive, CDM failure of the gated diode **14** can occur from the gate to the anode. Hence, this can lead to failure of the fuse **1** and a false reading of whether the fuse **1** is open or shorted. Hence, the introduction of SOI ESD networks to improve the robustness of the SOI fuse network leads to two new failure issues due to a CDM event. Also by connection of the gates of element **14** and element **16** leads to electrical overstress of the gates in overshoot or undershoot conditions.

Reference now being made to FIG. 3, a schematic diagram is shown illustrating the fuse network of FIG. 1 modified to avoid ESD, EOS, and CDM failures according to the teachings of the present invention. In this embodiment, the gated diode elements **14** and **16** are isolated from the power rails VDD and VSS to avoid CDM failures. This is accomplished by connected the gates of the diode elements **14** and **16** to the control circuit logic **12**. Alternatively, the gates of the diode elements **14** and **16** could be controlled by other circuitry such as a gate biased network (e.g. invertor tied to ground).

Reference now being made to FIG. 4, a schematic diagram is shown illustrating the fuse network of FIG. 1 modified to avoid the introduction of ESD, EOS, and CDM failures according to the teachings of the present invention. In this embodiment, additional measures are provided to avoid the misreading of the fuse **1** state. Specifically, PFET **18** is placed below NFET **8** to avoid having NFET element **8** undergo MOSFET second breakdown and leading to false reading of the fuse **1**. The PFET **18** will not undergo MOSFET second breakdown from a positive pulse. Additionally, using the control circuit **10**, the PFET **18** can be shutoff to insure that the NFET **8** does not have a path to ground even if it did have a MOSFET second breakdown failure. Thus, PFET **18** serves as a means to prevent electrical shorting, missreading, and as a means to isolate NFET **8** logically from the evaluation of the fuse **1**. For example, to verify a low, the PFET gate **18** can be set low, and the NFET **8** can be set high via control circuit **10**.

In order to avoid the pinning of the well of the pullup PFET **6**, a second PFET **20** is placed such that the well of PFET **6** is not connected to VDD. For example, when the output of the half pass NFET **4** is high, PFET **20** is off leading the well of pullup PFET **6** to float, thus, preventing the pinning of the potential across the hall pass NFET **4**. From this example, it can be seen that the addition of PFET **20** avoids the electrical overstress of the pullup PFET **6** and/or failure of the half pass NFET **4** due to EOS or ESD events (e.g. from the pad signal).

To further improve the fuse network, a NFET **22** and resistor **24** have been coupled to the receiver inverter **2**. The gate of the NFET **22** is connected to the control circuitry to avoid electrical overstress of the gate structure (e.g. avoiding the grounding of the gate to a VSS or ground rail), and to ensure that a CDM mechanism does not lead to failure. With the addition of the NFET **22** and resistor **24**, the HBM and MM ESD robustness will improve (e.g. They form a resistor divider when the NFET **22** is on, reducing the voltage at the receiver inverter **2** input). Additionally, to avoid the CDM failure mechanisms observed in SOI technology a resistor element **26** is placed in series with the gate of the half pass NFET **4**. This provides a robust pass transistor and avoids CDM failures.

The discussed modifications to the fuse network of FIG. 1 are applicable to CMOS, RF CMOS, BiCMOS, BiCMOS SiGe, BiCMOS SiGeC, strained Si, and other technologies which construct semiconductor products that require fuses. The fuse **1** can be, for example, constructed from aluminum, titanium/aluminum/titanium, copper, refractory metals, silicides, polysilicon and silicon elements.

Various modifications may be made to the structures of the invention as set forth above without departing from the spirit and scope of the invention as described and claimed. Various aspects of the embodiments described above may be combined and/or modified.

What is claimed is:

1. In an integrated circuit, a fuse network comprising:

a fuse;

ESD circuitry capable of providing ESD protection; and

control circuitry capable of controlling the activation of the ESD circuitry and

being protected by ESD events by the ESD circuitry, the control circuitry including an inverter capable of reading the status of the fuse, and a pass through transistor to gate the inverter.

2. The fuse network of claim **1** wherein the ESD circuitry includes:

a cascoded set of diodes each having a gate activated by the control circuitry.

3. The fuse network of claim **2** wherein the control circuitry includes:

a pfet coupled to the input of the inverter and the output of the pass through transistor.

4. The fuse network of claim **3** wherein the control circuitry includes a Nfet coupled in parallel with the input of the pass through transistor and cascoded set of diodes.

5. In an integrated circuit, a fuse network comprising:

a fuse

a cascoded pair of lubistors coupled to the fuse, each one of the lubistors having a gate for activation

a pass through NFET coupled to the pair of lubistors;

a NFET coupled to the source of the pass through NFET;

a first PFET coupled to the drain of the pass through NFET;

an inverter coupled to the drain of the pass through NFET; and

control circuitry coupled to the gates of the lubistors.

Figure 4.A.3 (*Continued*)

US 6,762,918 B2

6. The fuse network of claim 5 wherein the control circuitry is also coupled to the gates of the NFET and the first PFET.

7. The fuse network of claim 6 further comprising:
a second PFET coupled to the drain of the NFET for providing ESD protection to the NFET, the gate of the second PFET being activated by the control circuit.

8. In an integrated circuit, a fuse network comprising:
a fuse;
an inverter for reading the status of the fuse;
a pass through transistor for gating the inverter;
a PFET coupled to the input of the inverter and the output of the pass through transistor;
a control circuit for controlling the activation of the pass through transistor;
a cascoded set of diodes capable of providing ESD protection to the control circuit, pass through transistor, PFET, and inverter, the gate of the diodes being activated by the control circuit;
a NFET coupled in parallel with the input of the pass through transistor and cascoded set of diodes; and
a PPET coupled to the drain of the NFET and ground.

9. In an integrated circuit, a fuse network comprising:
a fuse;
control circuitry capable of controlling the reading circuitry;
an ESD circuit for protecting the reading circuitry and control circuitry from high voltage events, the ESD circuit being activated by the control circuitry;
reading circuitry capable of reading the value of the fuse, the reading circuitry including:
an inverter for reading the status of the fuse;
a pass through transistor for gating the inverter;
a PFET coupled to the input of the inverter and the output of the pass transistor; and
an NFET coupled in parallel with the input of the pass through transistor and the ESD circuit.

10. The fuse network of claim 9 wherein the control circuitry comprises:
a first control circuit capable of activating the NFET; and
a second control circuit capable of activating the PFET, and activating the ESD circuit.

11. The fuse network of claim 10 wherein the ESD circuit comprises a cascoded pair of lubistors.

* * * * *

Figure 4.A.3 (Continued)

第 5 章
专利说明书附图

5.1 引言

在本章中，重点讨论专利的说明书附图，展示了不同类型的图的例子[1]，讨论了附图与说明书和权利要求相关的规则和要求[2-21]。

根据《美国法典》第 35 篇第 113 条要求推荐专利附图[1]：

"申请人应在必要时提供附图，以便了解要申请专利的主题。"根据 35 USC 113，大多数专利需要专利附图。

对于可以用文字清楚描述的发明，不需要专利附图。该描述必须足够清楚。以下主题不能用附图来描述[1]：

- 过程
- 混合物
- 涂层制品
- 层压结构
- 其主要区别特征是特定材料的发明

专利附图视图可以分为正交视图和透视图。正交视图是观察者在其中物体的特定一侧的位置观察到的视图。正交视图包含以下视图[1]：

- 前视图或正视图

- 后视图
- 左视图
- 右视图
- 俯视图
- 仰视图

透视图是把三维物体显示在二维平面上的一个视图[1]，透视图包含以下视图[1]：

- 前部透视图
- 后部透视图
- 右侧透视图
- 左侧透视图
- 顶部透视图
- 底部透视图

透视图可以根据透视缩短进行修改，透视缩短使用"灭点"，可以使视图看起来更逼真。透视缩短可以在人造地平线上引入多个灭点。

5.1.1 绘图技巧——手绘

可以使用钢笔和其他工具进行绘图，也可以从草图开始或使用描摹技术进行绘图。手工绘图时，通常先用铅笔绘制图形然后再用墨水画线。视图应采用可擦除草图线来绘制。最终提交前应擦除铅笔线。

5.1.2 绘图技术——通过计算机绘图

附图可以通过计算机生成。计算机辅助设计（CAD）工具可为专利申请生成要提交的附图。CAD 工具专为生成专利附图而设计，可以通过描摹照片的边缘并生成专利的线条图来实现。

5.1.3 绘图技术——通过照相机生成附图

附图可以使用照相机来生成。但是除了植物专利，照片不能作为其他专利附图提交。通过照相机拍照的图片可用于生成附图。

5.2 专利附图——发明专利

发明专利包括结构、装置和方法。对于不同类型的专利，附图是不同的。在发明专利中，附图中包括现有技术和本发明公开的内容。发明专利可以包含以下内容：

- 机器
- 制造物品
- 方法
- 混合物

专利附图可以是不同的类型。专利附图的标准包含在专利审查指南（MPEP）第 608.02 部分中，具体包括以下类型的附图[22]：

- 框图
- 流程图
- 二维图
- 透视和三维（3-D）视图
- 剖面或剖视图
- 分解图

图 5-1 展示了不同类型的发明专利。发明专利可能包含多种不同类型的附图。

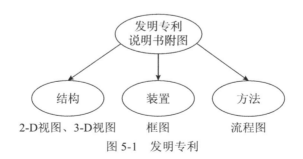

图 5-1　发明专利

5.3 专利附图——结构

结构图使发明可视化，以此支持专利申请[1]。这些图应该充分向公众公开，以便专利的读者可以重现本发明。结构图包括以下类型：

- 框图
- 二维视图
- 透视图和三维（3-D）视图
- 剖面图或剖视图
- 分解图

专利说明书中包含现有技术的附图和与本发明相关的附图。现有技术的附图用于提供现有技术中过去已知的图与新交底书之间的对比，这样便于对比并突出差异。

通常在专利申请的背景技术部分讨论现有技术附图。因此，这些附图通常放在附图清单的最前面。现有技术附图通常在其图号下标明"现有技术"，并在其下方指定为"图××"。专利申请的具体实施方式在附图的范围内讨论专利的实施示例。因此，这些附图通常放在附图清单中的现有技术附图之后，这些图在其下方被指定为"图××"。

图中显示的所有物理形状需要加以标识。通过一条连接数字和物理性状的数字线或"波浪线"来对这些形状加以标识[1-5]。

图 5-2 是一个结构图的示例。

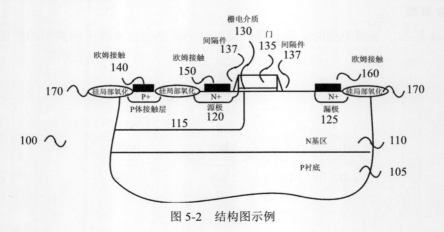

图 5-2　结构图示例

结构图绘制规则如下所述。

在美国专利商标局（USPTO），结构图有特定的规则。如果认为对理解本发明有必要，则需要结构图。结构图包括流程图和示意图。

附图不应包含文字内容，除非某个字或词对理解附图是必须的。如果结构图包含文字，在确保不干扰附图中任何线的基础上，可以将文字放置在图中[1]。

对附图的特殊要求包括以下内容：

- 草图绘制：先用铅笔绘制草图，然后在最终的铅笔标记上涂上墨水线。
- 导尺：所有的墨水线都应该借助导板、标尺、模板和曲线板绘制。
- 线条：用耐久的黑色线条来绘制附图。线条应当密实、均匀清晰、足够深，不得着色。形成锐角的线条必须相交且不重叠。还要为这些线条上墨。
- 擦除：避免擦墨水线。使用白色修正液来覆盖不需要的墨水标记。
- 横截面：横截面采用斜的平行线来交叉地画出阴影。剖面不会干扰或阻碍阅读附图标记。
- 比例：如果尺寸线性减小，附图应清晰，可以分辨出图中各个细节。
- 图形比例尺：可将比例尺置于附图上，比例尺用图形表示。
- 数字、字母和附图标记：所有数字、字母和附图标记都要足够简单和清晰，让读者能够识别结构中的形状，附图中不可以使用括号。
- 线的质量：通过制图工具或类似设备来绘制附图，以保证线的质量。
- 图中的元素：每一个元素与图中其他元素的比例关系要正确。
- 数字和字母的高度：高度不得小于0.32厘米。
- 图的文字：文字应采用拉丁字母或希腊字母。
- 图形符号：可使用流程图和电子符号的图形符号。
- 形状创建工具：形状创建工具，生成附图的圆、椭圆、线、多边形。
- 多个图：多个附图可以绘制在同一张纸上。
- 图编号：不同的图应当用阿拉伯数字顺序编号。
- 附图标记：说明书中未出现的附图标记不得出现在附图中。

5.4 专利附图——装置

装置图使发明可视化，以此支持专利申请[1]。这些图应该充分向公众公开，以便专利的读者可以重现该装置。装置图包括以下类型的图：

- 框图

- 二维视图
- 透视图和三维（3d）视图
- 剖面图或剖视图
- 分解图

图 5-3 展示了一个装置附图的示例。

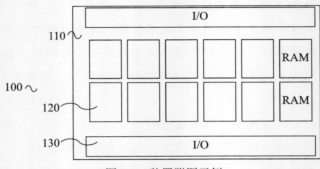

图 5-3　装置附图示例

装置图绘制规则如下所述。

装置图不应包含文字内容，除非确定一个字或词对理解附图是必须的。如果装置图包含文字，确保在不干扰图中任何线的基础上，可以将文字放置在图中[1]。

对附图的特殊要求如下：

- 线条：用耐久的黑色线条来绘制附图。线条应当密实、均匀清晰、足够深，不得着色。形成锐角的线条必须相交且不重叠。还要为这些线条上墨。
- 横截面：横截面采用斜的平行线来交叉地画出阴影。剖面不会干扰或阻碍阅读附图标记。
- 比例：如果尺寸线性减小，附图应清晰，可以分辨出图中各个细节。
- 图形比例尺：可将比例尺置于绘图上，比例尺用图形表示。
- 数字、字母和附图标记：所有数字、字母和附图标记都要足够简单和清晰，让读者能够识别装置中的形状，附图中不可以使用括号。
- 线的质量：通过制图工具或类似装置来绘制附图，以保证线的质量。
- 图中的元素：每一个元素与图中其他元素的比例关系要正确。
- 数字和字母的高度：高度不得小于 0.32 厘米。

- 图的文字：文字应来自拉丁字母或希腊字母。
- 多个图：多个附图可以绘制在同一张纸上。
- 图编号：不同的图应当用阿拉伯数字顺序编号。
- 附图标记：说明书中未出现的附图标记不得出现在附图中。

5.5 专利说明书附图——电路图

专利说明书中可以添加电路图，以显示元件和它们之间的电气连接[3-5]。电路图使发明可视化，以此支持专利申请[1]。这些图应该充分向公众公开，以便专利的读者可以重现该电路，从而展示块和元件的连接性。电路图包括以下类型的图：
- 框图
- 单元素视图
- 框图和单元素视图

电路图可以是代表电路或装置中的功能块的框图，需要显示功能块之间的电气连接和信号线。

电路图也可以是电路中所有有源和无源的单个元件的视图。无源元件可以是电阻器、电容器和电感器。有源元件可以包括双极结晶体管（BJT），金属氧化物半导体场效应晶体管（MOSFETS），和其他晶体管类型（例如，节型场效应管 JFETS），电路也可以提供其他内容表示，如逻辑门（如与 AND、或 OR、或非 NOR、与非 NAND）和更高级别的电路块（如复用器 MUX）。

电路图还可以用至少一个框图和至少一个单一元素的混合图表示。混合图可以提供对电路进行比较的一般表示。

图 5-4 展示了一个电路图的示例，图中有一个高级描述和一些元件。通过提供高级图，可以扩大权利要求的保护范围。这种方式允许对权利要求进行概括。在图中的元件和权利要求之间可以建立一种一对一的对应关系。对于独立权利要求来说，对图像进行概括是一个很好的做法。

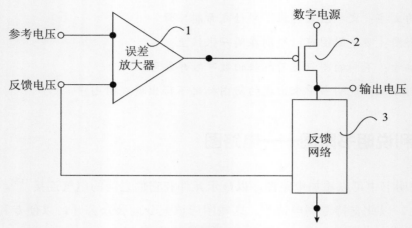

图 5-4 电路图——高级

图 5-5 展示了在电路元件层面上电路图的示例。权利要求对图中的元件进行标识。权利要求也将陈述各种元件是如何集成和相互关联的。这里允许对权利要求进行概括。图中的元件和权利要求之间可以建立一一对应关系。通常，元件的符号是 IEEE 标准符号，但这不是必须的。因此，可以建立新的符号系统。许多年以来，IBM 都使用自己的符号来表示晶体管元素。

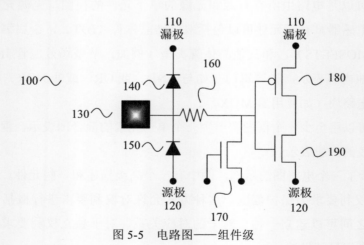

图 5-5 电路图——组件级

电路图的特殊要求如下[1-5]。

- 电路元件：标准符号可用于单个无源或有源元件的电路元件，以及逻辑门和其他确定的符号表示，可以包括电阻器、电容器、电感、晶体管（如 BJT、

MOSFET、JFET、IGFET）和逻辑电路（如 NAND、NOR、OR、XOR），电源（如电流源、电压源、振荡源）。

- 线条：用耐久的黑色线条来绘制附图。线条应当密实、均匀清晰、足够深，不得着色。
- 比例：如果尺寸线性减小，附图应清晰，可以分辨出图中的各个细节。
- 数字、字母和附图标记：所有数字、字母和附图标记都要足够简单和清晰，让读者能够识别装置中的形状，附图中不可以使用括号。
- 线的质量：通过制图工具或类似设备来绘制图，以保证线的质量。
- 数字和字母的高度：高度不应小于 0.32 厘米。
- 图的文字：文字应采用拉丁字母或希腊字母。
- 多个图：多个附图可以绘制在同一张纸上。
- 图编号：不同的图应当用阿拉伯数字顺序编号。
- 附图标记：说明书中未出现的附图标记不得出现在附图中。

5.6　专利附图——系统

系统的专利图可以表示为结构、装置或方法。机械系统可以表示为结构图。电气系统可以表示为装置图。图 5-6 是一个系统图的例子。

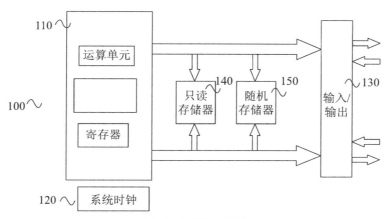

图 5-6　系统级图示例

系统图绘制规则如下所述。

系统级图与结构、装置甚至方法的发明专利中使用的规则相同。例如，系统级图可以是一个框图（见前文）。

5.7 专利附图——方法

方法图采用流程图的格式，其中流程图由方框和方框箭头组成[1-5]。每个框表示方法流程中的一个步骤。图 5-7 是一个方法图的示例说明。在方法权利要求的步骤和图中方框内包含的文本之间建立一对一的对应关系。

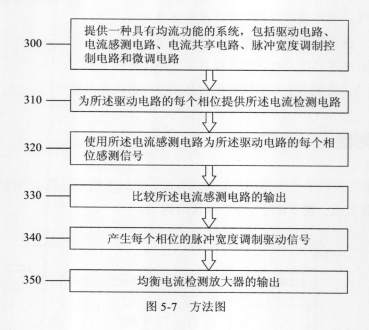

图 5-7　方法图

5.7.1　方法图规则

方法图包含文本内容，其中流程图中的每个步骤都表示方法中的一个步骤。方法图的特殊要求包括以下几项。

- 流程图形状：流程图由表示方法步骤的形状组成。
- 流程图文本：流程图中形状中的每一步都包含方法的步骤。

- 流程图连接：流程图中的每个连接都用方框箭头表示，以展示流程图中的流向。
- 线条：用耐久的黑色线条来绘制附图。线条应当密实、均匀清晰、足够深，不得着色。
- 数字、字母和附图标记：所有数字、字母和附图标记都要简单明了，以便读者在流程图中识别所示方法的步骤。
- 线的质量：通过制图工具或类似设备来绘图，以保证线的质量。
- 数字和字母的高度：高度不得小于 0.32 厘米。
- 图的文字：文字采用拉丁字母或希腊字母。
- 图编号：不同的图应当用阿拉伯数字顺序编号。

5.7.2 方法图与权利要求间的对应关系

在方法权利要求的步骤和图中方框内包含的文本之间需要一对一的对应关系。在权利要求中，每个步骤单独成行，后面要跟逗号。

5.8 专利附图——外观设计

图是外观专利中最重要的部分。外观专利申请应包含所有的视图，对外观设计进行完整描述，以便充分向公众公开。图可以是正式的和非正式的。正式的图纸必须符合一套严格的标准。手绘草图被视为非正式的图。使用非正式图提交的申请可以提交并获得申请日。在审查专利申请之前，应当提交正式的图。虽然非正式图不是最终的正式图，但它们仍然需要足够高的质量以便理解本发明。注意，工程图纸，如蓝图，不适用于外观专利。如果工程图清楚地显示了发明，可以作为非正式图纸提交。

外观设计图可以包括后视图、俯视图、仰视图、左侧和右侧视图。对于圆形或对称的对象，可以采用俯视图、仰视图和侧视图来全面描述专利设计。

外观专利图必须准确地展示物体的形状、比例和表面轮廓。它们还可以显示特殊的材料属性或纹理。图纸必须显示正常使用过程中的对象的每个可见的特征。因此，外观设计的任何部分都不能遗漏。外观专利必须加阴影来描绘表面特征，如透明度，并区分开放区域和实心区域。例如，编号 USD712 405S、标题为电子设备[23]、发明人为 Akana 等人的专利图，是手机的外观专利。在这个专利中，有 10 幅图提供了不同的透视图。图 5-8（a）～（d）提供了外观专利的不同视图示例。

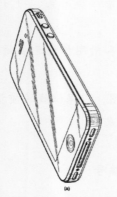
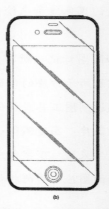

（a）外观设计图——底前立体图　　（b）外观设计图——前视立体图

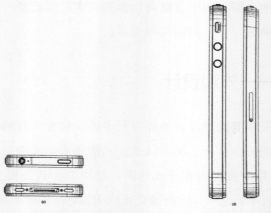

（c）外观设计图——顶视图和底视图　　（d）外观设计图——左侧视图和右侧视图

图　5-8

专利 US D712 405S 的说明书中的描述如下[23]：

- 图 1 是我们新设计的电子设备的底前立体图；
- 图 2 是其正面立体图；
- 图 3 是其后顶立体图；
- 图 4 是其底后立体图。
- 图 5 是其主视图；
- 图 6 是其后视图；

- 图 7 是其俯视图；
- 图 8 是其仰视图；
- 图 9 是其左侧视图；
- 图 10 是其右侧视图。

外观设计图绘制的特殊要求包括以下内容。

- 实线：用耐久的黑色线条来绘制附图。线条应当密实、均匀清晰、足够深，不得着色。
- 边界线：边界线将一个实施例中所声明的区域与放弃的区域分开。该线采用点画线的样式。
- 假想线：假想线表示图的放弃部分。
- 投影线：投影线代表与另一实施例分离的部件或设备。该线采用点画线的样式。
- 虚线：虚线可在发明专利中使用。在专利申请中，当一条线被另一个部分或实体挡住时，使用虚线来表示要披露的这条线。
- 阴影：阴影可以清楚地显示外观设计的三维的所有面的特征和轮廓。
- 数字、字母和附图标记：所有数字、字母和附图标记都要足够简单和清晰，让读者能够识别结构中的形状，附图中不可以使用括号。
- 线的质量：要用绘图工具或同等工具绘制线，以保证线的质量。
- 图中的元素：每个元素与图中其他元素的比例应当合适。
- 数字和字母的高度：高度不得小于 0.32 厘米。
- 绘图的文字：文字应采用拉丁字母或希腊字母。
- 多个图：多个附图可绘制在同一张纸上。
- 图编号：不同的图应当用阿拉伯数字顺序编号。
- 附图标记：说明书中未出现的附图标记不得出现在附图中。

5.9　专利附图——植物图

凡发明或发现，并以无性生殖方式繁殖一个独特的植物新品种，都可以被授予植物专利。这一授权保护发明人的权利，排除他人通过无性繁殖、销售或使用该专利中的植物新品种的权利[1-5]。

植物图与发明专利或外观专利有明显不同。植物专利的图必须要为授权的植物提供描述信息。植物专利中的图必须显示植物最显著的特征，以便能够与之前已有植物进行区别。图 5-9 给出了一个植物图的示例。注意，多个图可以放在同一页上。图 5-10 给出了植物照片的一个示例，彩色照片是允许使用的。

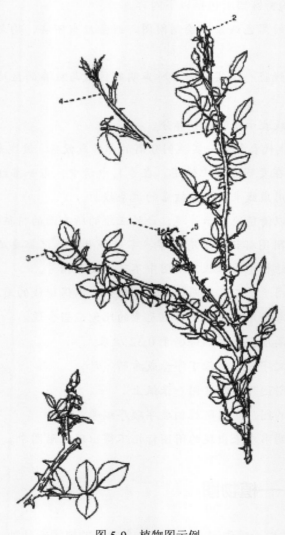

图 5-9　植物图示例

图 5-10　植物照片

植物专利图绘制的规则如下[1-5]。

- 植物专利中的图必须以足够的尺寸显示植物的最显著的特征，以便在缩小图形比例 50% 后仍能够识别。
- 植物专利图的尺寸应足够大，以区分其与亲本植物的不同特征。
- 附图必须如照片般清晰。
- 如果颜色具有区别性的特征，那么必须是彩色附图。
- 植物专利的图必须提供与重要的亲本植物进行区分的手段。如果叶子、树皮、花朵和果实，其特征与其他植物有明显区别，应附一幅以上的视图并清楚指明。
- 除非专利审查员明确要求，否则附图不需要页码。
- 裱装的植物附图，必须与发明申请中的图的要求相同。

5.10　独特的专利附图——计算机可读介质权利要求

计算机可读介质权利要求——Beauregard 权利要求，是这样一种权利要求：计算机程序按照权利要求的形式，写到某一实体——计算机存储介质上，其中计算机程序通过编码的指令来完成某一过程。这些权利要求考虑的计算机可读介质通常是软盘或 CD-

ROM，因此这种类型的权利要求有时被称为"软盘"（floppy-disk）权利要求。过去，纯指令的权利要求通常被认为是不可授予专利权的。用于计算机存储介质类权利要求的图是一个框图，代表与权利要求相关的系统。图 5-11 是一个用于计算机存储介质类权利要求的图的示例。

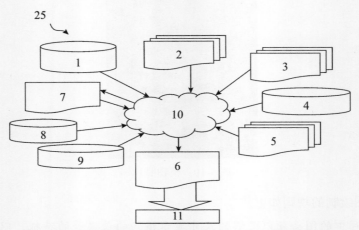

图 5-11　计算机可读介质权利要求（Beauregard claims）

5.11　专利图和审查意见

在专利审查过程中，要审查专利附图以验证其是否符合专利规则[1]。美国专利商标局的两个部门对图进行审查：图审查部门和审查部门。在收到专利受理通知书之后或随着第一次审查意见通知书，可以收到图审查部门的结果。细则 85（a）（37 CFR.1.85（a），其中 CFR 为美国联邦法规）规定，在将申请分配给专利审查员之前，附图必须符合所有手续。专利受理通知书发送后，附图被发送到图审查部门。在图审查部门，美国专利商标局的初步审查人员将对图纸进行研究，并发送一份名为"专利图审查通知"（PTO 948）的审查通知。审查表 PTO 948 将包含所有不合规范和所需的更正结果。如果附图中没有错误，则不发送 PTO 948 表格。如果存在不合规范的情况，新的图纸就需要发送更正。

5.11.1 初步审查专利图

"专利图初步审查通知"包含一系列要求,包括以下内容:

- 附图 -37 CFR 1.84(a)可接受的图纸种类
- 照片 -37 CFR 1.84(b)
- 纸的类型 -37 CFR 1.84(e)
- 纸的尺寸 -37 CFR 1.84(f)
- 边距 -37 CFR 1.84(g)
- 视图 -37 CFR 1.84(h)
- 剖面图 -37CFR 1.84(h)(3)
- 视图安排 -37 CFR 1.84(j)
- 尺寸比例 -37 CFR 1.84(k)
- 线条、数字及文字的特质 -37 CFR 1.84(i)
- 阴影 -37 CFR 1.84(m)
- 数字、字母和附图标记 -37 CFR 1.84(p)
- 导引线 -37CFR1.84(q)
- 纸的编号 -37 CFR 1.84(t)
- 图的编号 -37 CFR 1.84(u)
- 更正 -37 CFR 1.84(w)
- 设计图纸 -37 CFR 1.152

5.11.1.1 图 37 CFR 1.84(a)可接受的图的种类

除非请求被批准,否则彩色图是不被接受的。只有批准了请求,才允许使用彩色图。其次,铅笔和非黑色墨水也是不允许的。图必须用黑色书写工具或喷墨打印机来完成。

如果请求未获批准,或者使用了铅笔和非黑色墨水,专利审查员会提出异议。对此的补救办法是启动请求书或用适当的打印机或黑色墨水笔以纯黑色线条重新作图。

5.11.1.2 照片 37 CFR 1.84(b)

在下列情况发生时,照片将被提出异议:

- 除非请求被批准，否则照片是不被接受的。
- 需要三种不同色调。
- 照片未正确裱装（必须是布里斯托纸板双张照片纸）。
- 质量差（中间色）。

5.11.1.3　纸的类型 37 CFR 1.84（e）

对于图所用的纸的类型也有要求。纸的质量和类型要求如下：

- 纸张必须是白色的。
- 纸张必须耐用。
- 纸张必须柔韧。
- 不接受删除、更改、覆盖、交错、折叠和复印机标记。
- 聚酯薄膜和牛皮纸不可使用。

5.11.1.4　纸的尺寸 37 CFR 1.84（f）

对于图所用的纸的大小也有要求。可接受的尺寸如下：

- 21.0 厘米乘 29.7 厘米（DIN A4）
- 21.6 厘米乘 27.9 厘米（8.5 英寸 ×11 英寸）
- 所有图用的纸必须是相同的尺寸

5.11.1.5　边距 37 CFR 1.84（g）

对于图，有可接受的页边距。对于 A4 纸的可接受页边距为顶部 2.5 厘米，左侧 2.5 厘米，右侧 1.5 厘米和底部 1.0 厘米。尺寸为 8.5×11 英寸的纸张，可接受页边距分别是顶部 2.5 厘米，左侧 2.5 厘米，右侧 1.5 厘米和底部 1.0 厘米。如果没有遵循边距的限制，则会被提出异议。

5.11.1.6　视图 37 CFR 1.84（h）

对于图，有可接受的视图。若图改变，则需要对说明书进行相应的修改。这些包括以下内容：

- 通过投影线或引线连接的视图。
- 视图没有单独标记或正确标记。
- 放大的视图没有单独标记或正确标记。

- 需要括号将图作为一个整体显示。

5.11.1.7 剖面图 37 CFR 1.84（h）（3）

对剖面图会有一定的要求，截面的名称应该用阿拉伯数字或罗马数字标明。此外，还有对剖面线的要求，物体的截面部分需要使用剖面线。

5.11.1.8 视图安排 37 CFR 1.84（j）

合理安排视图是很重要的。下列情况可导致对图的异议：
- 无论纸为直立还是将顶端移转至右侧的情形，文字皆须由左到右横向书写，但图表除外。
- 视图不在纸上的同一平面上。

5.11.1.9 尺寸比例 37 CFR 1.84（k）

为了避免图纸产生异议，尺寸比例应大到足以显示其构造而不拥挤，且将其缩小至三分之二重制时仍能清晰显示。

5.11.1.10 线条、数字和文字的特质 37 CFR 1.84（i）

线条、数字和文字的特性必须是高质量的。它们应该是耐久、整洁、黑色的，粗细一致，界线分明。

5.11.1.11 阴影 37 CFR 1.84（m）

对于阴影，有质量上的要求。阴影线不能苍白、粗糙或模糊。纯黑色阴影是不允许的，纯黑色区域不能变淡。

5.11.1.12 数字、文字和附图说明 37 CFR 1.84（p）

对于图，数字、文字和附图说明应符合下列要求：
- 数字和附图说明应该清晰易读。
- 附图说明必须清楚。
- 数字和附图说明必须与视图方向一致。
- 不能使用英文字母。
- 数字、文字和附图说明的高度必须至少为 0.32 厘米。

5.11.1.13 导引线 37 CFR 1.84（q）

图中必须有导引线，导引线不能交叉，也不能漏掉。

5.11.1.14 纸的编号 37 CFR 1.84（t）

附图纸应用阿拉伯数字从 1 起依序编号。

5.11.1.15 图的编号 37 CFR 1.84（u）

附图中的图应用阿拉伯数字连续编号。

5.11.1.16 更正 37 CFR 1.84（w）

根据美国专利商标局（USPTO）-948 要求更正。

5.11.1.17 设计图 37 CFR 1.152

对于设计图，不得使用表面阴影。实心黑色阴影不用于颜色对比。

5.11.2 异议或驳回

在后续章节中，我们将深入讨论审查意见通知书。在本章中，我们讨论了与附图相关的一些异议或驳回。

基于 35 USC 112 提出的异议如下：

"依据 35 USC 112 提出的异议，本发明的说明书未能提供足够的描述，以对本发明充分公开。"

假定用于专利申请的附图未提供"可实现的公开"，该申请的附图对于本领域技术人员而言不够详细或不够清晰，便无法应用本发明。因此，如果图不符合附图的规则和要求，可能会被提出基于 35 USC 112 的异议。对于发明专利来说，附图和说明书必须配合提供可实现的公开。对于外观专利，图可能不完整、精确或清晰。

对于发明或外观专利，如果图中未显示所有特征，则可能会出现基于 37 CFR 1.83（a）的异议。附图必须显示权利要求中指定的本发明的所有特征。如果权利要求中包含未在附图中显示的特征，则可能会发生异议或驳回。

5.12 专利示例

在本节中,将展示两项植物专利和一项外观专利。植物专利是先前讨论过的专利。

5.13 总结

本章讨论了不同类型的专利的图的要求。发明专利的图与外观专利、植物专利的要求有显著不同。本章给出了结构、装置和方法的图的例子,还给出了外观专利图和植物专利图。本章讨论了专利图的规则和要求,以及与图的驳回相关的一些问题。专利图对专利申请的提交起着关键作用。

第 6 章将着重论述权利要求和权利要求撰写。将介绍不同类型的权利要求结构,突出权利要求的独特风格。该章将涵盖结构权利要求、方法权利要求、混合权利要求、方法加功能权利要求、计算机可读介质权利要求(Beauregard)、马库什(Markus)权利要求、两分法(Jepson)权利要求、方法特征限定的产品权利要求、综合权利要求、信号权利要求和第二医药用途(Swiss-Type)权利要求。

问题

1. 所有专利都需要图吗?请解释相关要求。
2. 专利是否会因为图不合格而被异议?
3. 专利是否会因为图不正确而被驳回?
4. 被异议的图有哪些问题?
5. 图上的所有项目都要标识吗?如果标识有误会怎样?
6. 对于外观专利来说,为使专利完整所需要的图的数目有什么规则?若不完整会怎样?
7. 外观专利要被允许,需要的最少的视图数是多少?
8. 对于电路图,可以使用原理图的屏幕截图吗?
9. 对于电路图,能否使用 IEEE 符号作为装备的标识?
10. 对于电路图,能发明自己的符号系统吗?
11. 对于电路图,可以混合使用高级元素和设备元素吗?为什么要这样做?

12. 对于方法专利图，给出一个流程图的例子。

13. 对于方法专利图，是否要求为方法中的各个元素提供"活跃动词"？

14. 对于植物图，可以有照片吗？规则是什么？

15. 对于植物图，可以有颜色吗？规则是什么？

16. 植物图中应该标记什么？

17. 什么用于系统图？

18. 系统图的规则是什么？

19. 画一个系统图。

20. 画一个植物专利图。

21. 画一个发明专利图。

22. 画一个计算机可读介质权利要求（Beauregard）的图，并标记各个元素。

案例研究

案例研究 A

画一个可乐瓶的图，需要多少个视图？

案例研究 B

绘制符合专利要求的手机的图，需要多少个视图？现有专利采用多少个视图？

案例研究 C

画出计算机可读介质权利要求（Beauregard）所需的图。标识图中的所有元素。

案例研究 D

使用高级电路模块和低级电路元件绘制电路。使用最低等级的电路元件来绘制同样的电路。使用高级电路模块的优点是什么？给图中的元素编号。

案例研究 E

用电路元件（包括二极管和金属氧化物半导体场效应晶体管）绘制电路图，并给图中的元素编号。

参考文献

1. Lo, J. and Pressman, D. (2015). *How to Make Patent Drawings*, 7e. Nolo. ISBN: 978-1-4133-2156-2.

2. Amernick, B.A.(1991). *Patent Law for the Nonlawyer: A Guide for the Engineer, Technologist, and Manager*, 2e. Von Nostrand Reinhold. ISBN: 13978-0442001773.

3. Voldman, S. (2014). Short Course, *Innovating, Inventing, and Patenting*, Dr. Steven H. Voldman LLC, Ministry of Science Technology and Innovation (MOSTI), Putrajaya, Malaysia, May 2014.

4. Voldman, S. (2015).Short Course, *Innovating, Inventing, and Patenting*, Dr. Steven H. Voldman LLC, FITIS，Sri Lanka, February 2015.

5. Voldman, S. (2016). Short Course, *Writing and Generating Patents*, Dr. Steven H. Voldman LLC, FITIS, Sri Lanka, February 2016.

6. D.O. Adams. *Patents Demystified: An Insider's Guide to Protecting Ideas and Invention*, American Bar Association, ISBN-13978-163425679, December 7,2015.

7. D. Pressman, T. Tuytschaevers. *Patent It Yourself: Your Step-by-Step Guide to filing at the U.S. Patent Office*, American Bar Association, 2016.

8. H.J.A. Charmsson, and J. Buchaca. *Patents, Copyrights, and Trademarks for Dummies*, American Bar Association, 2008.

9. U.S.Patent Office (USPTO). https://www.uspto.gov(accessed 20 December 2017).

10. European Patent Office (EPO). https://www.epo.org (accessed 20 December 2017).

11. Japan Patent Office. https://www.jpo.go.jp (accessed 20 December 2017).

12. Malaysian Patent Office (MyIPO). https://www.myipo.gov/my (accessed 20 December 2017).

13. World Intellectual Property Organization (WIPO). https://www.wipo.int(accessed 20 December 2017).

14. Organisation Africaine de la Propriete Intellectuelle (OAPI). https://www.oapi.int (accessed 20 December 2017).

15. African Regional Intellectual Property Organization (ARIPO). https://www.airpo.org (accessed 20 December 2017).

16. U.S. Department of Commerce (2000). *Patents and How to Get One: A Practical Handbook*. Department of Commerce, Courier Corporation.

17. Mueller, J. M. (2016).*Patent Law*, 5e. Wolter Kluwer Publications.

18. Stim, R. (2016).*Patent, Copyright, and Trademark: An Intellectual Property Desk Reference*. NOLO.ISBN:978-1-4133-2221-7.

19. Durham, A.L.(2013). *Patent Law Essentials: A Concise Guide*,4e.Oxford: Praeger, ABC-CLIO, LLC. ISBN:13 978-1440828782.

20. DeMatteis, B., Gibb, A., and Neustal, M. (2006). *The Patent Writer: How to Write Successful Patent Applications*. Garden City Park, NY: Patents for Commerce, Square One Publishers.

21. Sutton, E. *Software Patents: A Practical Perspective*, 2016. CreateSpace Independent Publishing Platform.

22. Brown and Michaels. https://www.bpmlegal.com(accessed 20 December 2017).

23. Akana et al.(2014). Electronic device. US Patent D712,405S, issued 2 September 2014.

附录 5.A

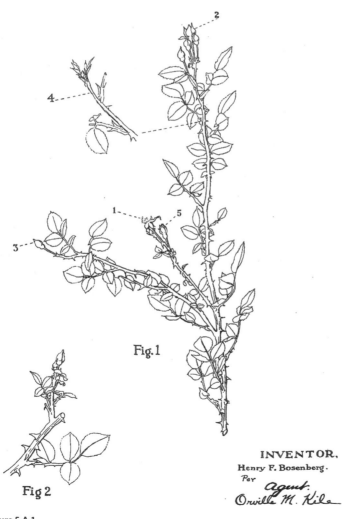

Figure 5.A.1

Patented Aug. 18, 1931

Plant Pat. 1

UNITED STATES PATENT OFFICE

HENRY F. BOSENBERG, OF NEW BRUNSWICK, NEW JERSEY, ASSIGNOR TO LOUIS C. SCHUBERT, OF NEW BRUNSWICK, NEW JERSEY

CLIMBING OR TRAILING ROSE

Application filed August 6, 1930. Serial No. 473,410.

My invention relates to improvements in roses of the type known as climbing or trailing roses in which the central or main stalks acquire considerable length and when given moderate support "climb" and branch out in various directions.

In roses it is very desirable to have a long period of blooming. This has been acquired in non-climbing roses of the type ordinarily called monthly roses or everblooming roses. My invention now gives the true everblooming character to climbing roses.

The following description and accompanying illustrations apply to my improvements upon the well known variety Dr. Van Fleet, with which my new plant is identical as respects color and form of flower, general climbing qualities, foliage and hardiness, but from which it differs radically in flowering habits —but the same everblooming habits may be attained by breeding this new quality into other varieties of climbing roses.

Figure I shows (1) a flower that is just dropping its petals, (2) a bud about to open, (3) a terminal bud just forming on a large side shoot, and (4) a new shoot which has not yet finished its growth and formed buds at its terminus. This shoot would not appear on the branch illustrated until several weeks later than the stage of development shown, when it would grow out ordinarily from the axil of the first or second leaf below the bloomed-off flower. (5) shows a second way in which new flowering shoots form, by branching off on a short stem immediately or closely adjacent to the blossom that has just finished blooming. Figure II shows a further method of branching and bud formation in cases where the bloom has been cut off, but the formation of new flowering shoots is not dependent upon pruning off the old blossoms. It is evident that this succession of blooms continuously or intermittently supplied by new shoots branching out throughout the summer and fall gives the true everblooming character. When grown in the latitude of New Brunswick, New Jersey, my new climbing rose named "The New Dawn" and illustrated herewith in exact drawings from photographs, provides a succession of blossoms on a single plant from about the end of May to the middle of November, or until stopped by frost.

No claim is made as to novelty in color or other physical characteristics of the individual blossoms, nor as to the foliage or growing habits of this rose other than as described above.

I claim:

A climbing rose as herein shown and described, characterized by its everblooming habit.

In testimony whereof I affix my signature hereunto.

HENRY F. BOSENBERG.

Figure 5.A.1 (*Continued*)

United States Patent [19]
Lemon

[11] Patent Number: **Plant 10,000**
[45] Date of Patent: **Aug. 12, 1997**

[54] GERANIUM PLANT NAMED 'LOIS'

[75] Inventor: **David Lemon**, Lompoc, Calif.

[73] Assignee: **John Bodger and Sons Company**, South El Monte, Calif.

[21] Appl. No.: **571,168**

[22] Filed: **Dec. 12, 1995**

[51] Int. Cl.^{6} A01H 5/00
[52] U.S. Cl. Plt./87.12
[58] Field of Search Plt./87.12

[56] **References Cited**
U.S. PATENT DOCUMENTS

P.P. 8,909 9/1994 Walters Plt./87.12

Primary Examiner—Howard J. Locker
Attorney, Agent, or Firm—Fulwider Patton Lee and Utecht, LLP

[57] **ABSTRACT**

The cultivar is characterized by its distinctive multi-color presentation. Its strong growth habit and continuous flowering provides for superb cuttings and also for quick recovery following wet weather conditions. The blooms exhibit relatively longer post-harvest production compared to that of other Regal varieties.

1 Drawing Sheet

BACKGROUND OF THE NEW PLANT

The present invention comprises a new and distinct cultivar of *Pelargonium×domesticum* known by the varietal name Lois (Oglevee No. 551, Bodger No. 6 PD 23–3). The new variety was discovered in a selective breeding program by David Lemon at Bodger Seeds, Ltd., Lompoc, Calif. The new variety is a selection from the self-pollenation of Hazel Ripple (unpatented). Hazel Ripple is distinguished from Lois primarily by its smaller flower.

The new cultivar was first asexually reproduced by cuttings at Oglevee Ltd., Connellsville, Pa., and has been repeatedly asexually reproduced by cuttings at Oglevee Ltd. in Connellsville, Pa. It has been found to retain its distinctive characteristics through successive propagations.

The new cultivar, when grown in a glass greenhouse in Connellsville, Pa., using full light, 60° Fahrenheit night temperature, 68° Fahrenheit day temperature, 72° Fahrenheit vent temperature and grown in a soilless media of constant fertilizer 150–200 parts per million of nitrogen and potassium has a response time of fourteen weeks from the rooted cutting to a flowering plant in a 6.0 inch pot.

DESCRIPTION OF THE DRAWING

The accompanying drawing illustrates the new cultivar, the color being as nearly true as possible with color illustrations of this type.

DESCRIPTION OF THE NEW PLANT

The following detailed description sets forth characteristics of the new cultivar. The data which defines each characteristic was collected from asexual reproductions carried out by Oglevee Ltd. in Connellsville, Pa. The plant histories were taken on rooted cuttings potted on Nov. 4, 1994 which flowered on Feb. 10, 1995 under full light and greenhouse, and colorings were taken indoors under 200–220 foot candles of fluorescent cool white light using The R.H.S. Colour Chart of The Royal Horticultural Society of London.

THE PLANT

Classification:
  Botanical.—*Pelargonium×domesticum.*

Form: Upright mound.
  Height.—23.0–25.0 cm above the media surface.
  Growth.—Strong; controlled, continuous flowering.
  Strength.—Free standing; quick rooting.
Foliage: Stalked leaf attachment.
Leaves:
  Size.—8.0–11.0 cm across; fully expanded leaf.
  Shape.—Depressed ovate; truncate base.
  Margin.—Serrately lobed.
  Texture.—Slightly pubescent; leathery.
  Color.—Top: Yellow/Green Group 146B; Zone: Not Present. Bottom: Yellow/Green Group 146C.
  Ribs and veins.—Palmate venation; Color: Yellow/Green group 147C.
Petioles:
  Length.—4.5–6.5 cm.
  Color.—Yellow/Green group 146B.
Stem:
  Color.—Yellow/green group 146C.
  Internodes.—2.0–3.5 cm in length.

THE BUD

Shape: Elliptical.
Size: 2.5–3.0 cm across; bud just showing color.

INFLORESCENCE

Blooming habit: Continuous blooming; many large florets opening into a full flower head.
Size: 9.0–11.5 cm across.
Borne: Umbel; florets on pedicel; pedicel on peduncle.
Number: About 10 inflorescences per plant.
Florets:
  Closed.—Bud size: 1.5–2.0 cm in length, 0.5–1.0 cm in width; elliptical; bud showing no color.
  Open.—Form — Cupped. Color — Top: Outer Edge Red/Purple Group 69D; Upper two petals have a blotch of Red/Purple Group 59A; Upper petal marking is large; Color between margin and marking is RHS 59A; Lower petals: Red/Purple Group 74C with veining of Red/Purple Group 59A; Moderate sized markings of RHS 74C and margin color is RHS 69D; Bottom of petals: Red/Purple Group 69D with veining and blotch showing through faintly; General

---

Figure 5.A.2

Plant 10,000

3

appearance is bi-color with magenta purple upper petals and light lavender lower petals. Petals — 5–6 in number; single, not united; margin entire. Size — 5.0–6.0 cm across. Texture and appearance — Smooth, satiny; appearance from a distance is a medium purple with dark purple blotch present on upper two petals and dark purple veining present on all petals above medium green foliage.

Petaloids:
   Quantity.—None.
Pedicel:
   Length.—2.5–3.0 cm.
   Color.—Yellow/Green Group 146B.
Peduncle:
   Length.—5.0–6.5 cm.
   Color.—Yellow/Green Group 146B.
Persistence:
   Disease resistance.—Not Known.
   Lasting quality.—Slow to shatter; continuous blooming.

REPRODUCTIVE ORGANS

Stamens:
   Anthers.—Length: 4.0–5.0 mm.
   Filaments.—Length: 1.4–1.9 cm; Color: White Group 155D.

Figure 5.A.2 (*Continued*)

4

   Pollen.—Golden Brown.
Pistils:
   Number.—1.
   Length.—2.2–2.5 cm.
   Stigma.—5 parted; Color: Red/Purple Group 60C.
   Style.—Length: 1.4–1.6 cm; Color: Red/Purple Group 60C.
   Ovaries: Pubescent; Length: 6.0–7.0 mm; Color: Light Green; superior.
Fruit: None observed.

GENERAL CHARACTERISTICS

Lois adds a new color pattern to the Regal geranium group. The interesting blotching and veining gives Lois' flowers a multi-color presentation. Lois exhibits ease and quickness of rooting, good cutting production, a controlled plant habit, rich medium green foliage and longer post-harvest production, all of which are unique to Lois among Regal geraniums. Given all these positive properties, this variety should please both the grower and the gardener.

I claim:
1. A new and distinct variety of Geranium plant, substantially as shown and described.

* * * * *

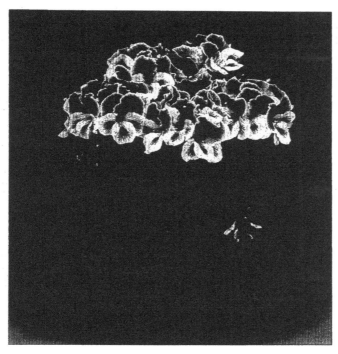

Figure 5.A.2 (*Continued*)

(12) United States Design Patent
Akana et al.

(10) Patent No.: **US D762,208 S**
(45) Date of Patent: ** Jul. 26, 2016

(54) **PORTABLE DISPLAY DEVICE WITH GRAPHICAL USER INTERFACE**

(71) Applicant: **Apple Inc.**, Cupertino, CA (US)

(72) Inventors: **Jody Akana**, San Francisco, CA (US); **Bartley K. Andre**, Palo Alto, CA (US); **Daniel J. Coster**, San Francisco, CA (US); **Elizabeth Caroline Cranfill**, San Francisco, CA (US); **Daniele De Iuliis**, San Francisco, CA (US); **M. Evans Hankey**, San Francisco, CA (US); **Richard P. Howarth**, San Francisco, CA (US); **Mikio Inose**, Cupertino, CA (US); **Jonathan P. Ive**, San Francisco, CA (US); **Steven P. Jobs**, Palo Alto, CA (US); **Duncan Robert Kerr**, San Francisco, CA (US); **Stephen O. Lemay**, San Francisco, CA (US); **Shin Nishibori**, Kailua, HI (US); **Matthew Dean Rohrbach**, San Francisco, CA (US); **Peter Russell-Clarke**, San Francisco, CA (US); **Christopher J. Stringer**, Woodside, CA (US); **Eugene Antony Whang**, San Francisco, CA (US); **Rico Zörkendörfer**, San Francisco, CA (US)

(73) Assignee: **Apple Inc.**, Cupertino, CA (US)

(**) Term: **14 Years**

(21) Appl. No.: **29/491,128**

(22) Filed: **May 16, 2014**

Related U.S. Application Data

(63) Continuation of application No. 29/454,965, filed on May 15, 2013, now Pat. No. Des. 705,223, which is a continuation of application No. 29/365,379, filed on Jul. 8, 2010, now Pat. No. Des. 683,730.

(51) LOC (10) Cl. 14-02
(52) U.S. Cl.
USPC **D14/341**; D14/493
(58) **Field of Classification Search**
USPC .... D14/341–347, 137, 138, 138.1, 138 AA, D14/138 R, 138 AB, 138 AC, 138 AD, 138 C, D14/138 G, 203.1, 203.3, 203.4, 203.5, D14/203.6, 203.7, 426, 427, 428, 432, 429, D14/127, 130, 420, 437–441, 448, 125, 147, D14/156, 218, 247–248, 250, 389, 336, D14/485–495; 715/700–867, 961–978; D10/65, 104.1; D18/6–7; D21/29, 686, D21/33, 324, 329, 330; D6/596, 601, 605; 345/169. 901, 905; 348/169; 361/814; 341/122; 346/173; 379/433.04, 433.07, 379/433.11, 433.12, 433.13, 910, 916, 379/433.01, 433.06

See application file for complete search history.

(56) **References Cited**

U.S. PATENT DOCUMENTS

805,678 A   11/1905   Smith
(Continued)

FOREIGN PATENT DOCUMENTS

AU   315078   7/2007
(Continued)

OTHER PUBLICATIONS

U.S. Appl. No. 29/365,378, filed Jul. 8, 2010 (not published).
(Continued)

Primary Examiner — Barbara Fox
(74) *Attorney, Agent, or Firm* — Sterne, Kessler, Goldstein & Fox P.L.L.C.

(57) **CLAIM**

The ornamental design for a portable display device with graphical user interface, as shown and described.

DESCRIPTION

FIG. 1 is a front view of a portable display device with graphical user interface showing our new design;
FIG. 2 is a rear view thereof;
FIG. 3 is a top view thereof;
FIG. 4 is a left side view thereof;
FIG. 5 is a right side view thereof; and,
FIG. 6 is a bottom view thereof.
The broken lines shown in the drawings represent unclaimed Figures show portions of the portable display device with graphical user interface and-that form no part of the claimed design.
The shade lines in FIG. 1 show transparency and not surface ornamentation.

1 Claim, 3 Drawing Sheets

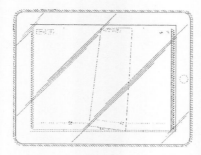

Figure 5.A.3

US D762,208 S
Page 2

(56) **References Cited**

U.S. PATENT DOCUMENTS

2,424,630 A	7/1947	Perez
3,877,729 A	4/1975	Friedman
D262,151 S	12/1981	Sussman
D265,326 S	7/1982	Sugiyama
D270,061 S	8/1983	Ackeret
D270,062 S	8/1983	Ackeret
D270,063 S	8/1983	Ackeret
D270,066 S	8/1983	Ackeret
4,420,112 A	12/1983	Cline
D278,276 S	4/1985	Bakic
4,545,023 A	10/1985	Mizzi
D284,084 S	6/1986	Ferrara, Jr.
4,610,392 A	9/1986	DaRosa
D289,873 S	5/1987	Gemmell et al.
4,860,217 A *	8/1989	Sasaki H04N 5/2628 345/653
D306,583 S	3/1990	Krolopp et al.
D308,055 S	5/1990	Tedham et al.
D317,609 S	6/1991	Wei
D321,215 S	10/1991	Shamis
5,123,676 A	6/1992	Donnelly et al.
D332,328 S	1/1993	Lombardi, Jr.
D333,574 S	3/1993	Ackeret
D337,322 S	7/1993	Yang
D337,569 S	7/1993	Kando
5,237,651 A	8/1993	Randall
D340,701 S	10/1993	Takeuchi
D340,917 S	11/1993	Sakaguchi et al.
D345,346 S	3/1994	Alfonso et al.
D346,589 S	5/1994	Andros
D346,793 S	5/1994	Iino et al.
5,345,543 A	9/1994	Capps et al.
5,347,295 A	9/1994	Agulnick et al.
5,398,310 A	3/1995	Tchao et al.
D361,552 S	8/1995	Iino
5,463,725 A	10/1995	Henckel et al.
D364,153 S	11/1995	Umaba et al.
5,488,204 A	1/1996	Mead et al.
5,495,566 A	2/1996	Kwatinetz
5,499,330 A	3/1996	Lucas et al.
D377,169 S	1/1997	Chida
D384,050 S	9/1997	Kodosky
5,703,626 A	12/1997	Itoh et al.
D395,639 S	6/1998	Ham et al.
D396,215 S	7/1998	Inukai
D396,452 S	7/1998	Naruki
D398,299 S	9/1998	Ballay et al.
D408,372 S	4/1999	Ota et al.
D408,794 S	4/1999	Ogasawara
D409,185 S	5/1999	Kawashima
D409,654 S	5/1999	Harris
D410,440 S	6/1999	Carnell
D412,157 S	7/1999	Stevenson
D415,136 S	10/1999	Newton et al.
D420,354 S	2/2000	Morales
D422,582 S	4/2000	Bright et al.
D422,583 S	4/2000	Herceg et al.
D422,991 S	4/2000	Regan, Jr. et al.
D424,535 S	5/2000	Peltola
D425,887 S	5/2000	Edwards et al.
6,061,063 A	5/2000	Wagner et al.
6,067,068 A	5/2000	Hussain
D438,201 S	2/2001	Ogawa et al.
6,217,443 B1	4/2001	Green, Jr.
D444,465 S	7/2001	Do
D448,031 S	9/2001	Goetz
D448,764 S	10/2001	Marsalka et al.
6,297,824 B1	10/2001	Hearst et al.
6,310,610 B1	10/2001	Beaton et al.
D451,505 S	12/2001	Iseki et al.
D452,687 S	1/2002	Yeh
D453,166 S	1/2002	Ording
6,337,698 B1	1/2002	Keely, Jr. et al.
D453,333 S	2/2002	Chen
D453,938 S	2/2002	Graham
D454,348 S	3/2002	Yeh
D455,433 S	4/2002	Alviar et al.
D456,023 S	4/2002	Andre et al.
D456,805 S	5/2002	Ono et al.
D457,530 S	5/2002	Amron
D458,252 S	6/2002	Palm et al.
6,407,757 B1	6/2002	Ho
D461,175 S	8/2002	Yokota
D461,802 S	8/2002	Tu
D466,096 S	11/2002	Takada
D466,501 S	12/2002	Hirota
D467,890 S	12/2002	Lai et al.
6,501,967 B1	12/2002	Mäkelä et al.
D469,413 S	1/2003	To et al.
6,509,907 B1	1/2003	Kuwabara
D469,762 S	2/2003	Iwama et al.
D474,163 S	5/2003	Araki
D480,721 S	10/2003	Hsieh
D483,021 S	12/2003	Shih
D483,032 S	12/2003	Bertrand et al.
D483,037 S	12/2003	Whitehorn et al.
D483,359 S	12/2003	Lin et al.
D484,471 S	12/2003	Lin et al.
D485,265 S	1/2004	Sato et al.
D486,149 S	2/2004	Kawami et al.
6,690,387 B2	2/2004	Zimmerman et al.
D488,469 S	4/2004	Batsikas
D489,731 S	5/2004	Huang
D490,420 S	5/2004	Solomon et al.
D491,936 S	6/2004	Jao
D492,684 S	7/2004	Ozolins et al.
D494,164 S	8/2004	Wu et al.
6,773,195 B2	8/2004	Tims
D496,040 S	9/2004	Jobs et al.
D496,362 S	9/2004	Ozolins et al.
D496,363 S	9/2004	Ozolins et al.
D497,364 S	10/2004	Ozolins et al.
D497,910 S	11/2004	Huang et al.
D498,763 S	11/2004	Totten et al.
6,822,635 B2	11/2004	Shahoian et al.
D500,037 S	12/2004	Ozolins et al.
D502,179 S	2/2005	Cha et al.
D502,944 S	3/2005	Kawami et al.
D504,889 S	5/2005	Andre et al.
D506,195 S	6/2005	Leveridge et al.
6,924,822 B2	8/2005	Card et al.
D511,169 S *	11/2005	Totten D14/492
6,989,815 B2	1/2006	Liang et al.
D514,558 S	2/2006	Nagel et al.
D514,568 S	2/2006	Huang et al.
D515,082 S	2/2006	Moskaluk et al.
7,009,596 B2	3/2006	Seet et al.
7,079,111 B2 *	7/2006	Ho G06F 3/03547 345/156
D526,661 S	8/2006	Arai
D528,517 S	9/2006	Izumi
7,107,522 B1	9/2006	Morgan et al.
D532,791 S	11/2006	Kim
7,139,982 B2 *	11/2006	Card G06F 3/0483 715/757
D534,516 S	1/2007	Lheem
D534,517 S	1/2007	Cho et al.
D535,281 S	1/2007	Yang
D535,657 S	1/2007	Ording
D535,661 S	1/2007	Gusmorino et al.
7,165,039 B2 *	1/2007	Seet G06F 3/0483 705/13
7,171,630 B2	1/2007	O'Leary et al.
D540,203 S	4/2007	Jeon
D543,183 S	5/2007	Cho et al.
D543,979 S	6/2007	Lee
D548,713 S	8/2007	Lee et al.
D548,747 S	8/2007	Andre et al.
D555,663 S	11/2007	Nagata et al.
D555,664 S	11/2007	Nagata et al.
D556,211 S	11/2007	Howard
D557,259 S	12/2007	Hirsch
7,304,635 B2	12/2007	Seet et al.
D558,756 S	1/2008	Andre et al.

Figure 5.A.3 (Continued)

US D762,208 S
Page 3

(56) **References Cited**

U.S. PATENT DOCUMENTS

Patent No.		Date	Inventor(s)	
D558,757 S		1/2008	Andre et al.	
D558,758 S		1/2008	Andre et al.	
D559,858 S		1/2008	Gusmorino et al.	
D560,227 S		1/2008	Bennett et al.	
D561,782 S		2/2008	Kim	
D563,424 S		3/2008	Gusmorino et al.	
D565,596 S		4/2008	Kim	
D567,819 S		4/2008	Devericks et al.	
D568,309 S		5/2008	Cebe et al.	
D568,892 S		5/2008	Stabb et al.	
7,386,804 B2		6/2008	Ho et al.	
D572,694 S		7/2008	Park	
D573,143 S		7/2008	Park et al.	
D575,760 S		8/2008	Kim et al.	
D577,703 S		9/2008	Lee et al.	
7,437,005 B2		10/2008	Drucker et al.	
D580,387 S		11/2008	Andre et al.	
D580,431 S		11/2008	Morita	
D581,922 S		12/2008	Andre et al.	
D583,346 S		12/2008	Jung et al.	
D583,373 S		12/2008	Chen	
D584,711 S		1/2009	Kim et al.	
D584,738 S		1/2009	Kim et al.	
D586,800 S		2/2009	Andre et al.	
D592,211 S		5/2009	Ichise et al.	
D592,212 S		5/2009	Hamada et al.	
D593,059 S		5/2009	Kim et al.	
D593,132 S		5/2009	Kim	
7,536,654 B2		5/2009	Anthony et al.	
D594,863 S		6/2009	Ichise et al.	
D597,067 S		7/2009	Oh et al.	
D598,016 S		8/2009	Han et al.	
7,581,186 B2		8/2009	Dowdy et al.	
D599,342 S		9/2009	Andre et al.	
D600,241 S		9/2009	Andre et al.	
D600,690 S		9/2009	Miyaji	
D601,170 S		9/2009	Pell et al.	
D601,353 S		10/2009	Sadler et al.	
D601,558 S		10/2009	Andre et al.	
D602,014 S		10/2009	Andre et al.	
D602,015 S		10/2009	Andre et al.	
D602,016 S		10/2009	Andre et al.	
D602,017 S		10/2009	Andre et al.	
D602,486 S		10/2009	Andre et al.	
D602,488 S		10/2009	Jiang et al.	
D604,289 S		11/2009	Andre et al.	
D604,290 S		11/2009	Andre et al.	
D604,291 S		11/2009	Andre et al.	
D604,292 S		11/2009	Andre et al.	
D604,293 S		11/2009	Andre et al.	
D604,297 S		11/2009	Andre et al.	
D604,733 S		11/2009	Andre et al.	
D606,129 S		12/2009	Ben-Moshe	
D606,559 S		12/2009	Kocmick	
D606,988 S		12/2009	Andre et al.	
D606,989 S		12/2009	Andre et al.	
D606,991 S		12/2009	Liu et al.	
D606,992 S		12/2009	Liu et al.	
D607,428 S		1/2010	Kim et al.	
D607,889 S		1/2010	Poling et al.	
D609,705 S		2/2010	Andre et al.	
D609,715 S		2/2010	Chaudhri	
D610,161 S		2/2010	Matas	
7,658,675 B2		2/2010	Hotta	
D611,045 S		3/2010	Andre et al.	
D611,469 S		3/2010	Andre et al.	
7,688,574 B2		3/2010	Zadesky et al.	
D613,300 S		4/2010	Chaudhri	
D613,735 S		4/2010	Andre et al.	
D613,736 S		4/2010	Andre et al.	
D613,753 S	*	4/2010	Barcheck	D14/492
D615,083 S		5/2010	Andre et al.	
D615,554 S		5/2010	Andre et al.	
D620,004 S		7/2010	Andre et al.	
D621,825 S		8/2010	Andre et al.	
D622,718 S		8/2010	Andre et al.	
D622,719 S		8/2010	Andre et al.	
D622,720 S		8/2010	Andre et al.	
D623,057 S		9/2010	Kletz	
D624,072 S		9/2010	Andre et al.	
D624,536 S		9/2010	Andre et al.	
D624,914 S		10/2010	Wang	
D624,935 S		10/2010	Umezawa	
D625,307 S		10/2010	Cheng	
D625,326 S		10/2010	Allen	
D627,343 S		11/2010	Andre et al.	
D627,777 S		11/2010	Akana et al.	
D627,778 S		11/2010	Akana et al.	
D628,592 S		12/2010	O'Donnell et al.	
D628,593 S		12/2010	O'Donnell et al.	
D628,595 S		12/2010	Danhope-Smith	
D629,010 S		12/2010	O'Donnell et al.	
D629,413 S		12/2010	Sriver	
D629,799 S		12/2010	Andre et al.	
D630,207 S		1/2011	Seong	
D630,630 S		1/2011	Andre et al.	
D633,523 S		3/2011	Trabona et al.	
D633,908 S		3/2011	Akana et al.	
D633,917 S		3/2011	Poling et al.	
D634,319 S		3/2011	Andre et al.	
D634,742 S		3/2011	Andre et al.	
7,898,541 B2		3/2011	Hong et al.	
D635,540 S		4/2011	Kim et al.	
D635,952 S		4/2011	Park et al.	
D635,989 S		4/2011	Bright et al.	
D636,390 S		4/2011	Andre et al.	
D637,596 S		5/2011	Akana et al.	
D638,027 S		5/2011	Towbin et al.	
D638,439 S		5/2011	Cavanaugh et al.	
D638,835 S		5/2011	Akana et al.	
D647,911 S		11/2011	Allen et al.	
D648,347 S		11/2011	Chaudhri	
D658,195 S		4/2012	Cranfill	
D669,069 S		10/2012	Akana et al.	
D669,468 S		10/2012	Akana et al.	
D669,906 S		10/2012	Cranfill et al.	
D670,692 S		11/2012	Akana et al.	
D670,713 S		11/2012	Cranfill et al.	
D673,974 S		1/2013	Sepulveda	
8,358,280 B2	*	1/2013	Li	G06F 3/0483 345/173
D677,659 S		3/2013	Akana et al.	
D681,630 S		5/2013	Akana et al.	
D681,631 S		5/2013	Akana et al.	
D682,262 S		5/2013	Akana et al.	
D683,345 S		5/2013	Akana et al.	
D683,346 S		5/2013	Akana et al.	
D683,730 S		6/2013	Akana et al.	
8,499,251 B2	*	7/2013	Petschnigg	G06F 3/0483 715/773
8,539,384 B2	*	9/2013	Hinckley	G06F 3/04883 715/763
D693,833 S		11/2013	Inose et al.	
D700,191 S	*	2/2014	Faenza	D14/485
D701,234 S		3/2014	Cranfill et al.	
D702,727 S	*	4/2014	Abratowski	D14/492
D705,223 S		5/2014	Akana et al.	
D707,706 S	*	6/2014	Cranfill	D14/492
8,977,977 B2	*	3/2015	Chong	G06F 3/0483 345/156
D727,924 S	*	4/2015	Yu	D14/485
D729,818 S	*	5/2015	Bae	D14/485
D734,765 S	*	7/2015	Lee	D14/485
D735,219 S	*	7/2015	Young-Ri	D14/486
9,092,118 B2	*	7/2015	Gunji	G06F 3/0483
D735,734 S	*	8/2015	Lee	D14/485
D736,782 S	*	8/2015	Lee	D14/485
D737,280 S	*	8/2015	Lee	D14/485
D738,903 S	*	9/2015	Lee	D14/485
2003/0020687 A1	*	1/2003	Sowden	G06F 3/03547 345/157
2003/0117425 A1		6/2003	O'Leary et al.	
2003/0125079 A1		7/2003	Park et al.	
2003/0125094 A1		7/2003	Hyun et al.	

Figure 5.A.3 *(Continued)*

US D762,208 S
Page 4

(56) **References Cited**

U.S. PATENT DOCUMENTS

2003/0125959 A1	7/2003	Palmquist	
2003/0184525 A1	10/2003	Tsai	
2003/0206202 A1	11/2003	Moriya	
2003/0210949 A1	11/2003	Tims	
2004/0021676 A1	2/2004	Chen et al.	
2004/0039750 A1	2/2004	Anderson et al.	
2004/0041504 A1	3/2004	Ozolins et al.	
2004/0145603 A1	7/2004	Soares	
2004/0255254 A1	12/2004	Weingart et al.	
2005/0144565 A1	6/2005	Hemmings	
2005/0156902 A1	7/2005	McPherson et al.	
2005/0283742 A1	12/2005	Gusmorino et al.	
2006/0066510 A1	3/2006	Takahashi	
2007/0168883 A1	7/2007	Sugimoto	
2008/0004083 A1	1/2008	Ohki et al.	
2009/0247244 A1	10/2009	Mittleman et al.	
2011/0167369 A1	7/2011	van Os	

FOREIGN PATENT DOCUMENTS

CA	72548	5/1993
CA	89155	3/2000
CA	101263	1/2004
CN	200430011774.1	6/2005
CN	1695105 A	11/2005
CN	200630030574.X	8/2007
CN	200630174414.2	12/2007
CN	200730008464.8	3/2008
CN	200730058820.7	6/2008
CN	200930166104.X	2/2010
CN	200930166516.3	5/2010
CN	200930166757.8	6/2010
CN	301499241 S	3/2011
DE	40104198-0001	9/2001
DE	40301867-0001	9/2003
DE	40705936-0001	5/2008
EM	00048061-0001	8/2003
EM	00046198-0001	11/2003
EM	000181607-0001	8/2004
EM	000317490-0001	5/2005
EM	000328265-0001	6/2005
EM	000257621-0004	7/2005
EM	000375191-0001	9/2005
EM	000401898-0003	11/2005
EM	000493721-0002	5/2006
EM	30-0424148	9/2006
EM	000569157-0005	9/2006
EM	000614565-0001	12/2006
EM	00066789-0001	3/2007
EM	000718770-0007	6/2007
EM	000748280-0001	8/2007
EM	000748280-0006	8/2007
EM	000767959-0001	8/2007
EM	000748314-0001	10/2007
EM	000891809-0003	4/2008
EM	000939731-0001	6/2008
EM	001005839-0007	3/2009
EM	001098149-0001	4/2009
EM	001098149-0005	4/2009
EM	001143010-0005	7/2009
EM	001594771-0004	7/2009
EM	001677816-0002	4/2010
EM	001694712-0001	4/2010
EM	001214290-0003	6/2010
EM	001222905-0001	10/2010
EM	001222905-0002	10/2010
EM	01772252-0001	10/2010
EM	001000152-0001	11/2010
EM	001140115-0001	11/2010
EP	0 701 220 B1	7/2001
ES	10147763	3/2001
ES	D0505712-01	4/2008
FI	4897	7/1980
FR	990052-001	5/1999
FR	985598-001	10/2001
FR	985598-002	10/2001
FR	985598-003	10/2001
FR	985598-004	10/2001
GB	1042780	12/1986
GB	1058720	10/1988
GB	1058721	10/1988
GB	2030050	10/1992
GB	2033245	8/1993
GB	2101448	10/2000
GB	2103809	2/2001
GB	3009998	1/2003
GB	3010002	1/2003
GB	3014024	8/2003
GB	3022447	10/2005
JP	61-69284	4/1986
JP	4-370868	12/1992
JP	886431	12/1993
JP	887388	12/1993
JP	6-102341	4/1994
JP	000921403	3/1995
JP	949744	3/1996
JP	953136	5/1996
JP	9-325180	12/1997
JP	001009317	5/1998
JP	11-112918	4/1999
JP	11-126149	5/1999
JP	11-289169	10/1999
JP	1056385	12/1999
JP	2000-163031	6/2000
JP	2000-322495	11/2000
JP	2001-22471	1/2001
JP	2001-22704	1/2001
JP	D1098145	1/2001
JP	001104685	3/2001
JP	D1124750	10/2001
JP	D1133939	2/2002
JP	001142127	5/2002
JP	D1143615	6/2002
JP	D1145748	7/2002
JP	2002-254614	9/2002
JP	D1155168	10/2002
JP	D1155176	10/2002
JP	D1159881	12/2002
JP	D1178470	7/2003
JP	001188041	10/2003
JP	2003-330613	11/2003
JP	D1194485	1/2004
JP	D1204221	5/2004
JP	2004-290256	10/2004
JP	001235888	4/2005
JP	001241638	6/2005
JP	D1241383	6/2005
JP	D1247215	8/2005
JP	D1250487	9/2005
JP	D1263649	2/2006
JP	001280315	9/2006
JP	D1285057	10/2006
JP	D1295003	2/2007
JP	D1302929	6/2007
JP	D1325539	3/2008
JP	D1351273	2/2009
KR	30-0110034	11/1980
KR	30-0037851	8/1982
KR	30-0109896	11/1990
KR	30-0213519	4/1998
KR	30-0222985-1	2/1999
KR	30-0237041	6/1999
KR	30-0241312	7/1999
KR	30-0243584	8/1999
KR	30-0288858	1/2002
KR	30-0290725	2/2002
KR	30-0291915	2/2002
KR	30/0294002	3/2002
KR	30-0296928	5/2002
KR	30-0297741	5/2002
KR	30-0304213	8/2002
KR	30-0307008	9/2002
KR	30-0307235	9/2002

Figure 5.A.3 (*Continued*)

(56) **References Cited**

FOREIGN PATENT DOCUMENTS

KR	30-0307307	9/2002
KR	30-0309692	10/2002
KR	30-0312151	11/2002
KR	30-0312152	11/2002
KR	30-0327126	6/2003
KR	30-0330831	8/2003
KR	30-0332851	9/2003
KR	30-0336260	10/2003
KR	30-0338781	11/2003
KR	2003-0088374	11/2003
KR	30-0338886	12/2003
KR	30-0326632-2	1/2004
KR	30-0398307	11/2005
KR	30-0394921-1	12/2005
KR	30-0418422-1	7/2006
KR	30-0418547	7/2006
KR	30-0422221	8/2006
KR	30-0422222	8/2006
KR	30-0441230	2/2007
KR	30-0452432	6/2007
KR	30-0452985	8/2007
KR	30-0529167	5/2009
KR	30-0533504	7/2009
KR	30-0546031	11/2009
KR	30-0598018	5/2011
KR	30-0598021	5/2011
SE	55044	10/1993
SE	2001/0343	9/2002
TW	584453	4/2004
TW	D106137	8/2005
TW	D137754	11/2010
WO	WO 00/74240 A1	12/2000
WO	WO 01/88679	11/2001
WO	WO 2004/023272 A2	3/2004

OTHER PUBLICATIONS

U.S. Appl. No. 29/365,381, filed Jul. 8, 2010 (not published).
U.S. Appl. No. 29/366,479, filed Jul. 26, 2010 (not published).
U.S. Appl. No. 29/384,918, filed Feb. 4, 2011 (not published).
Axiotron, Macworld Expo 2009 Wrap-Up, 4 pages, http://www.axiotron.com/index.php?id=events.
Arrington, Michael, CrunchPad: The Launch Prototype, 17 pages, http://techcrunch.com/2009/06/03/crunchpad-the-launch-prototype/, Jun. 3, 2009.
Arrington, Michael, TechCrunch Tablet Update: Prototype B, 25 pages, http://techcrunch.com/2009/01/19/techcrunch-tablet-update-prototype-b/, Jan. 19, 2009.
Chubb, Daniel, Amtek's iTablet T221: new to the tablet and UMPC scene, 4 pages, http://www.product-reviews.net/2008/01/11/amteks-itablet-t221-new-to-the-tablet-and-umpc-scene/, Jan. 11, 2008.
Coldeway, Devin, TechCrunch Tablet makes an early debut, 26 pages, http://www.crunchgear.com/2009/04/09/crunchtablet-hits-the-net-a-little-early/, Apr. 9, 2009.
Davies, Chris, Are Dell planning a keyboard-less UMPC?, 15 pages, http://www.slashgear.com/are-dell-planning-a-keyboard-less-umpc-096130/, Jul. 9, 2007.
Dunn, Jason, Ultra-Portable Concept PC Shown at WinHEC 2005, 1 page, http://www.digitalhomethoughts.com/new/show/27459/ultra-portable-concept-pc-shown-at-winhec-2005.html, Apr. 26, 2005.
Lerg, Andreas, Billig-Tablet-Computer von Grabbeltisch, 2 pages, http://digitalleben.t-online.de/tablet-pc-von-jay-tech-fuer-140-euro-bei-rossmann/id_43326132/index, Nov. 3, 2010.
Loyola, Roman, Macworld Video: Axiotron Modbook Pro first look, 4 pages, http://www.macworld.com/article/137878/2009/01/mwvodcast86.html, Jan. 5-9, 2009.
Newgadgets.de, Mysteriöser Tablet-PC?, 6 pages, http://www.newgadgets.de/6072/mysterioser-tablet-pc/, Nov. 10, 2009.
Radioshack, Motorala DEFY with MOTOBLUR Black, 1 pages, http://radioshackwireless.com/eCommerce/SpecialOffer.aspx?cid=25578_47f08265355e4a8a92dcd57eed6cbe69.

Receipt of an Invalidity Application Notification to the HOlder of the Contested RCD Community Design No. 001222905-0002, File No. ICD 8453, 1 page, mailed May 23, 2011.
Sierra, Amtek Releases the iTablet T221 Tablet PC, Laptops Arena, 5 pages, http://www.laptopsarena.com/amtek-releases-the-itablet-t221-tablet-pc/, Dec. 22, 2007.
Sze, Amtek has five iTablets for the Computex 2010, iTech News Net, 13 pages, http://www.itechnews.net/2010/05/27/amtek-has-five-itablets-for-the-computex-2010/.
Topolsky, Joshua, Amtek intros the iTablet T221 UMPC, 3 pages, http://www.engadget.com/2007/12/18/amtek-intros-the-itablet-t221/umpc/, Dec. 18, 2007.
Wikipedia, Axiotron Modbook, 4 pages, http://en.wikipedia.org/wiki/Axiotron_Modbook.
Wikipedia, Datei:Modbook pro.jpg, 4 pages, http://de.wikipedia.org/wiki/Datei;Modbook_pro.jpg, Feb. 21, 2009.
Wikipedia, Datei:VIA Tablet PC Reference Design.jpg, 3 pages, http://de.wikipedia.org/wiki/Datei:VIA_Tablet_PC_Reference_Design.jpg, Mar. 10, 2009.
Wikipedia, Dynabook, 2 pages, http://de.wikipedia.org/wiki/Dynabook, last updated Apr. 2, 2011.
Wikipedia, Tablet-Computer, 3 pages, http://de.wikipedia.org/wiki/Tablet-Computer, Apr. 7, 2011.
Wikipedia, Tablet-PC, 2 pages, http://de.wikipedia.org/wiki/Tablet-PC, Apr. 14, 2011.
Request for declaration of invalidity of design No. 001222905-0002, 7 pages, May 5, 2011.
Search Report for ROC (Taiwan) Design Patent Application No. 099303076, completed May 25, 2011, 1 page.
Search Report for ROC (Taiwan) Design Patent Application No. 099303077, completed May 25, 2011, 1 page.
HP Compaq Tablet PC TC1000-TM5800 1GHz-10.4" TFT, CNET Reviews, http://reviews.cnet.com/tablet-pcs/hp-compaq-tablet-pc-4505-1707_7-20627295.html. Downloaded Apr. 20, 2010. 15 pages.
HP Compaq Tablet PC TC1000, http://www.tc-one-thousand.com. Downloaded Apr. 20, 2010, 2 pages.
HP Compaq Tablet PC TC1000, http://www.hardwarezone.com.au/reviews/view.php?cid=45&id=533&pg=2. Downloaded Apr. 20, 2010, 1 page.
Bookeen—Cybook ePaper—the eBook reading device, http://www.bookeen.com/ebook/ebook-reading-device.aspx. Downloaded Apr. 26, 2010. 2 pages.
BeBook Neo e-reader launches with WiFi and WACOM capabilities—Engadget, http://www.engadget.com/2010/01/22/bebook-neo-e-reader-launches-with-wifi-and-wacom-capabilities/. Downloaded Apr. 26, 2010, 2 pages.
CES: Plastic Logic's Que E-Reader Revealed—PCWorld, http://www.pcworld.com/article/186224/ces_plastic_logics_que_ereader_revealed.html. Downloaded Apr. 26, 2010, 1 page.
The Best of CES 2010, http://www.pcworld.com/article/186511/another_excellent_ereader.html. Downloaded Apr. 26, 2010. 1 page.
iRex Digital Reader 800 with 8-inch touchscreen and wireless | Gadget Folder, http://www.gadgetfolder.com/irex-digital-reader-800-with-8-inch-touchs . . . Downloaded Jun. 29, 2010. 3 pages.
Qarchive.org, "Book Of Time 3D Screensaver 3.1," (http://book-of-time-3d-screensaver.kryptile-screensavers.qarchive.org), released Dec. 13, 2004, 3 pages.
Amazon, "Developing Tablet PC Applications—(Charles River Media Programming)," (http://www.amazon.co.uk/Developing-Tablet-Applications-Charles-Programming/dp/1584502525), retrieved Jul. 20, 2011, 4 pages.
Association for Computing Machinery. The Open Video Project, "ACM CHI 1995—The Tablet Newspaper: A vision for the Future," (http://www.open-video.org/details.php?videoid=8315&surrogate=storyboard), 3 pages (1994).
Aumente, J., American Journalism Review, "Panel Vision," (http://www.ajr.org/Article.asp?id=1257), published Oct. 1994, 7 pages.
Blickensterfer, C., PenComputing Magazine, "Tablet PCs: HP Compaq Tablet PC TC1000", (http://pencomputing.com/frames/tpc_compaq.html), 3 pages (Dec. 2002).
Blickensterfer, C., PenComputing Magazine, "Tablet PCs: Motion Computing M1200," (http://pencomputing.com/frames/tpc_motion1200.html), 3 pages (Dec. 2002).

Figure 5.A.3 (*Continued*)

US D762,208 S
Page 6

(56) **References Cited**

FOREIGN PATENT DOCUMENTS

KR	30-0307307	9/2002
KR	30-0309692	10/2002
KR	30-0312151	11/2002
KR	30-0312152	11/2002
KR	30-0327126	6/2003
KR	30-0330831	8/2003
KR	30-0332851	9/2003
KR	30-0336260	10/2003
KR	30-0338781	11/2003
KR	2003-0088374	11/2003
KR	30-0338886	12/2003
KR	30-0326632-2	1/2004
KR	30-0398307	11/2005
KR	30-0394921-1	12/2005
KR	30-0418422-1	7/2006
KR	30-0418547	7/2006
KR	30-0422221	8/2006
KR	30-0422222	8/2006
KR	30-0441230	2/2007
KR	30-0452432	6/2007
KR	30-0452985	8/2007
KR	30-0529167	5/2009
KR	30-0533504	7/2009
KR	30-0546031	11/2009
KR	30-0598018	5/2011
KR	30-0598021	5/2011
SE	55044	10/1993
SE	2001/0343	9/2002
TW	584453	4/2004
TW	D106137	8/2005
TW	D137754	11/2010
WO	WO 00/74240 A1	12/2000
WO	WO 01/88679	11/2001
WO	WO 2004/023272 A2	3/2004

OTHER PUBLICATIONS

U.S. Appl. No. 29/365,381, filed Jul. 8, 2010 (not published).
U.S. Appl. No. 29/366,479, filed Jul. 26, 2010 (not published).
U.S. Appl. No. 29/384,918, filed Feb. 4, 2011 (not published).
Axiotron, Macworld Expo 2009 Wrap-Up, 4 pages, http://www.axiotron.com/index.php?id=events.
Arrington, Michael, CrunchPad: The Launch Prototype, 17 pages, http://techcrunch.com/2009/06/03/crunchpad-the-launch-proto-type/, Jun. 3, 2009.
Arrington, Michael, TechCrunch Tablet Update: Prototype B, 25 pages, http://techcrunch.com/2009/01/19/techcrunch-tablet-update-prototype-b/, Jan. 19, 2009.
Chubb, Daniel, Amtek's iTablet T221: new to the tablet and UMPC scene, 4 pages, http://www.product-reviews.net/2008/01/11/amteks-itablet-t221-new-to-the-tablet-and-umpc-scene/, Jan. 11, 2008.
Coldeway, Devin, TechCrunch Tablet makes an early debut, 26 pages, http://www.crunchgear.com/2009/04/09/crunchtablet-hits-the-net-a-little-early/, Apr. 9, 2009.
Davies, Chris, Are Dell planning a keyboard-less UMPC?, 15 pages, http://www.slashgear.com/are-dell-planning-a-keyboard-less-umpc-096130/, Jul. 9, 2007.
Dunn, Jason, Ultra-Portable Concept PC Shown at WinHEC 2005, 1 page, http://www.digitalhomethoughts.com/new/show/27459/ultra-portable-concept-pc-shown-at-winhec-2005.html, Apr. 26, 2005.
Lerg, Andreas, Billig-Tablet-Computer von Grabbeltisch, 2 pages, http://digitalleben.t-online.de/tablet-pc-von-jay-tech-fuer-140-euro-bei-rossmann/id_43326132/index, Nov. 3, 2010.
Loyola, Roman, Macworld Video: Axiotron Modbook Pro first look, 4 pages, http://www.macworld.com/article/137878/2009/01/mwvodcast86.html, Jan. 5-9, 2009.
Newgadgets.de, Mysteriöser Tablet-PC?, 6 pages, http://www.newgadgets.de/6072/mysterioser-tablet-pc/, Nov. 10, 2009.
Radioshack, Motorola DEFY with MOTOBLUR Black, 1 page, http://radioshackwireless.com/eCommerce/SpecialOffer.aspx?cid=25578_47f08265355e4a8a92dcd57eed6cbe69.

Receipt of an Invalidity Application Notification to the HOlder of the Contested RCD Community Design No. 001222905-0002, File No. ICD 8453, 1 page, mailed May 23, 2011.
Sierra, Amtek Releases the iTablet T221 Tablet PC, Laptops Arena, 5 pages, http://www.laptopsarena.com/amtek-releases-the-itablet-t221-tablet-pc/, Dec. 22, 2007.
Sze, Amtek has five iTablets for the Computex 2010, iTech News Net, 13 pages, http://www.itechnews.net/2010/05/27/amtek-has-five-itablets-for-the-computex-2010/.
Topolsky, Joshua, Amtek intros the iTablet T221 UMPC, 3 pages, http://www.engadget.com/2007/12/18/amtek-intros-the-itablet-t221/umpc/, Dec. 18, 2007.
Wikipedia, Axiotron Modbook, 4 pages, http://en.wikipedia.org/wiki/Axiotron_Modbook.
Wikipedia, Datei:Modbook pro.jpg, 4 pages, http://de.wikipedia.org/wiki/Datei;Modbook_pro.jpg, Feb. 21, 2009.
Wikipedia, Datei:VIA Tablet PC Reference Design.jpg, 3 pages, http://de.wikipedia.org/wiki/Datei:VIA_Tablet_PC_Reference_Design.jpg, Mar. 10, 2009.
Wikipedia, Dynabook, 2 pages, http://de.wikipedia.org/wiki/Dynabook, last updated Apr. 2, 2011.
Wikipedia, Tablet-Computer, 3 pages, http://de.wikipedia.org/wiki/Tablet-Computer, Apr. 7, 2011.
Wikipedia, Tablet-PC, 2 pages, http://de.wikipedia.org/wiki/Tablet-PC, Apr. 14, 2011.
Request for declaration of invalidity of design No. 001222905-0002, 7 pages, May 5, 2011.
Search Report for ROC (Taiwan) Design Patent Application No. 099303076, completed May 25, 2011, 1 page.
Search Report for ROC (Taiwan) Design Patent Application No. 099303077, completed May 25, 2011, 1 page.
HP Compaq Tablet PC TC1000-TM5800 1GHz-10.4" TFT, CNET Reviews, http://reviews.cnet.com/tablet-pcs/hp-compaq-tablet-pc-4505-1707_7-20627295.html. Downloaded Apr. 20, 2010, 15 pages.
HP Compaq Tablet PC TC1000, http://www.tc-one-thousand.com. Downloaded Apr. 20, 2010, 2 pages.
HP Compaq Tablet PC TC1000, http://www.hardwarezone.com.au/reviews/view.php?cid=45&id=533&pg=2. Downloaded Apr. 20, 2010, 1 page.
Bookeen—Cybook ePaper—the eBook reading device, http://www.bookeen.com/ebook/ebook-reading-device.aspx. Downloaded Apr. 26, 2010. 2 pages.
BeBook Neo e-reader launches with WiFi and WACOM capabilities—Engadget, http://www.engadget.com/2010/01/22/bebook-neo-e-reader-launches-with-wifi-and-wacom-capabilities/. Downloaded Apr. 26, 2010, 2 pages.
CES: Plastic Logic's Que E-Reader Revealed—PCWorld, http://www.pcworld.com/article/186224/ces_plastic_logics_que_ereader_revealed.html. Downloaded Apr. 26, 2010, 1 page.
The Best of CES 2010, http://www.pcworld.com/article/186511/another_excellent_ereader.html. Downloaded Apr. 26, 2010. 1 page.
iRex Digital Reader 800 with 8-inch touchscreen and wireless | Gadget Folder, http://www.gadgetfolder.com/irex-digital-reader-800-with-8-inch-touchs . . . Downloaded Jun. 29, 2010. 3 pages.
Qarchive.org, "Book Of Time 3D Screensaver 3.1," (http://book-of-time-3d-screensaver.kryptile-screensavers.qarchive.org), released Dec. 13, 2004, 3 pages.
Amazon, "Developing Tablet PC Applications—(Charles River Media Programming)," (http://www.amazon.co.uk/Developing-Tablet-Applications-Charles-Programming/dp/1584502525), retrieved Jul. 20, 2011, 4 pages.
Association for Computing Machinery. The Open Video Project, "ACM CHI 1995—The Tablet Newspaper: A vision for the Future," (http://www.open-video.org/details.php?videoid=8315&surrogate=storyboard), 3 pages (1994).
Aumente, J., American Journalism Review, "Panel Vision," (http://www.ajr.org/Article.asp?id=1257), published Oct. 1994, 7 pages.
Blickensторfer, C., PenComputing Magazine, "Tablet PCs: HP Compaq Tablet PC TC1000," (http://pencomputing.com/frames/tpc_compaq.html), 3 pages (Dec. 2002).
Blickensторfer, C., PenComputing Magazine, "Tablet PCs: Motion Computing M1200," (http://pencomputing.com/frames/tpc_motion1200.html), 3 pages (Dec. 2002).

US D762,208 S
Page 7

(56) **References Cited**

OTHER PUBLICATIONS

Justia, "*Apple v. Samsung Electronics Co. Ltd. et al*, Filing: 427 Attachment: 44," (http://docs.justia.com/cases/federal/district-courts/california/candce/5:2011cv01846/239768/427/44.html, dated Nov. 28, 2011, 3 pages.
Focus Online, "Rollei DF8 DVB-T," (http://www.focus.de/digital/foto/rollei-df8-dvb-t-fersehen-im-bilderrahmen__did__27152.html), published Feb. 18, 2010, 3 pages.
Foto-Branche.de, "Kodak EasyShare S730: Digitaler Bilderrahmen mit integriertem Akku," (http://www.foto-branche.de/news/singleview/browse/2/article/kodak-easyshare-s730-digitaler-bilder-rahmen-mit-integriertem-akku/2.html), published Jul. 13, 2009, 2 pages.
Fotofabrikas, "AgfaPhoto AF 5087 MS 8," http://www.fotofabrikas.lt/data/images/catalog_pics/n_preview/Skaitmeninian%20remeliai%20AgfaPhoto%20AF%205087%20MS%208%20skaitmeninis%20r%20melis_01.jpg), retrieved Nov. 22, 2011, 1 page.
Fotofabrikas, "AgfaPhoto AF 5087 MS 8," http://www.fotofabrikas.lt/en/items/Digital-Frames/Digital-photo-frames/2636ee4.2-AgfaPhoto-AF-5087-MS-8.html), retrieved Jul. 21, 2011, 5 pages.
Funamizu, M., Petitinvention, "(iPod Touch + iMac + Macbook Air) / 3," (http://petitinvention.wordpress.com/2008/05/03/ipod-touch-imac-macbook-air-3/), published May 3, 2008, 4 pages.
Kahney, L., Cult of Mac, "Mac Air Tablet Mockup From Isamu Sanada," (http://www.cultofmac.com/1952/mac-air-tablet-mockup-from-isamu-sanada/), published May 6, 2008, 13 pages.
KingVision, "Digital Photo Frame," (http://www.sz-kingvision.com/product.php?type_id=1&weblan=e), retrieved Nov. 21, 2011, 2 pages.
KingVision, "DPF-1902," (http://www.sz-kingvision.com/product_disp.php?productid=24&weblan=e), retrieved Nov. 21, 2011, 3 pages.
Kowalski, M., "Review: Kodak EasyShare S730," Digital Picture Frame Review, (http://www.digitalpictureframereview.com/2009/07/review-kodak-easyshare-s730), published Jul. 17, 2009, 14 pages.
Lazar, L., Cult of Mac, "Mac Pad Mock Up," (http://www.cultofmac.com/7921/mac-pad-mock-up/), published Nov. 19, 2008, 12 pages.
My Burmese Blog, "Apple Touch Screen Netbook Prototypes Hit The Web," (http://www.htootayzat.com/myblog/2009/03/apple-touch-screen-netbook-prototypes-hit-the-web/), published Mar. 12, 2009, 10 pages.
Netzwelt, "Rollei DF8 DVB-T," (http://www.netzwelt.de/news/81537-rollei-df8-dvb-t-test-digitaler-bilderrahmen-mini-fernseher.html), retrieved Nov. 21, 2011, 10 pages.
Samsung, "Samsung Introduces UltraThin Touch of Color Photo Frame," (http://www.samsung.com/us/news/newsPreviewRead.do?news_seq=13462), published May 23, 2009, 2 pages.
Samsung, "Samsung's Digital Photo Frames," (http://www.samsung.com/hk_en/consumer/detail/productPreviewRead.do?model_cd=LP08IPLSB/XK&group=photography&type=photo-frame&subtype=photo-frame), retrieved Nov. 21, 2011, 4 pages.
Savov, V., Engadget, "NVIDIA Tegra tablet prototype hands-on," (http://www.engadget.com/2009/11/27/nvidia-tegra-tablet-prototype-hands-on/), published Nov. 27, 2009, 5 pages.
Savov, V., Engadget, "NVIDIA Tegra tablet prototype hands-on," (http://www.engadget.com/photos/nvidia-tegra-tablet-prototype/#2485983), published Nov. 27, 2009, 2 pages.
Savov, V., Engadget, "NVIDIA Tegra tablet prototype hands-on," (http://www.engadget.com/photos/nvidia-tegra-tablet-prototype/#2485984), published Nov. 27, 2009, 2 pages.
Savov, V., Engadget, "NVIDIA Tegra tablet prototype hands-on," (http://www.engadget.com/photos/nvidia-tegra-tablet-prototype/#2485985), published Nov. 27, 2009, 2 pages.
Savov, V., Engadget, "NVIDIA Tegra tablet prototype hands-on," (http://www.engadget.com/photos/nvidia-tegra-tablet-prototype/#2485986), published Nov. 27, 2009, 2 pages.
Savov, V., Engadget, "NVIDIA Tegra tablet prototype hands-on," (http://www.engadget.com/photos/nvidia-tegra-tablet-prototype/#2485987), published Nov. 27, 2009, 2 pages.
Savov, V., Engadget, "NVIDIA Tegra tablet prototype hands-on," (http://www.engadget.com/photos/nvidia-tegra-tablet-prototype/#2485988), published Nov. 27, 2009, 2 pages.
Savov, V., Engadget, "NVIDIA Tegra tablet prototype hands-on," (http://www.engadget.com/photos/nvidia-tegra-tablet-prototype/#2485989), published Nov. 27, 2009, 2 pages.
Savov, V., Engadget, "NVIDIA Tegra tablet prototype hands-on," (http://www.engadget.com/photos/nvidia-tegra-tablet-prototype/#2485990), published Nov. 27, 2009, 2 pages.
Savov, V., Engadget, "NVIDIA Tegra tablet prototype hands-on," (http://www.engadget.com/photos/nvidia-tegra-tablet-prototype/#2485991), published Nov. 27, 2009, 2 pages.
Savov, V., Engadget, "NVIDIA Tegra tablet prototype hands-on," (http://www.engadget.com/photos/nvidia-tegra-tablet-prototype/#2485992), published Nov. 27, 2009, 2 pages.
Schulze, S. and Gratz, I., "1999 | Studio Display," in Apple Design, pp. 144-145, Hatje Cantz Verlag, Ostfildern, Germany (2011).
Softgate, "Fujifilm FinePix Real 3D VI," (http://www.softgate.ch/ger_details_31383/Fujifilm_FinePix_Real_3D_VI_Viewer_8_TFT.html), published Apr. 16, 2010, 2 pages.
Testberichte.de, "Fuji FinePix Real 3D VI," (http://www.testberichte.de/p/philips-tests/digital-photoframe-sph8008-testbericht.html), retrieved Nov. 21, 2011, 3 pages.
Testberichte.de, "Philips Digital PhotoFrame SPH8008," (http://www.testberichte.de/p/philips-tests/digital-photoframe-sph8008-testbericht.html), retrieved Nov. 21, 2011, 4 pages.
Touch User Guide, "What will the new Apple Tablet look like? Apple tablet images leaked?," (http://www.touchuserguide.com/2009/07/29/what-will-the-new-apple-tablet-look-like/), published Jul. 29, 2009, 29 pages.
Toxel.com, "12 Cool Apple Tablet Concepts," (http://www.toxel.com/tech/2009/08/31/12-cool-apple-tablet-concepts/), published Aug. 31, 2009, 16 pages.
Verkehrsrundschau, "Details zum Digitalen Bilderrahmen," (http://www.verkehrunschau.de/fotowettbewerb-mein-tag-der-logistik-835123.html?_apg=2), published Apr. 27, 2009, 1 page.
Vilas-Boas, R., "Nokia I.D.," (http://rdvb-designshowcase.blogspot.com/p/cheddar-1-2009.html), retrieved Nov. 7, 2011, 6 pages.
Yatego Shopping, "Lenco Digitaler Bilderrahmen DF-911 B," (http://www.yatego.com/index.htm?cl=spotshopping&popup=media&pid=4a13d0cc891ff), retrieved Nov. 21, 2011, 1 page.
Ballmer, S., Spiegel Online, "Bill Gates, Tablet PC: "Mira" war 2002 die große Vision des Microsoft-Gründers—kam aber wohl zu früh für den Markt," (http://www.spiegel.de/netzwelt/gadgets/bild-670427-46770.html), product announced 2002, 4 pages.
Ballmer, S., Spiegel Online, "Tablet PC," (http://www.spiegel.de/netzwelt/gadgets/0,1518,670427,00.html), published Jan. 6, 2010, 6 pages.
Cesweb, "MPIO," (http://www.cesweb.org/shared_files/innovations/innovations_2003/1616/mainphoto1616.jpg), retrieved Feb. 24, 2012, 1 page.
Chip Online, "VIA: Zwitter—PC für Schöngeister," (http://www.chip.de/ii/202812385_78f68afa8e.jpg), published May 8, 2002, 1 page.
Chip Online, "VIA: Zwitter—PC für Schöngeister," (http://www.chip.de/news/VIA-Zwitter-PC-fuer-Schoengeister_34204096.html), published May 8, 2002, 3 pages.
Cnet, "Sony Clie," (http://reviews.cnet.com/se/30733801-2-440-overview-1.gif), retrieved Feb. 26, 2012, 1 page.
Computer Museum, "Toshiba Dynapad T100X (1992)," (http://www.computermuseum.li/Testpage/Toshiba-Dynapad-1992.htm), product announced Nov. 1992, 1 page.
De Herrera, C., tabletpctalk, "Tatung Tablet PC Picture," (http://www.tabletpctalk.com/pictures/tatung.shtml, 2003, 3 pages.
Dg2000, Flickr, "Sony CLIE PEG-TH55/E1," http://www.flickr.com/photos/derekwilkinson/1323901259/), photo taken Sep. 4, 2007, 2 pages.
Digibarn Computer Museum, "Image of the GO tablet," http://www.digibarn.com/collections/systems/go/DSC02917.JPG), retrieved Feb. 24, 2012, 1 page.

Figure 5.A.3 (*Continued*)

US D762,208 S
Page 8

(56) **References Cited**

OTHER PUBLICATIONS

Digibarn Computer Museum, "Image of the GO tablet," http://www.digibarn.com/collections/systems/go/DSC02942.JPG), retrieved Feb. 24, 2012, 1 page.
Falcone, J., Cnet, "Digitalway MPIO-FL100 (128 MB)," (http://reviews.cnet.com/mp3-players/digitalway-mpio-fl100-128/4505-6490_7-20763548.html), published Apr. 18, 2003, 8 pages.
Gilder, G. "Digital Dark Horse—Newspapers," Forbes ASAP, published Oct. 25, 1993, 2 pages.
Gollwitzer, M., et al., Chip Online, "Microsofts CeBIT-Neuheiten 2002," (http://www.chip.de/artikel/c_druckansicht_9668117.html), published Mar. 13, 2002, 5 pages.
Heimi, "MPIO," http://www.heimi.net/FlashPlayer/MPIOFL100/pics/01a%20FL100Backlight.jpg), retrieved Feb. 24, 2012, 1 page.
Heimkino Markt, "CeBIT 2004 News: FLATRON Tablet—TFT-Monitor von LG mit Stifteingahemöglichkeit / Ideal für Industrie— und Designanwendungen," (http://www.heimkinomarkt.de/News/643/LG/CeBIT%202004%20News%20FLATRON%20TabletTFT%20-Monitor%20von/), published Mar. 25, 2004, 3 pages.
Justia, "*Apple v. Samsung Electronics Co. Ltd. et al.*, Filing 166 Attachment: 10," (http://docs.justia.com/cases/federal/district-courts/california/candce/5:2011cv01846/239768/166/10.html), dated Aug. 22, 2011, 2 pages.
Justia, "*Apple v. Samsung Electronics Co. Ltd. et al.*, Filing 166 Attachment: 11," (http://docs.justia.com/cases/federal/district-courts/california/candce/5:2011cv01846/239768/166/11.html), dated Aug. 22, 2011, 2 pages.
Justia, "*Apple v. Samsung Electronics Co. Ltd. et al.*, Filing 166 Attachment: 14," (http://docs.justia.com/cases/federal/district-courts/california/candce/5:2011cv01846/239768/166/14.html), dated Aug. 22, 2011, 2 pages.
Justia, "*Apple v. Samsung Electronics Co. Ltd. et al.*, Filing 166 Attachment: 15," (http://docs.justia.com/cases/federal/district-courts/california/candce/5:2011cv01846/239768/166/15.html), dated Aug. 22, 2011, 2 pages.
Justia, "*Apple v. Samsung Electronics Co. Ltd. et al.*, Filing 166 Attachment: 16," (http://docs.justia.com/cases/federal/district-courts/california/candce/5:2011cv01846/239768/166/16.html), dated Aug. 22, 2011, 2 pages.
Justia, "*Apple v. Samsung Electronics Co. Ltd. et al.*, Filing 166 Attachment: 9," (http://docs.justia.com/cases/federal/district-courts/california/candce/5:2011cv01846/239768/166/9.html), dated Aug. 22, 2011, 2 pages.
Justia, "*Apple v. Samsung Electronics Co. Ltd. et al.*, Filing 456 Attachment: 5," (http://docs.justia.com/cases/federal/district-courts/california/candce/5:2011cv01846/239768/456/5.html), dated Dec. 17, 2011, 3 pages.
Justia, "*Apple v. Samsung Electronics Co. Ltd. et al.*, Filing 456 Attachment: 6," (http://docs.justia.com/cases/federal/district-courts/california/candce/5:2011cv01846/239768/456/6.html), dated Dec. 17, 2011, 3 pages.
Justia, "*Apple v. Samsung Electronics Co. Ltd. et al.*, Filing 456 Attachment: 7," (http://docs.justia.com/cases/federal/district-courts/california/candce/5:2011cv01846/239768/456/7.html), dated Dec. 17, 2011, 3 pages.
Long, B., Macworld, "PL500 LCD Pen Tablet System," (http://images.macworld.com/images/legacy/2001/02/images/content/wacom.gif), published Feb. 1, 2001, 1 page.
Long, B., Macworld, "PL500 LCD Pen Tablet System," http://www.macworld.com/article/1938/2001/02/wacomtablet.html), published Feb. 1, 2001, 1 page.
Markoff, J., "A Media Pioneer's Quest: Portable Electronics Newspapers," The New York Times published Jun. 28, 1992, p. 11.
Microsoft, "Bill Gates Highlights the Magic of Consumer Software And the Coming Digital Decade at Consumer Electronics Show," (http://www.microsoft.com/presspass/press/2002/jan02/01-07ces2002keynotepr.mspx), published Jan. 7, 2002, 3 pages.

Monitor4u, "Flatron L1530TM," (http://www.monitor4u.co.kr/review/review4u/content.asp?idx=580&leftcode=16), published Mar. 17, 2004, 7 pages.
Old Computers, "Toshiba T100-X Dynapad," (http://www.old-computers.com/museum/computer.asp?st=1&c=801), product released 1993, 2 pages.
Oullette, D., "Dynapad weighs in at just 3.3 pound," Info World, vol. 15, Issue 31, Aug. 2, 1993, 2 pages.
PDA Museum, "Sony Clie,"(http://www.pdamuseum.com/sony_clie/clie_pics/th-55j2.jpg), retrieved Feb. 26, 2012, 1 page.
Presstime, "Life After Viewtron?Enter the E-Paper," Presstime, published Aug. 1993, p. 18-19.
Serapin_frame, s10forum, "fs: 2 lg flatron tablet monitors," (http://www.s10forum.com/forum/f296/fs-2-lg-flatron-tablet-monitors-483210/), published Nov. 27, 2011, 5 pages.
Sony, "The Sony Clié PEG-TH55 heralds The Next Generation of Digital Organisers," (http://dpnow.com/595.com.html), published Feb. 13, 2004, 2 pages.
Tatung, "Comdex Fall 2003 Tatung Booth# 102448," (http://www.tatung.com/en/news/comdex2003.htm), published Nov. 17, 2003, 2 pages.
Toshiba, "Dynapad T100X," product announced Nov. 1992. 2 pages.
Wacom, "Company History, PL-500," (http://wacom.jp/en/company/history/), product released Jun. 2000, 1 page.
Wikimedia, "MPIO," (http://upload.wikimedia.org/wikipedia/commons/e/e9/M-Pio_FL100.jpg), retrieved Feb. 24, 2012, 1 page.
Wiley, M.. IGN, "MPIO FL100 Review," http://uk.ign.com/article/2003/01/21/mpio-fl100-review), published Jan. 21, 2003, 4 pages.
"A Day in the Life of InfoLink," Stanford University Libaries, published May 1, 2003.
"The Oracle: The Personal Computer of the Year 2000," Stanford University Libraries.
"The Organizer: The Personal Computer of the Year 2000," Stanford University Libraries.
Aliexpress.com, "WACOM PL-500 15" LCD Tablet Mint," (http://www.aliexpress.com/product-fm/430172068-WACOM-PL-500-15-LCD-Tablet-Mint-Worldwide-Free-Shipping-wholesalers.html), accessed Nov. 14, 2011, 3 pages.
Amazon.com, "Digitalway MPIO FL100 128 MB MP3 Player" (http://www.amazon.com/Digitalway-MPIO-FL100-128-Player/dp/B00008V6NI), accessed Nov. 17, 2011, 7 pages.
Amazon.com, "Whirlpool Gold : GH7208XRS 2.0 cu. ft. Velos Speedcook Over the Range Microwave Oven," (http://www.amazon.com/Whirlpool-Gold-GH7208XRS-Speedcook-Microwave/dp/B000ZIPHM8), accessed May 23, 2012, 5 pages.
AndroidTabletFanatic.com "16 Old-Skool Tablets Rocking it Before the iPad," (http://www.androidtabletfanatic.com/android-tablet-news/16-old-skool-tablets-rocking-it-before-the-ipad/), published May 16, 2011, 10 pages.
Blandford, "Nokia N92 Preview," (http://www.allaboutsymbian.com/features/item/Nokia_N92_Preview.php), published Nov. 23, 2005, 10 pages.
Block, "Olympus @ CES Hand-on with the m:robe MR-500i and MR-100," Endadget.com, (http://www.engadget.com/2005/01/10/olympus-ces-hands-on-with-the-m-robe-mr-500i-and-mr-100), published Jan. 10, 2005, 3 pages.
Cell Phones, About.com, "More pictures of the LG KG800," (http://cellphones.about.com/library/b1-pi-lg_kg800.htm), accessed Feb. 13, 2012, 1 page.
Clarke, "Newspad by Authur C. Clarke from 2001: A Space Odyssey," (http://www.technovelgy.com/et/content.asp?Bnum=529), published 1968, 3 pages.
Computex.biz, "Tatung Tablet PC," (http://www.computex.biz/tatung/Default.aspx?pagetype=ProductDetail&pdt_id=36&cid=3 . . .). accessed Nov. 4, 2011, 3 pages.
Flickr.com, "Apple Graphics Tablet" (http://farm3.static.flickr.com/2691/4126733086_b24c987b0e_o.jpg), accessed Nov. 15, 2011, 1 page.
Fried, "Live-blogging Steve Ballmer," CNET, (http://www.cnet.com/8301-31045_1-10426723-269.html?tag=mncol%3btxt), published Jan. 6, 2010, 6 pages.

Figure 5.A.3 (*Continued*)

US D762,208 S
Page 9

(56) **References Cited**

OTHER PUBLICATIONS

Itbusiness.ca, "Microsoft Mira eHome Wireless Smart Display," (http://www.itbusiness.ca/images/articles/Jun10/clip_image003_0087.jpg), accessed Nov. 15, 2011, 1 page.
iTechNews.net, "Samsung K3 MP3 Player," (http://www.itechnews.net/2006/12/02/samsung-k3-mp3-player/), published Dec. 2, 2006, 9 pages.
Justia, "*Apple* v. *Samsung Electronics Co. Ltd. et al.*, Filing: 943 Attachment: 5," (http://docs.justia.com/cases/federal/district-courts/california/candce/5:2011cv01846/239768/943/5.html), dated May 18, 2012, 2 pages.
Knowyoumobile.com, "LG KG800 technical specifications," (http://knowyourmobile.com/lg/lgkg800chocolate/lgkg800/reviews/137/lg_kg800_technical_specifications.html), published Mar. 9, 2007, 3 pages.
McCracken, "The iPad Of 2000. As Envisioned in 1988," Technologizer.com, (http://technologizer.com/2010/05/14/the-ipad-of-2000-as-envisioned-in-1988/), published May 14, 2010, 12 pages.
MEL et al., "Tablet: Personal Computer in the Year 2000," *Communications of the ACM* 31:638-646, ACM (Jun. 1988).
Mobilegazette.com, "Nokia N92," (http://www.mobilegazette.com/nokia-n92-051102.htm), published Nov. 2, 2005, 2 pages.
Needleman, "Hands-on with the JooJoo," (http://news.cnet.com/8301-19882_3-10410960-250.html), CNET News, published Dec. 7, 2009, 31 pages.
Newlaunches.com, "Pidion BM-200 sleek Windows CE PDA phone from Bluebird," (http://www.newlaunches.com/archives/pidion_bm200_sleek_windows_ce_pda_phone_from_bluebird.php), published Dec. 15, 2005, 4 pages.
NYTimes.com "iRiver U10 (1GB)," (http://www.nytimes.com/technology/personaltech/music-players/iRiver-U10-1GB-/31432102.html), published Feb. 12, 2012, 2 pages.
NYTimes.com, "iRiver U10 (1GB)," (http://www.nytimes.com/technology/personaltech/music-players/iRiver-U10-1GB-/31432102.html), published Feb. 12, 2012, 2 pages.
Patel, "The HP Slate" (http://www.engadget.com/2010/01/06/the-hp-slate/), Engadget.com, published Jan. 6, 2010, 4 pages.
PDAdb.net, "Acer n20 Specs," (http://pdadb.net/index.php?m=specs&id=257&c=acern20), published May 13, 2005, 3 pages.
PDAdb.net, "Casio Cassiopeia IT-10 M20BR Specs," (http://pdadb.net/index.php?m=specs&id=95&c=casio_cassiopeia_it-10_m20br), published Dec. 30, 2004, 3 pages.
PDAdb.net, "Dell Axim X3 Basic Specs," (http://pdadb.net/index.php?m=specs&id=81&c=dell_axim_x3_basic), published Dec. 30, 2004, 3 pages.
PDAdb.net, "MiTAC Mio 338 Plus Digi-Walker Specs," (http://pdadb.net/index.php?m=specs&id=60&c=mitac_mio_338_plus_digi-walker), published Dec. 25, 2004, 3 pages.
PDAdb.net "O2 XDA (HTC Wallaby) Specs," (http://pdadb.net/index.php?m=specs&id=117&c=o2_xda_htc_wallaby), published Feb. 3, 2005, 3 pages.
PDAdb.net, "Sony Clie PEG-SL10 / PEG-SL10U / PEG-SL10E Specs," (http://pdadb.net/index.php?m=specs&id=1198&c=sony_clie_peg-sl10_peg-sl10u_peg-sl10e), published Mar. 4, 2008, 3 pages.
PDAdb.net, "Sony Clie PEG-T665C / PEG-T675C Specs," (http://pdadb.net/index.php?m=specs&id=1192&c=sony_clie_peg-t665c_peg-t675c), published Mar. 4, 2008, 3 pages.
PDAdb.net, "Toshiba Genio e330 Specs," (http://pdadb.net/index.php?m=specs&id=29&c=toshiba_genio_e330), published Dec. 9, 2004, 3 pages.
PDAdb.net, "Toshiba Genio e570," (http://pdadb.net/index.php?m=specs&id=229&c=toshiba_genio_e570), published May 4, 2006, 3 pages.
Ricker, "LG's KE850 PRADA official: iPhone says, wha?" Engadget.com, (http://www.engadget.com/2007/01/18/lgs-ke850-prada-official-iphone-says-wha/), published Jan. 18, 2007, 4 pages.
Ricker, "The Pidion BM-200: it's time to move to Seoul," (http://www.engadget.com/2005/10/04/the-pidion-bm-200-its-time-to-move-to-seoul/), Engadget.com, published Oct. 4, 2005, 4 pages.
Rojas, "Hands-on with the Nokia N92," Engadget.com, (http://www.engadget.com/2005/12/01/hands-on-with-the-nokia-n92/), published Dec. 1, 2005, 4 pages.
S. Dimitris, "Pidion BM200 Smartphone," PDASNews, (www.pdasnews.com/articles/pidion-bm200-smartphone.html), accessed Mar. 7, 2012, 2 pages.
Salesstores.com, "LG L1530TM Tablet LCD Monitor," (http://salesstores.com/lg11taledmos.html), accessed Nov. 17, 2011, 2 pages.
WebdesignerDepot.com, "The Evolution of Cell Phone Design Between 1983-2009" (http://www.webdesignerdepot.com/2009/05/the-evolution-of-cell-phone-design-between-1983-2009/), accessed May 6, 2011, 56 pages.
Wikipedia.org, "Clié," (http://en.wikipedia.org/wiki/CLI%C3%89), last modified Feb. 12, 2011, 3 pages.
Wikipedia.org, "E-Ten" (http://en.wikipedia.org/wiki/E-TEN), last modified Dec. 15, 2010, 2 pages.
Wikipedia.org, "File:LG KE850 Prada Hauptmenü.jpg," (http://en.wikipedia.org/wiki/File:LG_KE850_Prada_Hauptmen%C3%BC.jpg), accessed Feb. 12, 2012, 4 pages.
Wikipedia.org, "iPAQ" (http://en.wikipedia.org/wiki/IPAQ), last modified May 4, 2011, 7 pages.
Wikipedia.org, "Pocket viewer," (http://en.wikipedia.org/wiki/Pocket_viewer), last modified Apr. 10, 2011, 4 pages.
Wikipedia.org, "Zire Handheld," (http://en.wikipedia.org/wiki/Zire_Handheld), last modified Dec. 11, 2010, 4 pages.
Bhangal, S., O'Reilly Media, "The Page Turn Effect in Flash MX," (http://oreilly.com/javascript/archive/flashbacks.html), published Sep. 3, 2004, 3 pages.
FlippingBook, "Flipping Book Publisher 2.2 Demonstrations," (http://flippingbook.com/products/publisher/#demonstrations), retrieved Apr. 25, 2012, 2 pages.
Hosteasy Solutions, "Flipping Book Publisher 2.0," (http://www.hosteasysolutions.com/what-we-offer.html), retrieved Apr. 25, 2012, 2 pages.
Skyhool, naver.com, "FLEX," (http://blog.naver.com/PostView.ahn?blogId=skyhool&logNo=150027158023&widgetTypeCall=true), published Jan. 25, 2008, 3 pages.
Block, "Philips PET830 and PET1030 media players," Engadget, (http://www.engadget.com/2006/11/29/philips-pet830-and-pet1030-media-players/), published Nov. 29, 2006, 2 pages.
Boggs, "Electrovaya Scribbler SC4000 Review," Tablet PC Review, (http://www.tabletpcreview.com/default.asp?newsID=1130), published Apr. 15, 2008, 6 pages.
Fidler, "Mediamorphosis: Understanding New Media," pages 200, 222-225, and 236-252, Pine Forge Press, Thousand Oaks, CA (1997).
Fidler, "Predicting e-readers in 1981: A look back at the future," Associated Press Managing Editors, (http://www.snd.org/2010/01/predicting-e-readers-in-1981-a-look-back-at-the-future/) published Jan. 14, 2010, 5 pages.
Flatley, "BoEye MID700 unveiled with Android OS, vaguely familiar form factor," Engadget.com, (http://engadget.com/2009/11/11/boeye-mid700-unveiled-with-android-os-vaguely-familiar-form-fae/?icid=eng_Dubai_art), published Nov. 11, 2009, 2 pages.
Ganapati, "Hands on With the JooJoo, Formerly Known as CrunchPad," (http://www.wired.com/gadgetlab/2009/12/hands-on-joo-joo-crunchpad/all/1), published Dec. 7, 2009, 3 pages.
Justia, "*Apple* v. *Samsung Electronics Co. Ltd. et al.*, Filing: 166 Attachment: 2," (http://docs.justia.com/cases/federal/district-courts/california/candce/5:2011cv01846/239768/166/2.html), dated Aug. 22, 2011.
Justia, "*Apple* v. *Samsung Electronics Co. Ltd. et al.*, Filing: 166 Attachment: 3," (http://docs.justia.com/cases/federal/district-courts/california/candce/5:2011cv01846/239768/166/3.html), dated Aug. 22, 2011.
Justia, "*Apple* v. *Samsung Electronics Co. Ltd. et al.*, Filing: 166 Attachment: 4," (http://docs.justia.com/cases/federal/district-courts/california/candce/5:2011cv01846/239768/166/4.html), dated Aug. 22, 2011.
Miles, "56 Apple iTablet concept designs: Could any be real?," (http://www.pocket-lint.com/news/26500/56-apple-itablet-concept-designs), published Aug. 21, 2009, 14 pages.

Figure 5.A.3 (*Continued*)

US D762,208 S
Page 10

(56) **References Cited**

OTHER PUBLICATIONS

Miller, "MSI shows off 10-inch Android tablet running new Tegra chipset," Engadget.com, (http:///www.engadget.com/2010/01/08/msi-shows-off-10-inch-android-tablet-running-new-tegra-chipset/), published Jan. 8, 2010, 11 pages.
Perton, "Samsung digital picture frame stores pics, movies, music," Engadget, (http://www.engadget.com/2006/03/09/samsung-digital-picture-frame-stores-pics-movies-music/), published Mar. 9, 2006, 1 page.
Rothman, "HP's Windows 7 Slate Device Revealed by Steve Ballmer," Gizmodo.com, (http:/1 gizmodo.com/ 544 2200/hps-windows-7-slate-device-revealed-by-steve-ballmer), published Jan. 6, 2010, 2 pages.
Stevens, "ICD confirms Vega tablet, includes Android 2.0, Tegra processor, our hearts," Engadget.com, (http://www.engadget.com/2009/11/13/icd-confirms-vega-tablet-includes-android-2-0-tegra-processor/), published Nov. 13, 2009, 2 pages.
TechLahore.com, "China's Boeye releases "Apple tablet" before Apple can—TechLahore@GITEX," (http://www.techlahore.com/2009/11/08/chinas-boeye-releases-apple-tablet-before-apple-can-techlahoregitex/), published Nov. 8, 2009, 3 pages.
The Apple Collection, "Apple PowerPad G3 Pro—Apple Design & Prototype," (www.theapplecollection.com/design/macdesign/Apple_PowerPad_G3_Pro.shtml), updated Oct. 31, 2011, 2 pages.
The Apple Collection, "MyPal," (www.theapplecollection.com/design/macdesign/MyPal.shtml), updated Oct. 31, 2011, 2 pages.
The Apple Collection, "PowerPad," (www.theapplecollection.com/design/macdesign/PowerPad2.shtml), updated Oct. 31, 2011, 2 pages.
Tingle, "ICD Vega Android 2.0 Tablet Available in 2010 [Vega Android Tablet, Nvidia Tegra Chipset Touting 'Low Cost Tablet' Confirmed]," (http://nexus404.com/Blog/2009/11/13/icd-vega-android-2-0-tablet-available-in-2010-vega-android-tablet-nvidia-tegra-chipset-touting-low-cost-tablet-confirmed/), TFTS, published Nov. 13, 2009, 2 pages.
To, "Boeye MID 700 tablet runs Android with Apple-esque aesthetics," Pocketables.com, (http://www.pocketables.net/2009/11/boeye-mid-700-tablet-runs-android-with-applesque-aesthetics.html), published Nov. 12, 2009, 2 pages.
Topolsky, "Leaked pics of the CrunchPad make it look dangerously close to availability," Engadget.com, (http://www.engadget.com/2009/04/09/leaked-pics-of-the-crunchpad-make-it-look-dangerously-close-to-a/), published Apr. 9, 2009, 2 pages.
Topolsky, "NVIDIA CEO shows off mystery tablet, makes zero statements about mystery tablet," Engadget.com (http://www.engadget.com/2009/11/09/nvidia-ceo-shows-off-mystery-tablet-makes-zero-statements-about/), published Nov. 9, 2009, 1 page.
Weiss, "Hands-On: NVIDIA Tegra Tablet Prototyp ICD Vega (Video)," (http://de.engadget.com/2009/11/28/hands-on-nvidia-tegra-tablet-prototyp-icd-vega-video/), Engadget German, published Nov. 28, 2009, 13 pages.
Klass, C., "Plastic Logic Reader—erste Partner bekannt," (http://ww.golem.de/0902/65175.html), published Feb. 10, 2009, 9 pages.
JKKMobile.com, "Hanvon TouchPad BC10C tablet review," (http://ww.jkkmobile.com/2010/03/hanvon-touchpad-be10c-tablet-revew.html), published Mar. 13, 2010, 2 pages.
MacTablet, "More Pictures of MacTablet Revealed.," (http://www.mactablet.co.uk/2009/11/more-pictures-of-mactablet-revealed.html), published Nov. 27, 2009, 3 pages.
Stromer, G., "CES 2010: PLastic Logic zeigt superflachen und-teuren eBook-Reader Que," (http://www.cnet.de/41525409/ces-2010-plastic-logic-zeigt-superflachen-und-teuren-ebook-reader-que/), published Jan. 10, 2010, 2 pages.
Vantage Emea, "TPT1210-1 Web Tablet," (http://www.vantageemea.com/technical/User%20Interfaces/Touchscreens/Cutsheets/TPT1210-1%20Web%20Tablet_cutsheet.pdf), published 2008, Belgium, 2 pages.
Cranfill et al., "Display Screen or Portion Thereof with Animated Graphical User Interface," U.S. Control No. 90/012,756, filed Dec. 31, 2012 (pending ex parte reexamination of U.S. Pat. No. D670,713).
Cranfill et al., "Display Screen or Portion Thereof with Animated Graphical User Interface," U.S. Control No. 90/012,757, filed Dec. 31, 2012 (pending ex parte reexamination of U.S. Pat. No. D669,713).
Girvan, R., "Spitfire 40," http://www.rolandandcaroline.co.uk/html/spitfire_40.html, accessed Jan. 8, 2013, 4 pages.
Gorenje Kitchen 2008 Online Catalog, http://web.archive.org/web/20090301042519/http:/page-flip.com/new/demos/03-kitchen-gorenje-2008/index.html, dated Nov. 3, 2008, accessed Jan. 11, 2013.
Liesaputra et al., "Creating and Reading Realistic Electronic Books," *Computer*, vol. 42, No. 2, pp. 72-81, IEEE Computer Society (Feb. 2009).
United Video, http://cogentlogic.com/apple/, accessed Jan. 11, 2013.

* cited by examiner

Figure 5.A.3 (*Continued*)

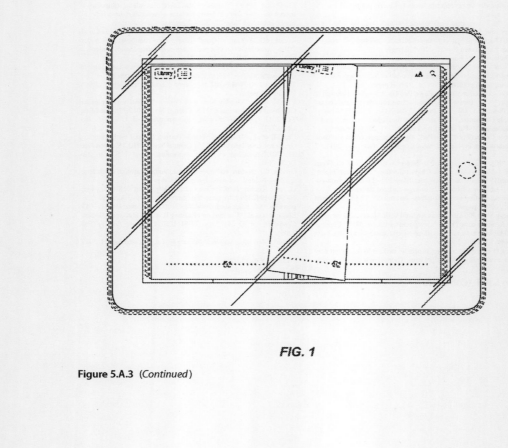

FIG. 1

Figure 5.A.3 (*Continued*)

FIG. 2

Figure 5.A.3 (*Continued*)

U.S. Patent　Jul. 26, 2016　Sheet 3 of 3　US D762,208 S

FIG. 3

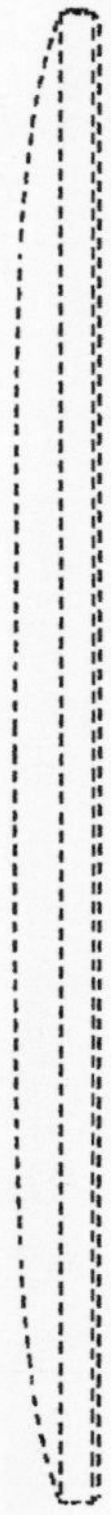

FIG. 4

FIG. 5

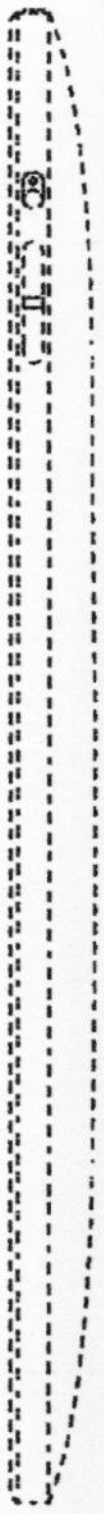

FIG. 6

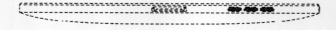

Figure 5.A.3 (*Continued*)

第 6 章 权利要求

6.1 引言

在本章中,讨论的是专利中的不同类型的权利要求。权利要求是专利中最关键的部分,会增大不同专利之间的价值差异[1-18]。

6.2 独立和从属权利要求

权利要求撰写中的一个重要概念是撰写独立权利要求和从属权利要求。

在专利权利要求部分构建权利要求间的逻辑联系,这两种类型的权利要求撰写都很重要(如图 6-1 所示)。

图 6-1 独立权利要求和从属权利要求

6.2.1 独立权利要求

"独立权利要求"是指不引用其他权利要求的权利要求[6]。在专利申请中,独立权利要求后面通常有从属权利要求跟随。通过允许从属权利要求和减少独立权利要求中要素的数量,独立权利要求可以不太具体或有更宽泛的保护范围。

6.2.2 从属权利要求

"从属权利要求"与独立权利要求相关联。从属权利要求在前序部分对独立权利要求(或另一个从属权利要求)进行引用[6]。通过添加其他元素或特征,从属权利要求比独立权利要求更具体。在专利申请中,从属权利要求引用独立权利要求,且两者成组出现。

从属权利要求可以存在于发明专利中,如一个结构、方法或组合物。从属权利要求的前序通常以"The"开头,而不是"a"。从属权利要求的前序部分如下:

"2.The structure of Claim 1,wherein XXXX."

或

"2.The method of claim 1, wherein said method of XXX, further comprising the steps of:"

独立权利要求和从属权利要求形成了"逻辑树",独立权利要求可以表示为根(独立权利要求),而分支是由从属权利要求组成的(如图6-2所示)。从属权利要求可以从独立权利要求中分支出来。图6-2展示了一个"逻辑树",展示了独立权利要求和从属权利要求之间的联系。每个独立权利要求都形成另一个"逻辑树"。所有的权利要求互相联系到"逻辑树"中。

图6-2 独立权利要求和从属权利要求"逻辑树"

从属权利要求和独立权利要求可以是结构权利要求、装置权利要求和方法权利要求(如图6-3所示)。

图 6-3　结构、装置和方法权利要求

6.3　结构权利要求

结构权利要求是描述设备、电路或系统的专利权利要求[1-6]。例如，结构权利要求可以描述半导体设备。在权利要求中，结构权利要求必须包括半导体内或描述半导体的图中的所有物理结构。在电路中，电路元件组成电路，并构成一个装置或系统。结构权利要求根据结构要素来定义发明。结构权利要求可以定义结构的元素以及它们是如何集成或相互连接的。结构权利要求不声明本发明的功能。

在权利要求书的前序中，可以使用短语"结构"或者发明的另一个名称。例如，对于一个晶体管而言，术语"晶体管"可以用于描述它是什么，而不是它的功能。

关于晶体管权利要求的一个例子如下：

（1）一种晶体管，包括：

衬底晶片；

所述衬底晶片的顶部平面对向的孔；

所述衬底晶片的所述顶部平面上的栅极电介质；

所述栅极电介质层上的栅极叠层；

与所述栅极叠层对齐的源极区域；

与所述栅极叠层对齐的漏极区域；

连接到所述源极、漏极和栅极叠层的至少一个触点。

6.4　装置权利要求

装置权利要求是描述通常具有有源元件的设备、电路或系统的专利权利要求[1-6]。

例如，在电路中，电路元件组成电路，形成装置或系统。装置权利要求根据本发明的部件来定义本发明。装置权利要求可以定义设备的元件以及如何将它们集成或相互连接，并不要求发明所实现的功能。装置权利要求不同于方法权利要求，因为装置权利要求通常根据物理或结构特性描述设备或系统的组件。实际设备或系统会对装置权利要求侵权。

装置权利要求可以通过前序部分中的术语来标识。与方法权利要求相比，尽管装置权利要求在前序部分中使用的术语有更多的变化，但是装置权利要求的前序部分经常包括以下术语："装置""设备"或"系统"。

在装置中，描述装置的所有组件或元件以及它们是相互连接的方式是很重要的。如果涉及之前提到的某一元素，则该权利要求必须称其为"所述"。

装置权利要求的独立权利要求示例如下：

（2）一种设备，包括：

第一元件 AAA；

电连接到所述第一元件 AAA 的第二元件 BBB；

连接到所述第一元件 AAA 和所述第二元件 BBB 的第三元件 CCC；

耦合到所述第三元件 CCC 的第四元件。

6.5　方法权利要求

方法权利要求包含在发明专利中，是描述如何提供本发明的一组步骤[1-6]。方法或过程属于发明专利的范畴。方法是用于执行本发明的功能的一系列步骤。

方法权利要求前序的格式如下所述：

"一种提供 XXXX 的方法，包括以下步骤："

接着是方法（或过程）的每个步骤，其中每个步骤是单独成行的项目，步骤中的第一个词应该是主动词，然后每一步都以分号结尾，倒数第二个步骤也以分号结尾，后跟"以及"，方法的最后一个步骤以句号结尾。

一些方法权利要求为这些步骤添加符号体系，例如数字（如 1.，2.）或带括号的字母［如（a）、（b）等］。

例如，方法的独立权利要求的步骤如下所示：

（3）一种提供 XXXX 的方法，包括以下步骤：

提供 AAA；

比较 BBB；

生成 CCC；

均衡 DDD。

当方法权利要求要描述某结构或装置时，该方法的第一步骤是对结构或装置的描述，其中装置内的每个元件都包含在该方法的第一步骤内。因此，对于具有元件 X、Y、Z 的放大器，可以写成：

（1）一种提供 XXXX 的方法，包括以下步骤：

提供包括 X、Y、Z 的放大器；

比较 BBB；

生成 CCC；

均衡 DDD。

方法的从属权利要求也可以通过添加额外的步骤来进一步扩展流程步骤。既然是从属权利要求，就必须在前序部分中引用独立权利要求。

（2）根据权利要求 1 所述的方法，其中，所述 EEE 还进一步包括以下步骤：

烘焙 FFF；

摇晃 GGG。

以前，只有当个人或实体实施了权利要求请求保护的所有步骤时，方法专利权利要求才可能被侵犯。这表明了化学方面的流程专利的历史背景，制造已知物质的新的工艺需要被保护。

6.6 混合权利要求

混合权利要求是不同类型的专利的组合，可以是以下类型的组合[1-6]：

- 结构
- 装置
- 方法
- 组成

6.7 方法加功能权利要求

"方法加功能"权利要求是一项包括用功能术语表示的技术特征的权利要求[1-6]。

方法加功能权利要求的示例如下：

"将数字电信号转换成模拟电信号的方法"；

方法加功能权利要求在某种意义上是一种混合权利要求，因为它可以描述一种结构和方法。许多专利代理人不喜欢这类权利要求，因为他们发现其价值很有限。

6.8 Beauregard 权利要求

Beauregard 权利要求是有关计算机可读介质的权利要求，它是以 Beauregard 案的裁决命名的。在 Beauregard 权利要求中，涉及一种计算机可读存储设备，其包含一组指令，这些指令使计算机执行一系列程序[1-6, 8]。

这种权利要求类型是由于"软盘"而产生的。什么是"软盘"？它是一种存储设备，用于存储插入到系统中并被系统读取的数据。因此，其格式必须与"机器可读介质"相关联。

Beauregard 权利要求的独立权利要求实例如下[1-6, 8]所述：

"一种体现在设计过程中使用的机器可读介质中的设计结构，该设计包括：……"

对于从属权利要求：

"根据权利要求1所述的设计结构，其中所述设计结构包括描述所述电路的网表；

根据权利要求1所述的设计结构，停留在所述存储介质上，其中，所述设计结构作为用于交换集成电路的布局数据的数据格式。"

这种权利要求类型在公司中用于软件的保护。许多人认为我们不能为软件申请专利或保护软件，因为我们可以很容易地修改代码。专利代理人用来保护软件代码的方法是在专利中包括 Beauregard 权利要求，从而提供另一层的专利保护。

在撰写软件专利时，专利代理人将这类权利要求添加到清单表中。通过这种方式，他们可以获得一个结构专利要求、装置专利要求、方法专利要求和一个额外的 Beauregard 权利要求的授权专利。

6.9 用尽的组合权利要求

用尽的组合权利要求是这样一种权利要求,具有新颖性和创造性的装置与常规的其他元件相结合,并且新的装置以传统的方式与其协作[6]。

6.10 替代权利要求

本节讨论了新的权利要求的类型。对于特定学科或领域,存在适合特定主题的新型权利要求类型。权利要求类型是为特殊行业的需求而开发的,如电气行业和制药行业(如图6-4所示)。

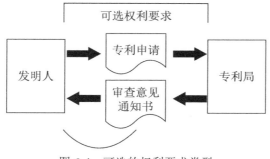

图6-4 可选的权利要求类型

6.10.1 马库什(Markush)权利要求

马库什权利要求或结构是这样一种权利要求,在化合物的一个或多个部分中允许有多个"功能等同"化学成分[6]。

根据美国专利商标局(USPTO)的规定,马库什权利要求的正确格式是:

"选自如下化合物的组合,包括A、B和C。"

6.10.2 两分法(Jepson)权利要求

在美国专利法中,两分法(Jepson)权利要求是一种方法或产品权利要求,其中一个或多个限定被特别标识为创新点,至少可以与前序部分的内容区别[6]。可以像下面这样撰写:

"一种用于存储信息的系统包括(……),其中改进包括:(……)"。

6.10.3 方法特征限定的产品权利要求

方法特征限定的产品权利要求是一种产品权利要求,其中产品是由其制造过程定义的[6]。

例如:

"通过权利要求 X 的方法获得的产品"。

6.10.4 计算机程序权利要求

"计算机程序"权利要求的形式如下:

"……一个通用数字计算机程序,被编程为执行这样和那样的步骤,其中的步骤属于一个方法,如计算……方法。"

这种权利要求的目的是避免判例法将这种特定类型的方法视为不符合专利资格[6]。

6.10.5 综合权利要求

综合权利要求是指包括引用说明书或附图,而没有明确说明所要求保护的产品或过程的任何技术特征的权利要求[6]。

6.10.6 信号权利要求

"信号权利要求"是用于电子信号的权利要求,例如,该电子信号可以具体化、可用于实现期望的结果或服务于一些其他有用目标的信息[6]。

下面是信号权利要求的一个例子:

"携带用于执行新方法的计算机可读指令的电磁信号……"

6.10.7 瑞士型权利要求

瑞士型权利要求或"瑞士型用途权利要求"是以前使用的权利要求形式,旨在涵盖已知物质或组合物[6]的第一、第二或随后发现的医疗用途(或用于功效指示)。

6.10.8 延展性权利要求

延展性权利要求是试图涵盖基础研究发明或发现的权利要求[6]。注意:它是试图在一项发现成为一项完整的发明之前,夺得它的价值。

具体而言，延展性权利要求是指"产品或产品用途的权利要求，其作为筛选方法或工具来提供实验数据，用于鉴别产品"。

6.11　总结

鉴于权利要求撰写的重要性，整个章节侧重于权利要求和撰写。本章涉及撰写独立权利要求和从属权利要求，展现了常见权利要求的结构以及不常见权利要求的结构。本章包括结构权利要求、方法权利要求、混合权利要求、方法加功能权利要求、Beauregard 权利要求、马库什权利要求、两分法（Jepson）权利要求、方法特征限定的产品权利要求、综合权利要求、信号权利要求和瑞士型权利要求。

在第 7 章中，我们将集中讨论审查意见，重点是如何答复专利局和专利审查员，并将侧重于美国专利商标局、欧盟专利商标局和德国专利商标局的审查意见通知书。这一章将包括对审查意见的阅读和权利要求的处理，重点介绍允许的权利要求、驳回的权利要求和异议的权利要求，并讨论详细地处理，撰写审查意见答复并修改权利要求。

问题

1. 独立权利要求和从属权利要求有什么区别？
2. 独立权利要求前序中的第一个词是什么？
3. 从属权利要求前序中的第一个词是什么？
4. 写一组权利要求，其中包括一个以上的独立权利要求，每一个独立权利要求有多个从属权利要求。画出独立和从属权利要求的"逻辑树"连接。
5. 权利要求什么时候使用包含或者组成？有什么区别？
6. 什么是混合权利要求？
7. 写一个自行车的权利要求。
8. 写一个独轮车的权利要求。
9. 写一个有多条腿的椅子的权利要求。
10. 写一个计算机的权利要求。
11. 写一个计算机的装置权利要求。
12. 写一个装置权利要求。

13. 写一个 Jepson 权利要求。
14. 谁会使用 Jepson 权利要求？它有什么作用？
15. 写一个 Markush 权利要求。
16. 谁使用 Markush 权利要求？它有什么作用？
17. 写一个 Beauregard 权利要求。
18. Beauregard 权利要求的目的是什么？它的重要性体现在哪里？
19. 写一组软件权利要求，并在其中添加一个 Beauregard 权利要求。
20. 写一个方法特征限定的产品权利要求。

案例研究

案例研究 A

写一个有四条腿的椅子的权利要求。写一个有一条腿的椅子的权利要求。怎样为有四条腿的椅子写一个权利要求，以涵盖有一条腿的椅子的这种情况？

案例研究 B

为自行车写一个独立的权利要求。为独轮车写一个权利要求。如何写一组权利要求，使权利要求的部分要素放到从属权利要求中？

案例研究 C

为包括磁盘驱动器、软盘、键盘和屏幕的计算机写一个独立权利要求。将针对软盘的 Beauregard 权利要求作为独立权利要求写入其中。

案例研究 D

为计算机写一个独立权利要求，其允许屏幕与计算机分离。为平板电脑写一个权利要求。如何撰写计算机权利要求，以涵盖平板电脑的这种情况？

案例研究 E

为一种允许在不同的配置下旋转屏幕的计算机写一个独立权利要求。你能以重新排列为理由争辩吗？

为"蝴蝶"电脑写一个权利要求，其可以折叠成更小的电脑。

案例研究 F

为一种用于心率监测的运动手表写一个独立权利要求和从属权利要求，这种手表在胸部有一个附件和一个给手表发送信号的发送器。

案例研究 G

为没有发射器、也未在胸部周围安装监听器的运动手表写一个独立权利要求。怎样写权利要求使这款运动手表获得专利权？

可以用简化作为争辩理由吗？移除呢？

案例研究 H

为安装了应用程序的蜂窝电话撰写蜂窝电话的结构、装置和方法权利要求。能为软件和应用程序撰写 Beauregard 权利要求吗？

案例研究 I

为一种嵌入物联网电子产品的服装分别撰写结构、装置和方法权利要求。

参考文献

1. Slusky, R. (2013).*Invention Analysis and claiming: A Patent Lawyer's Guide*, 2e. American Bar Association. ISBN: 13 978-1614385615.

2. Rosenberg, M. (2016). *Essentials of Patent Claim Drafting*, LexisNexis IP Law and Strategy Series. Matthew Bender.

3. Voldman, S. (2014). Short Course, *Innovating, Inventing, and Patenting*, Dr. Steven H. Voldman LLC, Ministry of Science Technology and Innovation(MOSTI), Putrajaya, Malaysia, May 2014.

4. Voldman, S. (2015). Short Course, *Innovating, Inventing,Patenting*, Dr. Steven H. Voldman LLC, FITIS, Sri Lanka, February 2015.

5. Voldman, S. (2016).Short Course, *Writing and Generating Patents*, Dr. Steven H. Voldman LLC, FITIS, Sri Lanka, February 2016.

6. U.S.Patent Office (USPTO). https://www.uspto.gov (accessed 21 December 2017).

7. European Patent Office (EPO). https://www.epo.org (accessed 21 December 2017).

8. Amernick, B.A. (1991). *Patent Law for the Nonlawyer: A Guide for the Engineer, Technologist, and Manager*, 2e. Von Nostrand Reinhold. ISBN: 13 978-0442001773.

9. Mueller, J. M. (2016). *Patent Law*, 5e. Wolter Kluwer Publications.

10. Stim, R. (2016). *Patent, Copyright, and Trademark: An Intellectual Property Desk Reference*. Nolo. ISBN: 978-1-4133-2221-7.

11. Adams, D. O. (2015). *Patents Demystifed: An Insider's Guide to Protecting Ideas and Invention*. American Bar Association. ISBN: 13 978-163425679.

12. Pressman, D. and Tuytschaevers, T. (2016). *Patent It Yourself: Your Step-by-Step Guide to filing at the U.S. Patent Office*. Nolo.

13. Charmsson, H.J.A. and Buchaca, J. (2008). *Patents, Copyrights, and Trademarks for Dummies*. Wiley.

14. Stim, R. and Pressman, D. (2015). Patent Pending in 24 Hours, 7e. Nolo. ISBN: 978-1-4133-2201-9.

15. Durham, A.L. (2013). *Patent Law Essentials: A Concise Guide*, 4e. Oxford: Praeger, ABC-CLIO, LLC. ISBN: 13 978-1440828782.

16. DeMatteis, B., Gibb, A., and Neustal, M. (2006). *The Patent Writer: How to Write Successful Patent Applications*. Garden City Park, NY: Patents for Commerce, Square One Publishers.

17. Sutton, E. *Software Patents: A Practical Perspective*, 2016. CreateSpace Independent Publishing Platform.

18. Jackson Knight, H. (2013). *Patent Strategy for Researchers and Research Managers*, 3e. Chichester, England: Wiley.

第 7 章
审查意见通知书

7.1 引言

本章讨论了审查意见通知书（OA）及其答复。本章首先讨论美国专利商标局（USPTO）的审查意见通知书。了解如何阅读专利审查员的审查意见通知书以及如何写答复，这是发明流程中的关键。本章还重点介绍了如何撰写美国专利商标局的审查意见通知书答复的基础知识，并讨论了欧盟审查意见通知书以及其与美国专利商标局的审查意见通知书的不同之处。

7.2 审查意见通知书——美国专利商标局

申请人向美国专利商标局（USPTO）提交专利申请后，会收到专利审查员的回复[1]。了解如何阅读专利审查员发出的审查意见通知书，对于准备该通知书的回复至关重要。根据审查意见通知书的内容，必须发送答复给美国专利商标局，以解决专利申请中专利审查员提出的问题（如图 7-1 所示）。

学会阅读审查意见通知书是了解专利处理的重要技能。通过了解如何阅读审查意见通知书，可以学习如何撰写审查意见通知书的答复[2-6]。

在本节中，将讨论审查意见通知书的内容（如图 7-2 所示）。对于美国专利申请，审查意见通知书文件使用 USPTO 的信头并包含 USPTO 的地址[1]。

从发明到专利——科学家和工程师指南. 第一版. 史蒂文·H. 沃尔德曼 ©2018 John Wiley & Sons 公司，2018 年由 John Wiley & Sons 公司出版。

图 7-1　审查意见通知书与美国专利商标局

图 7-2　USPTO 审查意见通知书的内容

1. 申请号

在提交专利申请后，该案件会被分配一个申请号。这一申请号作为该案例的指代，并在审查意见通知书文档中列出。

2. 代理人案卷号

代理人案卷号也列在专利申请中。在提交专利申请后，代理人案卷号列在审查意见通知书文件中，并作为处理此案专利代理人的指代。

3. 第一发明人

在审查意见通知书中，申请引用第一发明人。在首页上，申请会写出第一发明人的全名。在整个审查意见通知书中将使用发明人的姓氏。如果专利申请具有多个发明人（共同发明人），则专利申请将其他人记作"等"。

4. 律师事务所

在审查意见通知书中，专利申请与提交该申请的律师事务所相关联。审查意见通知书上有该律师事务所的名称和地址。审查意见通知书不显示律师姓名，但在审查意见通知书答复中会明确。

5. 审查员

给出审查意见通知书的专利审查员是专利局的雇员。专利审查员的姓名首先列出姓氏，然后是名和中间名。被指派的专利审查员来自指定专利被分配的"技术领域"（Art Unit）。这是为了确保专利审查员具有处理专利申请的相关教育背景或技能。

6. 技术领域（Art Unit）

"技术领域"是指专利局中分配专利申请的某个小组。专利申请被分配给一个小组，该小组的专利审查员的技能或教育背景适于处理该专利申请。

7. AIA（第一个申请的发明人）

在审查意见通知书中，将涉及美国发明法案（AIA）。美国发明法案改变了"先申请"和"先发明"相关的专利申请程序的处理。因此，将提到案件采用 AIA 还是 AIA 之前的方式处理。

8. 申请日

在审查意见通知书中，第一页列有申请日。申请日是专利递交到专利局的那一天。

9. 邮寄日

在审查意见通知书中，第一页列有邮寄日。邮寄日是将审查意见通知书邮寄到要处理的律师事务所的日期。这一点至关重要，因为律师事务所必须在有限的时间内作出回应。在审查意见通知书中，规定了答复期限，如几个月。例如，文本会指出：

缩短的法定答复期限设定为自本函邮寄日起，为期 3 个月。

10. 状态

在审查意见通知书的状态部分，列出了专利申请的状态。
有以下状态选项：
- 对本次沟通的回应。
- 最终审查意见通知书。
- 非最终审查意见通知书。

- 申请人在回应时针对限制进行选择。
- 申请有条件地被允许，但需要解决形式问题。
（如美国专利法规定的 Ex parte Quayle，1935 CD，11,45 OG 213）。

7.3 权利要求的处理

审查意见通知书最重要的部分之一是权利要求处理（如图 7-3 所示）。这部分列出了权利要求的状态和可专利性。

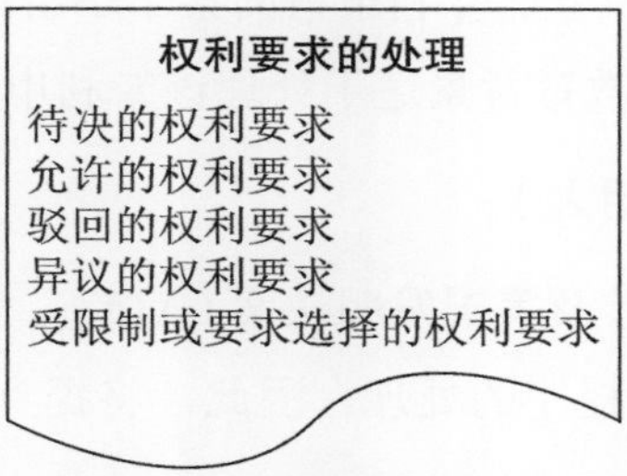

```
权利要求的处理
待决的权利要求
允许的权利要求
驳回的权利要求
异议的权利要求
受限制或要求选择的权利要求
```

图 7-3　权利要求的处理

7.3.1 待决的权利要求

在权利要求处理部分中，有一栏写明本申请中哪些权利要求是待决的。它还涉及已经撤回审议的一项或多项权利要求。选中此框，并写入文档中待决的权利要求的清单里（按编号）。例如，可做如下表述：

"本申请的权利要求 1-13 是待决的。"

7.3.2 允许的权利要求

在权利要求处理部分中，有一栏写明哪些权利要求是允许的。选中此框，并写入文档中允许的权利要求清单里（按编号）。例如，可做如下表述：

"本申请的权利要求 1-5 是允许的。"

7.3.3 驳回的权利要求

在权利要求处理部分中，有一栏写明哪些权利要求是被驳回的。选中该框，并写入文档中驳回的权利要求的清单里（按编号）。例如，可做如下表述：

"权利要求 <u>7-13</u> 被驳回。"

7.3.4 异议的权利要求

在权利要求处理部分中，有一栏写明哪些权利要求是有异议的。选中此框，并在文档中写入异议的权利要求的清单里（按编号）。例如，可做如下表述：

"权利要求 <u>6</u> 有异议。"

对于有异议的权利要求，有时提出异议是因为它是与被驳回的独立权利要求相关的从属权利要求。专利审查员可以声明，该异议的权利要求若写成独立权利要求的形式则可以被接受。

7.3.5 受限制或要求选择的权利要求

在权利要求处理部分中，有一栏写明哪些权利要求受到限制和/或要求选择的限制。选中此框，并在文档中写入权利要求的清单里（按编号）。例如，可做如下表述：

"权利要求 __ 受限制和/或要求选择。"

7.4 申请文件

在申请文件部分中，讨论了对说明书的异议和附图的接受问题。

7.4.1 说明书的异议

在申请文件部分中，审查员可以对说明书提出异议。

7.4.2 说明书附图状态——接受或反对

在申请文件部分，审查员可以对附图接受或提出异议。也会提及附图提交的日期，如下所示：

"审查员对 <u>2017 年 11 月 5 日</u>提交的说明书附图是：a）接受或 b）异议。"

7.5 具体审查意见

在具体审查意见部分中，给出了权利要求的异议、驳回和允许的详细说明。在本部分中，增加了来自专利审查员程序手册（MPEPs）的权利要求规则作为讨论的参考对象。

7.5.1 权利要求的异议

在具体审查意见部分中，列出了权利要求的异议。该部分首先列出申请中存在异议的所有权利要求，然后列出与每个权利要求相关的问题清单。

权利要求异议的原因之一是不合规范。常见的异议有以下原因：

- 缺乏在前的基础
- 缺乏变量定义

例如，它可以用如下方式列出：

由于如下不合规范的原因，对权利要求 8-10 提出异议：

权利要求 8，其中"输出"应为"输出终端"。

权利要求 9，需要定义公式中的变量 A。

权利要求 10，其中"所述输入终端"应为"输入终端"。

需要进行适当的修正。

7.5.2 权利要求的驳回

专利审查员将根据美国法典第 35 卷（35 USC）的规则来判断可专利性[1]。该规则包含在专利审查员程序手册（MPEPs）中，供专利审查员使用，来处理专利申请[1]。美国法典第 35 卷的关键要求[1]（如图 7-4 所示）。

> **常见USPTO驳回条款**
> 35 U.S.C. 101
> 35 U.S.C. 112
> 35 U.S.C. 102 新颖性
> 35 U.S.C. 103 显而易见性
> 附图异议

图 7-4 常见的美国专利商标局权利要求的驳回

7.5.2.1 35 U.S.C. 101

35 U.S.C. 101对可授权的发明进行了规定,是决定一项发明能否获得专利权的关键。35 U.S.C. 101可表述如下[1]:

任何人发明或发现新的有用的工艺、机器、制造或物质组成,或任何新的有用的改进,都可以在符合本条的条件和要求的情况下获得专利。

这可能看起来很宽泛,但某些主题不能被授予专利权,其被称为101司法例外。

7.5.2.2 35 U.S.C. 112

35 U.S.C. 112是专利申请中常见的驳回。当审查员认为说明书没有清楚地说明本发明时,通常会提出驳回。当说明书中存在妨碍审查员理解本发明的错误时,会触发此条款。

以下是对35 U.S.C. 112(b)款的引用:

(b)结论——说明书应当推导出一个或多个权利要求,这一个或多个权利要求特别地指出并清楚地要求发明人或共同发明人视其为他的发明的主题。

以下是35 U.S.C. 112(AIA之前)第二段的引用:

说明书应当推导出一个或多个权利要求,这一个或多个权利要求特别地指出并清楚地要求申请人视其为他的发明的主题。

7.5.2.3 35 U.S.C. 102 新颖性驳回

对于35 U.S.C. 102的情况,驳回是基于对权利要求中的所有内容都是现有技术这一事实的。这是基于这样一个事实的驳回,即权利要求中的所有内容都已获得过专利并且没有新的内容。因此,它是基于权利要求中没有任何新颖的内容这一事实的观点。

以下是对35 U.S.C. 102的适当段落的引用,其构成了审查意见通知书中驳回的基础[1]:

除非符合以下规定,否则任何人都有权获得专利:

(a)(1)请求保护的发明,在请求保护的发明的有效提交日期之前,已获得专利,在印刷出版物中描述过,或在公开使用中,出售,或社会公众可以通过其他方式获取。

在35 U.S.C. 102驳回中,现有技术陈述如下:

"基于35 U.S.C. 102,权利要求1-2被驳回,因为它相对于Voldman等人的文献(US XXXXXXX)不具备专利性"。

7.5.2.4　35 U.S.C. 103 显而易见性驳回

在本节中，讨论 35 U.S.C. 103。在这种情况下，专利权利要求的一些要素包含在一篇专利（或文件）中，剩余的要素可以在一篇或多篇其他的专利（或文件）中找到。这是基于这样一个事实的驳回：将这两篇专利结合起来实现该专利的所有要素是显而易见的。因此，把上述内容结合起来就是一种"显而易见"的陈述。

以下是 35 U.S.C. 103 的引文，其构成了审查意见通知书中所有因显而易见而拒绝的基础[1]：

要求保护的发明与 102 部分中规定的内容不相同，在要求保护发明的有效提交日期之前，如果要求保护的发明和现有技术之间的差异使得要求保护的发明对于要求保护的发明所属领域的普通技术人员来说整体上是显而易见的，那么要求保护的发明的专利不能被授权。发明创造的方式不得否定其可专利性。

在 35 U.S.C.103 的驳回中，现有技术表述如下[1]：

"基于 35 U.S.C.103，权利要求 108 将被驳回，因为相对于 Voldman 等人的文献（USXXXXXXX）不具备专利性"。

在 35 U.S.C. 103 驳回的情况下，引用了多篇专利。在这种情况下，将第一篇和第二篇专利的内容放在一起，陈述第一篇专利公开了一些内容，第二篇专利公开了部分内容。专利审查员使用这两篇（或多篇）内容，来主张这两篇专利的要素结合在一起使得专利申请显而易见，主张本领域的技术人员都可以将上述专利组合起来以获得专利申请内容。这是对没有新元素的论证。

在审查意见通知书中，表述如下：

"基于 35 U.S.C. 103，权利要求 21-25 被驳回，因为相对于 Mandelman 的文献（USXXXXXXX）和 Voldman 的文献（USXXXXXXX）不具备专利性"。

7.5.2.5　附图异议

鉴于图可能存在问题，申请文件部分中会注明"审查员异议"。异议意见的细节将在审查意见通知书中突出显示。对图的常见异议意见如下：

- 图例中缺少"现有技术"标签
- 未标识图形中的物理形状
- 缺少连接线

- 使用不允许的颜色

图例中缺少"现有技术"标签的示例如下：

图

图 1 应该指定图例，如现有技术借助旧的图例才能显示。参见专利审查员程序手册（MPEP）第 608.02（g）部分。为避免放弃申请，答复审查意见通知书修改后的图应符合 37 CFR 1.121（d）的要求。替换页应在页面页眉中标记为"替换页"（根据 37 CFR 1.84（c）），以免妨碍图中的任何部分。如果审查员不接受修改，则会在下一次审查意见通知书中通知，并指出需要改正的地方。对图的异议意见不会被搁置。

7.5.3 允许的主题

在"允许的主题"一节中，讨论允许的权利要求。此外，权利要求可能被异议，因为它们从属于被驳回的权利要求。在这种情况下，审查员指出如何修改才能使这一权利要求被允许。例如：

权利要求 2 被异议，因为其从属于被驳回的基础权利要求，但如果以独立权利要求的形式重写，包含基础权利要求和任何中间权利要求的所有限制的内容，则可以被允许。

7.5.4 结论部分

在结论部分中，可以包含以下话题：

- 现有技术的陈述
- 通知书的终局性
- 专利审查员的联系方式

在结论部分，可以通过以下内容开始关于现有技术的陈述：

"记录但不限于如下现有技术，被认为与本申请的披露相关。"

在该部分中，可以重复现有技术问题的简短概述。结论部分将包含需要进行答复的时间期限。

该通知书的终局性如下：

"本次通知书是最终的。提醒申请人注意 37 CFR 1.136（a）中规定的延长时间政策"。

与专利审查员的沟通可以讨论审查意见通知书的答复。审查意见通知书结论部分包含以下内容[1]：

- 专利审查员的电话号码
- 专利审查员的工作时间
- 传真电话号码
- 审查员的主管姓名和电话号码
- 专利申请信息检索（PAIR）系统
- 电子商务中心（EBC）的电话号码
- USPTO 客户服务代表的电话号码

7.6 撰写审查意见通知书答复

审查意见通知书的答复有特定的格式、风格和要求。本节将讨论这些内容（如图 7-5 所示）。

图 7-5　撰写审查意见通知书的答复

7.6.1 审查意见通知书答复的概要

答复的第一页必须包含图 7-5 所示的信息[1]。

以下是美国专利商标局审查意见通知书答复的第一页示例[1]。起始页上没有其他内容。

```
                                                    December 21, 2017

        TO:     Commissioner for Patents
                P.O. Box 1450
                Alexandria, VA 22313-1450

        FROM:   Steven H. Voldman, Reg. No. 57,575
                116 Patch Lane
                Lake Placid, N.Y. 12946

        SUBJECT: Serial #: 15/201,007
                File Date: 08/05/2017
                Inventor: Joseph Inventor
                Examiner: Michael Examiner
                Art Unit: 2842
                        Title: Circuit and Method for a Digital Circuit



                        RESPONSE TO OFFICE ACTION

        Dear Sir:

              In response to the office action dated July 19th, 2017, please amend
        the above-identified application for patent as follows:


                        CERTIFICATE OF MAILING
              I hereby certify that this correspondence is being deposited with the United
        States Postal Service as first class mail in an envelope addressed to:
        Commissioner for
        Patents, P.O. Box 1450, Alexandria, VA 22313-1450 on          .
                Signature_____    Date:_____
                Steven H. Voldman Reg: 57,575
```

在审查意见通知书答复的第二页上，列出了图的修改、权利要求的修改和评析／争辩所对应的页数。此页不包含其他内容。文本和格式的示例如下[1]：

权利要求的修改反映在从本答复第 3 页开始的权利要求清单中
评析／争辩从本答复第 7 页开始

审查意见通知书的答复必须由提交专利申请的委托人撰写。该专利申请通常委托给专利法律公司。在审查意见通知书答复中，必须处理许多手续。

- 图：必须解决所有图不合规范的问题。通常，替换的图需要附在专利申请上。
- 异议的权利要求：必须处理所有异议的权利要求，必须修改权利要求以解决异议。
- 驳回的权利要求：必须处理所有拒绝的权利要求，必须修改权利要求以解决驳回。

7.6.2 修改权利要求

异议或驳回的权利要求必须修改。在修改过程中，可以修改权利要求的内容以避免异议或驳回。权利要求的内容可以修改，修改权利要求的部分格式如下：

权利要求的修改：此权利要求清单将取代申请中的权利要求所有先前版本和清单：

权利要求清单：权利要求均按顺序编号。为了修改权利要求，权利要求编号后跟着如下内容（如图7-6所示）：

- 原始的（original）
- 新增的（new）
- 当前修改的（currently amended）
- 删除的（cancelled）

图7-6　前序中的词组

original：如果权利要求没有修改，则标签被指定为original。

new：如果一项新权利要求被添加到专利申请中，则将其指定为new。

currently amended：如果由于异议或驳回而修改了权利要求，这一权利要求是amended。

cancelled：如果撤回某权利要求，或者内容并入其他权利要求，然后这个权利要求就被标记为cancelled。在USPTO中，如果撤回或取消权利要求，则权利要求不会重新编号。

7.6.3 修改驳回的权利要求

在修改驳回的权利要求的过程中，可以删除或添加文本。程序如下：
- 指示：如果对权利要求进行修改，则在权利要求的前序之前添加"当前修改的（currently amended）"。
- 删除文本：如果删除了声明中的文本，则在文本中显示删除线。
- 添加文本：如果在权利要求中添加了新文本，则文本必须加下划线。

在审查意见通知书答复中，处理的所有权利要求都要列出。

REMARKS/ARGUMENTS

Examiner Michael Examiner is thanked for thoroughly reviewing the subject Patent Application.

CLAIMS

The Examiner Michael Examiner is thanked for thoroughly reviewing the subject Patent Application. All claims are now believed to be in condition for allowance, and allowance is so requested.

Claims 1-20 remain in the application. Claim 2 has been cancelled.

7.6.3.1　35 U.S.C. 112 驳回

针对基于 35 U.S.C. 112 驳回的答复，根据权利要求修改、评析和评论重新考虑，做出正式的请求[1]。答复的开始陈述如下所示：

请求重新考虑基于 35 U.S.C. 102（b）驳回的权利要求 4，6-9 和 15。

该陈述之后是修改后的权利要求，权利要求以单倍行距显示。重复权利要求的编号，并选择所采取的行动：当前修改的、原始的、删除的。对权利要求的修改可以是带有下划线的新增文本或带删除线的删除文本。

7.6.3.2　35 U.S.C. 102 驳回

针对基于 35 U.S.C. 102 驳回的答复，根据权利要求修改、评析和评论重新考虑，做出正式的请求[1]。答复的开始陈述如下所示：

请求重新考虑基于 35 U.S.C. 102 驳回的权利要求 12-16。

该陈述之后是修改后的权利要求，权利要求以单倍行距显示。重复权利要求的编号并选择所采取的行动：当前修改的、原始的、删除的。对权利要求的修改可以是带有下划线的新增文本或带删除线的删除文本。

7.6.3.3　35 U.S.C. 103 驳回

针对基于 35 U.S.C. 103 驳回的答复，根据权利要求修改、评析和评论重新考虑，做出正式的请求[1]。答复的开始陈述如下所示：

请求重新考虑基于 35 U.S.C. 103 驳回的权利要求 4，6-9 和 15。

该陈述之后是修改后的权利要求，权利要求以单倍行距显示。重复权利要求的编号并选择所采取的行动：当前修改的、原始的、删除的。对权利要求的修改可以是带有下划线的新增文本或带删除线的删除文本。

7.6.4　撤回权利要求

撤回权利要求的过程如下：

- 指定：如果权利要求被撤回或取消，则在权利要求的前序部分之前，添加短语（cancelled）。
- 删除文本：如果撤回或取消权利要求，则删除文本。

7.6.5　处理异议权利要求

在处理存在异议的权利要求过程中，可以删除或添加文本。

如果存在不合规范的情况，则必须解决这个问题。程序如下：

- 指定：如果修改了一项权利要求，则在权利要求的前序之前增加"当前修改的（currently amended）"短语。
- 删除文本：如果删除了权利要求中的文本，则该文本必须加删除线。
- 添加文本：如果在权利要求中添加了新文本，则该文本必须加下划线。

基于对不合规范信息的处理，请求重新考虑对权利要求 1,2 和 10 的异议。

7.6.6 审查意见通知书答复的结语

在审查意见通知书答复中有结束语。审查意见通知书答复的结束语的例子如下所示[1]：

> As noted above, all claims are believed to be in condition for allowance. Reconsideration of the above rejection is therefore respectfully requested.
>
> Applicant respectfully requests that a timely Notice of Allowance be issued in this case.
>
> It is requested that should there be any problems with this Amendment, please call the undersigned Attorney at (801) 922-1447.
>
> Respectfully submitted,
>
> Steven H. Voldman Reg. No. 57,5745

7.7 预先修正

为了解决说明书中的错误或对专利审查员意见进行回应，可以对说明书进行修改，也可以在专利申请提交过程中对专利申请进行修改。可以通过添加、删除或替换段落，替换部分或全部说明书来进行修改。这是通过用标记对全文中任意段落进行替代来实现的，以显示相对于原始段落的先前版本的所有更改。任何添加的文本必须对添加文本加下划线来显示。任何删除的文本必须用删除线显示。如果删除线不明显，则所有删除的文本必须通过置于双括号内来显示。

7.8 最终审查意见通知书

在与专利局的沟通中，审查意见可以通过一系列连续的审查意见通知书进行，其中最后一个被称为最终审查意见通知书。最终审查意见通知书（FOA）是专利审查员提供的最后一次审查意见。然后，申请人将针对 FOA 做出最后答复。

在最终审查意见通知书中，之前对审查意见通知书中的权利要求所做的所有修改都附在草案中。所有删除的文本都在最终审查意见通知书中删除。本草案不再通过下划线显示所有增加或修改。

此时,未能通过此最终审查意见通知书解决的所有权利要求问题,日后都可以向专利局提出上诉。

7.9 欧盟审查意见通知书

在欧盟专利申请中,欧盟审查意见通知书采用与其他专利局不同的格式[7]。

7.9.1 阅读欧盟审查意见通知书

在不同的专利局中,格式、专利审查员裁决和答复可能会有很大不同。在本节中,将简要讨论欧盟专利局(European Patent Office,EPO)的欧盟审查意见通知书[7]。

7.9.2 欧盟审查意见通知书起始语段

在欧盟审查意见通知书中,文档以各种文件的说明开始。

欧盟起始语段的一个例子如下:

> The examination is being carried out on the following applicant documents:
>
> Description, Pages
>
> 1-15 as originally filed
> Claims, Numbers
>
> 1-35 as originally filed
>
> Drawings, Sheets
>
> 1/8-8/8 as originally filed.

7.9.3 欧盟专利审查员裁决

欧盟审查意见通知书后面是每项权利要求的裁决。欧盟专利审查员遵循欧盟专利局的指导原则(如图7-7所示)。在下一部分中,举例说明了欧盟专利裁决的要求。在裁决中,它们被称为欧盟专利公约(EPC)第XX条,其中XX是参考编号。

```
Article 78 EPC
Article 83 EPC 发明披露
Article 84 EPC 权利要求
Article 52 EPC 可专利发明
Article 54 EPC 新颖性
Article 56 EPC 创造性高度
Article 42（1） EPC 说明书内容
Article 43（1） EPC 权利要求形式和内容
```

图 7-7　欧盟专利审查员裁决

7.9.3.1　EPC 第 78 条——欧盟专利申请的要求

欧盟专利申请要满足如下要求[7]：

a）请求授予欧盟专利的请求书

b）本发明的说明书

c）一项或多项权利要求

d）说明书或权利要求中提及的任何附图

e）摘要

并满足实施细则的要求。

7.9.3.2　EPC 第 83 条——发明披露

EPC 第 83 条指出，欧盟专利申请应以足够清楚和完整的方式披露本发明，以使本领域技术人员可以实现本发明[7]。

7.9.3.3　EPC 第 84 条——权利要求

EPC 第 84 条声明："权利要求应界定寻求保护的对象。它们应该清晰简洁，并得到说明书的支持[7]。

7.9.3.4　EPC 第 52 条第（1）款——可专利发明

EPC 第 52 条第（1）款规定："任何适用于工业应用的发明都应获得欧盟专利，这些发明是有新颖性和创造性的"[7]。

7.9.3.5　EPC 第 54 条——新颖性

EPC 第 54 条在处理新颖性问题时规定："如果发明不属于现有技术的一部分，则应视为新发明"[7]。

7.9.3.6　EPC 第 56 条——创造性高度

EPC 第 56 条，即创造性高度，规定："如果考虑到现有技术，一项发明对于本领域技术人员来说不是显而易见的，发明被认为具有创造性高度。如果现有技术水平，还包括第 54 条含义范围内的文件，这些文件在决定是否有创造性高度时不予考虑"[7]。

7.9.3.7　EPC 第 42 条第（1）款——说明书内容

EPC 第 42 条第（1）款规定了对说明书内容的要求[7]：

（1）说明书应：

（a）写明发明所属的技术领域；

（b）写明背景技术，包括就申请人所知的、可被视为有助于理解本发明的、起草欧盟检索报告和审查欧盟专利申请的，并且最好引证反映这些背景技术的文件；

（c）写明请求保护的发明所要解决的技术问题，即使没有明确说明，也可以理解其解决方案，并且参照现有技术说明本发明的有益效果；

（d）简要描述说明书附图中的图（如果有的话）；

（e）详细描述至少一种实施请求保护发明的方式，在适当的情况下使用实例并参考附图（如果有的话）；

（f）如果从本发明的说明书或性质不能明显看出的话，应明确地指出本发明的工业实用性。

（2）说明书应以上述方式和顺序撰写，除非本发明的性质以不同的呈现方式可以使他人准确理解发明或节省说明书的篇幅。

7.9.3.8　EPC 第 43 条第（1）款——权利要求形式和内容

EPC 第 43 条第（1）款规定了对权利要求形式和内容的要求。第 43 条第（1）款规定如下[7]：

（1）权利要求应根据本发明的技术特征确定寻求保护的主题。在适当情况下，权利要求应包括：

（a）前序部分，写明请求保护的发明主题名称和最接近的现有技术共有的必要技术特征；

(b) 特征部分，以"其特征是"或者类似的用语开头，并写明技术特征，与（a）部分的特征合在一起，限定保护范围。

（2）在不影响第 82 条的情况下，当申请的主题涉及下列之一时，欧盟专利申请就可以包含同一类别（产品、过程、装置或用途）中的一个以上的独立权利要求：

（a）多个相互关联的产品；

（b）产品或装置的不同用途；

（c）针对特定问题的替代解决方案，其中不适合通过单一权利要求来涵盖这些替代方案。

（3）一项陈述本发明基本特征的权利要求之后，可以是关于本发明的特定实施例的一个或多个权利要求。

（4）任何包括其他权利要求所有特征的权利要求（从属权利要求），尽可能在开头包含对另一权利要求的引用，然后写明附加特征。直接引用另一项从属权利要求的从属权利要求也可以被接受。所有从属权利要求引用单个在前权利要求，以及所有从属权利要求引用多个在前权利要求，应尽可能以适合的方式分组在一起。

（5）就所请求保护的发明的性质而言，权利要求的数目应是合理的。权利要求应以阿拉伯数字连续编号。

（6）除非绝对必要，否则权利要求在写明本发明的技术特征时不应依赖于对说明书或附图的引用。特别是不应包含诸如"如说明书的……部分中所描述的"或"如附图的图……所示"这样的表达。

（7）如果欧盟专利申请的附图包含有附图标记，可以帮助理解权利要求的方案，则附图标记可以放在括号中，并放在权利要求中相应的技术特征后面。这些附图标记不得解释为对权利要求保护范围的限制。

在权利要求中插入附图标记，来将附图内容和特定权利要求的文本相关联。

驳回可写成如下形式：

"在权利要求中没有插入带括号的附图标记以帮助理解权利要求（EPC 第 43 条第（7）款）。"

"如果申请人坚持在前序部分中包含从最接近的现有技术得到的更多特征，则应该接受这一点。如果没有其他现有技术可用，则可以使用这样的前序部分来提出缺乏创造性高度的异议意见。"[7]

"在审查一项权利要求是否符合第43条第（1）款第二句规定的形式时，重要的是要评估这种形式是否"适合"。在这方面应该记住，两部分形式的目的是允许读者清楚地看到所请求保护的主题的定义所必需的那些特征，与现有技术的一部分组合在一起。如果从说明书中对现有技术的说明已指示得足够清楚的话，能满足第42条第（1）款（b）项的要求，则无须坚持这两部分的形式"[7]。

7.9.4　撰写欧盟审查意见通知书答复

对于欧盟审查意见通知书答复，美国代理人需要与欧洲代理人就答复而沟通，首页包括以下内容：

> TO:　　　　Dr. European Attorney
>
> FROM:　　Joe US Attorney
>
> SUBJECT:　Application No. 141010101.0
> 　　　　　Inventor: Ralph First Inventor
> 　　　　　Title:　HIGH-VOLTAGE TO LOW-VOLTAGE CIRCUIT
>
> Dear European Attorney,
>
> 　　In response to the office action issued for the above-referenced case, attached please find our suggested amendments and remarks.
>
> Please feel free to contact us with any questions.
>
> Signed
>
> Joe US Attorney

第二页必须指明修改后的权利要求将在何处开始，以及评析。例如：

权利要求的修改反映在本文件第3页开始的权利要求清单中。

评析/争辩从本文件第10页开始。

接下来的页面将开始呈现修改后的权利要求以及权利要求清单。例如：

权利要求的修改：此权利要求清单将取代本申请中权利要求的所有先前版本和清单：

权利要求清单

权利要求

在权利要求部分，与 USPTO 的一些区别如下：

- 元素标签放在括号里，例如（QN1）。
- 元素标签可以放在中括号里，例如 {MPN，MP1}。
- 欧盟权利要求可引用多项权利要求，例如，"如权利要求 20 或 28 的方法，……"

对于评析和争辩，答复将突出显示欧盟审查意见通知书中的"点"。在欧盟审查意见通知书中，必须解决所有问题。例如，下面是答复的第一种形式：

REMARKS/ARGUMENTS

CLARITY

The Examiner is thanked for pointing out various issues of clarity in the claims. These have been addressed as follows.

Points 2.1, 2.2., 2.3, 2.4 and 2.5 have been addressed via amendments to the claims, and the claims are now believed to be clear.

Regards to 2.1, and 2.2 have been addressed by defining the interconnection between circuit functions.

Regards to 2.3, the circuit, "said" was removed and replaced by "a".

Regards to Point 3.6, claim 3 was amended to address the issue.

Claim 3 is amended as follows:

[CLAIM 3]

对于新颖性问题，审查意见通知书将写为：

NOVELTY

All claims are now believed to be in condition for allowance, and allowance is so requested.

Claims 1-25 remain in this application.

With regard to Point 4.0, and 4.1, reconsideration of the rejection of claim 1 for lacking novelty with regard to Document D1 is requested, based on the following.

Claim 1 is amended as follows:

[Claim 1]

请注意,在欧盟审查意见通知书答复中,以 D1、D2 等方式引用现有技术。在审查意见通知书答复的开头放置文档列表,文档记作文档 D1、文档 D2 等。

7.10 德国专利商标局(DPMA)审查意见通知书

第二个例子是德国专利商标局(DPMA)审查意见通知书,其格式与 EPO 不同。德国专利商标局以一个附件清单开头[8]。附件清单包括以下内容:
- 引用的出版物
- 审查报告
- 现有技术/参考文献清单

在 DPMA 审查意见通知书中,"考虑引用的文献"表格包括文件号、文件编号以及文件引用。这些专利根据国家、年份和专利号进行引用。

引用的出版物清单后面是一个表格,其中文档被编号[8] 如下:

本次交流中首次提及引用文献如下(其编号也适用于后续程序):

D1:US 2009/0 301 899 A1

D2:US 7 621 985 B1

D3:US 6 367 050

7.10.1 审查文本

本节后面是对"审查文本"的引用。

1)审查文本

审查是基于 2016 年 7 月 12 日提交的原始文件的说明书的第 1~10 页、附图 1~5、权利要求 1~18 以及 2016 年 1 月 20 日提交的译文。

7.10.2 申请的主题

在 DPMA 审查意见通知书答复部分,突出了本发明的领域。这通常是与申请主题相关的简要说明[8]。

7.10.3 本领域技术人员

在 DPMA 审查意见通知书答复部分,使用本领域普通技术人员作为示例。这与 USPTO 申请有很大不同。引用具体人员学位以及他/她的专业。例如:

"本领域普通技术人员是电气工程专业的研究生,具有丰富的数字电路开发经验。"

7.10.4 解释

在解释部分,讨论了异议或驳回权利要求的细节[8]。本节讨论的权利要求组织如下:
- 权利要求
- 新颖性
- 创造性高度

7.10.4.1 权利要求

该部分的标题引用权利要求[8]。接下来是复制权利要求,之后是以下陈述:

5 Claim 1

Claim 1 (main claim) organized according to features, otherwise literally reproduced, reads:

1.

1.1

1.2

在本节中,如专利申请中所示,该权利要求逐行复制。

7.10.4.2 新颖性

在本节中,突出了新颖性的主题[8]。引用了专利法的权利要求和参考文献。现有技术文献将根据前面引用的出版物中的数字(如 D1、D2)引用。以下是文本示例:

"The subject matter of the present claim 1 is not novel (No. 3 Patent Act).

Document D1 (cf. Fig. 1 with associated text) shows a digital circuit having all the features of claim 1; cf. in D1 in particular:

Fig.1 (digital circuit according to feature 1).

Fig.1 S1 (switch according to feature 1.1).

Thus claim 1 is not allowable for lack of novelty of its subject matter."

7.10.4.3 创造性高度

在本节中，必须明确定义创造性高度，以确定区别部分具备新颖性[8]。这包括在专利申请的解释中。

7.10.5 形式缺陷

在形式缺陷这一节中，提出了不符合专利法的问题[7,8]。在本节中，有一个与引用的技术和专利名称相关的形式缺陷的例子。以下是"形式缺陷"部分中内容类型的示例：

"权利要求没有提供附图标记。通知申请人，权利要求书中特定的特征优选应附有附图标记[专利条例§9（9）和专利法条§34（6）]。尽管引用了现有技术，若申请人被说服，必须根据以下几点进行修改，否则，因为这些缺陷，最终可能被驳回：

1）适当的情况下，说明书的引入应适用于新的权利要求。

2）该说明书不符合专利条例§10（1）和专利法§34（6）的要求，因为说明书第1页中给出的标题为"电路和一种数字电路方法"。

7.10.6 结论部分

在结论部分，向申请人陈述了状态。以下是结论部分中内容的示例。

"根据现有文件，由于所述缺陷而无法保证专利授权的前景，因为基于专利法第4条没有可专利性的发明；相反，如果本申请继续使用同样内容的文件或权利要求，本申请将被驳回（专利法第48条）。"

7.11 支持欧洲代理所和欧盟的答复

当一篇专利在多个国家提交时，在一个国家的专利代理所可在该文件提交的第二个国家使用第二家律师事务所。上面的例子说明了这一点。请注意，不同的专利代理人在不同国家的注册号不同。

7.12 总结

本章重点讨论了审查意见通知书。在这一章中，发明人必须学会如何阅读审查意见通知书，研究现有技术并评估权利要求。本章还讨论了如何撰写审查意见通知书的答复。涉及审查意见通知书，重点讨论了如何答复专利局和专利审查员。讨论了美国专利商标局、欧盟专利商标局和德国专利商标局的审查意见通知书。本章介绍了阅读审查意见通知书和处理权利要求，重点介绍了允许的权利要求、驳回的权利要求和异议的权利要求，并讨论了具体审查意见，撰写审查意见通知书答复来修改权利要求。

第 8 章将讨论如果通过创意解决问题（CPS）会议、智囊团和工业中使用系统发明的技术来产生更多发明。讨论什么是创意解决问题（CPS）会议，如何构建 CPS 会议、会议规则、主题、发明产生过程、投票过程、分组和结束，以及发明问题解决理论（TRIZ）、系统发明思维（SIT）和统一系统发明思维（USIT）。

问题

1. 什么是审查意见通知书？
2. 什么是审查意见通知书的答复？
3. 什么是 35 U.S.C. 112 驳回？
4. 什么是 35 U.S.C. 102 驳回？
5. 什么是 35 U.S.C. 103 驳回？
6. 对于 35 U.S.C. 112 驳回，欧盟相同的条款是什么？
7. 对于 35 U.S.C. 102 驳回，欧盟相同的条款是什么？
8. 对于 35 U.S.C. 103 驳回，欧盟相同的条款是什么？

9. USPTO 与欧盟审查意见通知书的要求有什么不同？

10. 欧盟审查意见通知书要求与德国专利审查意见通知书的要求有什么不同？裁决是相同还是不同？

11. 如果独立权利要求被驳回，那么相关的从属权利要求是否被驳回？

12. 允许的从属权利要求，是否可以添加到被驳回的独立权利要求中？

13. 如果权利要求是原始权利要求并且没有改变，则权利要求中的标签是什么？

14. 如果权利要求被修改，权利要求中的标签是什么？

15. 如果权利要求被撤销，权利要求中的标签是什么？

16. 当权利要求撤回，权利要求中的标签是什么？

17. 解释审查意见通知书中的创造性高度。

18. 解释审查意见通知书中的形式缺陷。

19. 撰写 USPTO 审查意见通知书附信。

20. 撰写欧盟审查意见通知书附信。

案例研究

案例研究 A

美国专利商标局发回审查意见通知书。所有权利要求在美国专利商标局的第一次回复中被驳回。必须采取哪些步骤来回应？

案例研究 B

美国专利商标局发回审查意见通知书。所有独立权利要求在美国专利商标局第一次回复中被驳回，但从属权利要求被允许。必须采取哪些步骤来回应？

案例研究 C

美国专利商标局发回审查意见通知书。独立权利要求在美国专利商标局第一次回复中被驳回，但是从属权利要求被异议。该意见称，如果以独立权利要求的形式撰写，则可以接受。怎样解决这个问题？

案例研究 D

美国专利商标局发回审查意见通知书。该申请有一个以上的独立权利要求，专利审查员评论该申请中有多项发明。这种情况如何处理？

案例研究 E

美国专利商标局发回审查意见通知书。该申请具有结构权利要求和方法权利要求。专利审查员要求撤回权利要求。这种情况如何处理？

参考文献

1. U.S.Patent Office (USPTO), https://www.uspto.gov (accessed 05 January 2018).

2. Amernick, B.A. (1991). *Patent Law for the Nonlawyer: A Guide for the Engineer*, *Technologist, and Manager*, 2e. Von Nostrand Reinhold. ISBN: 13 978-0442001773.

3. Mueller, J. M. (2016). *Patent Law*, 5e. Wolter Kluwer Publications.

4. R. Stim. *Patent, Copyright, and Trademark: An Intellectual Property Desk Reference*, NOLO, ISBN 978-1-4133-2221-7, January 29, 2016.

5. Durham, A.L. (2013). *Patent Law Essentials: A Concise Guide*, 4e. Oxford, England: Praeger, ABC-CLIO, LLC. ISBN: 13 978-1440828782.

6. Voldman, S. (2016).Short Course. In: *Writing and Generating Patents* (ed. D.S.H. Voldman). Sri Lanka, February: LLC,FITIS.

7. European Patent Office(EPO), https://www.epo.org (accessed 05 January 2018).

8. German Patent and Trade Mark Office(DPMA), https://www.dpma.de (accessed 05 January 2018).

9. U.S. Department of Commerce. *Patents and How to Get One: A Practical Handbook*, April 1, 2000.

第 8 章
发明产生的方法

8.1 引言

本章将讨论企业用于产生发明的不同方法[1-22]。实践中,一些企业用来产生专利的方法称为创造性问题解决(Creative Problem-Solving,CPS)会议。CPS 会议在诸如 IBM 等一些大型美国企业中有所应用。在马来西亚与斯里兰卡,CPS 会议的方法也被整合到发明课程中[3-5]。第二种实践中,常见做法是使用诸如 TRIZ、SIT 及 USIT 等系统化创造性思考的方法[7-22]。TRIZ 方法在以色列与马来西亚等国已引起关注。

8.2 创造性问题解决会议

CPS 可以用来处理特殊的问题、专利产生和战略性专利组合的拓展。图 8-1 列举了 CPS 会议的构建方式[3-5]。

建设创造性解决问题(CPS)会议
组建CPS会议参与团队
CPS会议的指导准则
CPS会议主题
CPS会议发明产生的过程
CPS会议发明投票环节
激发团队与发明研发
CPS会议结束

图 8-1 创造性问题解决(CPS)会议

从发明到专利——科学家和工程师指南. 第一版. 史蒂文·H. 沃尔德曼 ©2018 John Wiley & Sons 公司,2018 年由 John Wiley & Sons 公司出版。

8.2.1 建设创造性解决问题（CPS）会议

CPS 会议由主导团队发起[3-5]。该团队通常由技术工程师和发明人组成。10 人部门就可以制定计划并协调 CPS 会议。通过这种方式，可制定以每月开 2～3 次 CPS 会议的计划。10 人部门可分为由 3～4 人组成的多个小组。CPS 会议同样需要专利律师的参与。专利律师需要在会议结束时出席，记录发明处理过程与每项发明的发明人。

并非每位参与者都对发明专利有所贡献。因此，在 CPS 会议上，参与者需要陈述他们在发明中所做出的贡献。

8.2.2 组建 CPS 会议参与团队

CPS 会议需要参与者在场[3-5]。优秀参与团队的组建对 CPS 会议的成功与否至关重要。参与者的组成如图 8-2 所示。

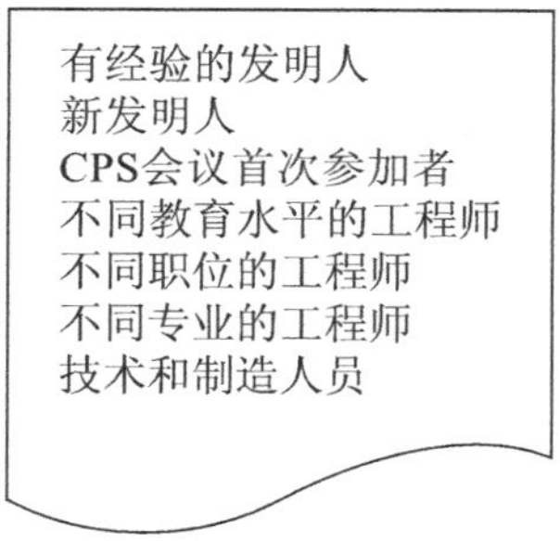

```
有经验的发明人
新发明人
CPS会议首次参加者
不同教育水平的工程师
不同职位的工程师
不同专业的工程师
技术和制造人员
```

图 8-2　CPS 会议出席者

有经验的发明人的参与对于激发探讨与发明过程十分重要。新人的参与也十分重要，这样他们可以学习如何发明并对发明做出贡献。学士、硕士及博士等不同教育水平的工程师的参与也是不错的选择。当工程师从事不同类型的工作时，他们思考的方式也不同。例如，在半导体企业中，半导体设备、半导体处理及半导体电路设计人员的参与对于获得不同的想法十分有益。

8.2.3 CPS 会议的指导准则

CPS 会议主导团队需要建立 CPS 会议的指导准则[3-5]（如图 8-3 所示）。例如，不允许批评别人的想法。第二条规则是允许全员参与。第三条规则是每次由一位发明人发言。一系列的指导准则对于保证全员参与、保证对话积极而富有成效非常重要。

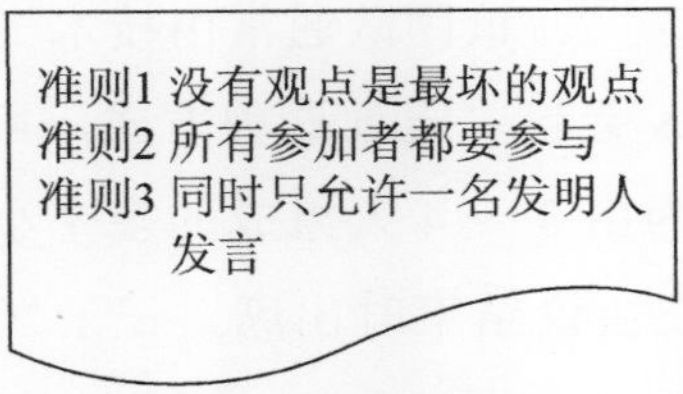

准则1 没有观点是最坏的观点
准则2 所有参加者都要参与
准则3 同时只允许一名发明人发言

图 8-3　CPS 会议准则

8.2.4　CPS 会议主题

CPS 会议中，CPS 主导或管理团队需要选择发明主题。主题的选择取决于目标。有些主题是为满足专利组合在特定领域的扩展需求而设定的；有些主题的设立是为处理某些未解决的问题；有些主题本质上是策略性的。此外，选择最适合特定主题的参与人也十分重要。

8.2.5　CPS 会议发明产生的过程

CPS 会议中发明过程包括以下步骤[3-5]：

- 向参与者公开主题。
- 为每名参与者提供"便利贴"。
- 每名参与者都将在便利贴上写下一个发明想法。
- 所有参与者都应独立工作。
- 发明相应的时间跨度取决于 CPS 会议主导团队的决定。

会议进尾声时，主导团队会在房间里四处走动，询问每位参与者的发明并让他们进行解释。然后参与者将他们的便利贴置于白板上。当所有的参与者完成所有的便利贴时，该过程停止。值得注意的是，一次积极的会议可以产生 50 至 150 项发明。

8.2.6　CPS 会议发明投票环节

接下来进行投票环节。为什么？因为并不是所有的发明都要进行研发，所以团队需要为自己支持或认为值得跟进的发明投票。投票的流程如下：

- 所有参与者持有三种颜色（红、黄与绿）的圆贴纸
- 红色代表最低分值，黄色代表中间分数，而绿色代表最高分值
- 然后所有参与者将贴纸贴于他们想跟进的发明上
- 当每位参与者投票完成后，选择分值最高的发明

通过这种方法选出 5～10 项要从事的发明。这些发明随后需要进行研发。

8.2.7 激发团队与发明研发

所选的发明将成为启动团队合作的焦点以进行进一步的研发[3-5]。启动团队组建后约花费小于一小时的时间来制定发明。启动团队准备了向其他团队宣读的报告。

8.2.8 CPS 会议结束

结束 CPS 会议时，有必要记录发明者所提出的发明[3-5]。专利律师将带走贴纸并记录不同发明的发明者。法律意义上，每个人必须为至少一项专利权利要求做出贡献才能成为法定的共同发明人。专利律师随后与发明团队举行会议以研发发明。

该过程一天可产生 5～10 项发明。在一天的 CPS 会议中产生 10 项发明的情况并不常见。

8.3 系统思考

接下来一节将讨论实践中所使用的基于系统思考的方法。

8.3.1 TRIZ

当前已有用于问题解决与发明的结构化方法，其中一种常用的方法是由 Genrikh Saulovich Altshuller 在 1948 年提出的 TRIZ（Teoriya Resheniya Izobetaltelskikh Zadatch，TRIZ）[7-11]。意为解决创造性问题的理论或创造性问题解决的理论[7]。

该方法通过消除矛盾得到发明。Genrikh Altshuller 声称，"发明就是在特定规则的帮助下消除矛盾"。

TRIZ 衍生出了另外两种系统创造性思考的方法，即 SIT 与 USIT（如图 8-4 所示）。

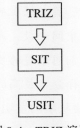

图 8-4 TRIZ 演进

8.3.2 TRIZ—Altshuller 方法

Altshuller 方法为发明提供了途径。他认为一定要浏览大量的发明，发现发明中的矛盾，并制定发明者用来消除矛盾的规则。Altshuller 方法注重发明的"消除"的概念[7-11]。

8.3.3 TRIZ——消除矛盾

Altshuller 方法聚焦消除或简化。移除或简化就是发明。TRIZ 方法是一种简化的系统化方法[7-11]。TRIZ 的 40 项原则可表述如下[11]：

- 分割
- 去除
- 局部质量
- 不对称
- 合并
- 通用
- 嵌套
- 防重
- 初步反行动
- 初步行动
- 预先缓冲
- 等势
- 逆向思维
- 曲面
- 动态

- 部分或过度行动
- 另一维度
- 机械振动
- 周期行动
- 持续有效行动
- 跳过
- 因祸得福
- 反馈
- 中间人
- 自助
- 复制
- 低开销短期目标
- 机制替代
- 气压与液压
- 柔性壳体与薄膜
- 多孔材料
- 颜色变化
- 均匀性
- 丢弃与回复
- 参数改变
- 相变
- 热膨胀
- 强氧化剂
- 惰性氛围
- 合成材料

8.4 系统化创造性思维（SIT）

系统化创造性思维（Systematic Inventive Thinking，SIT）是由 TRIZ 方法扩展出来

的新方法[12-22]。

该方法的关键是快速建立问题境况的独特认知能力。同时使用产生逻辑与创造性概念的两个大脑半球的启发式思维能获得创新成果。

SIT方法[11-22]（如图8-5所示）包括以下部分：

- 联合
- 倍增
- 分配
- 打破对称性
- 目标移除

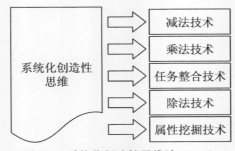

图8-5　系统化创造性思维法（SIT）

8.5　整合式系统化创造性思维（USIT）

整合式系统化创造性思维（Unified Systematic Inventive Thinking，USIT）是一种工程设计类、创新解决概念的、结构化问题解决方法[23]。USIT的目标是问题解决方能在短时间内找到多种问题解决路径。

8.6　数据挖掘

数据挖掘是一种主动从已有的数据中寻找发明的方法。该技术用于从专利中发现隐含的模式。专利挖掘是数据挖掘技术的扩展。

8.7 预测下一步的发明

一种竞争性分析技术或方法能在专利和已授权专利中获得专利灵感。可以根据以下评估来预测发明的途径[3-5]：

- 研究单个发明人的模式
- 研究共同发明人的模式
- 研究发明人的技术公开资料
- 研究企业的技术公开资料
- 研究企业的技术交底简报

通过研究单个发明人多项发明的路径，可以预测单个发明人下一步的计划。

8.8 总结

本章讨论了用于提出专利的发明方法，包括 CPS 会议、TRIZ、SIT 与数据挖掘。这些方法是产生大量专利的途径。本章也讨论了工业界使用的通过 CPS 会议、思维库与系统化发明技术产生更多专利的技术。本章还讨论了什么是 CPS 会议、如何组织 CPS 会议、会议规则、主题、发明产生过程、投票步骤、启动团队与结束，并讨论了 TRIZ、SIT 与 USIT 方法。这些方法已在大型企业中使用，并在全球引起关注。

第 9 章将讨论企业专利战略。第 8 章中讨论的方法也可作为企业专利战略中用到的工具。第 9 章将回顾企业专利申请流程，如审核流程、电子数据库、文档与提交过程。企业战略将包括追溯专利产品、软件工具与目标的方法。第 9 章还将讨论企业目标与个人目标及企业激励项目。

问题

1. 什么是 CPS 会议？
2. 团队中多少人需要参与 CPS 会议？
3. CPS 会议的参与人数量是多少？
4. 如果很少的人参与 CPS 会议会有什么问题？
5. 参与 CPS 会议的人数过多会有什么问题？

6. CPS 会议的步骤是什么？

7. 参会人的规则是什么？

8. 描述一下投票过程。

9. 在 CPS 会议中你用什么来识别发明？

10. CPS 会议中额外的发明将发生什么？

11. 对于 CPS 会议来说多少突破小组是合适的？

12. 为什么 CPS 会议中突破小组很适合头脑风暴？

13. 如何决定谁对发明进行贡献？

14. 律师如何决定谁应参与到发明中？

15. 什么是 TRIZ？

16. 什么是 SIT？

17. 什么是 USIT？

18. 描述一下 TRIZ 的核心思想。

19. TRIZ 是怎样工作的？

20. SIT 是怎样工作的？

案例研究

案例研究 A

一个技术团队开展了 CPS 会议。列出对所有参与者尽力参加有帮助、有价值的规则。

案例研究 B

一个技术团队开展了 CPS 会议。列出对所有参与者尽力参加有帮助、有价值的人选。列出 CPS 会议参与者预期的贡献。

案例研究 C

一个 CPS 团队开展 CPS 会议。列出 CPS 监督组有价值的贡献。

案例研究 D

CPS 会议主题确定瞄准一个特定领域以扩展企业的专利组合，针对 CPS 会议主题提供与你工作领域相关的案例。

案例研究 E

CPS 会议主题确定瞄准先进技术以扩展企业未来的专利组合。针对先进技术的 CPS 会议主题提供与你工作领域相关的一些例子。

参考文献

1. U.S. Patent Office (USPTO). https://www.uspto.gov (accessed 22 December 2017).
2. European Patent Office (EPO). https://www.epo.org (accessed 22 December 2017).
3. Voldman, S. (2014). Short Course, *Innovating, Inventing, and Patenting,* Dr. Steven H. Voldman LLC, Ministry of Science Technology and Innovation (MOSTI), Putrajaya, Malaysia, May 2014.
4. Voldman, S. (2015). Short Course, *Innovating, Inventing, and Patenting,* Dr. Steven H. Voldman LLC, FITIS, Sri Lanka, February 2015.
5. Voldman, S., Short Course, *Writing and Generating Patents,* Dr. Steven H. Voldman LLC, FITIS, Sri Lanka, February 2016.
6. Amernick, B.A. (1991). *Patent Law for the Nonlawyer: A Guide for the Engineer, Technologist, and Manager,* 2e. Von Nostrand Reinhold. ISBN: 13 978-0442001773.
7. Wikipedia. https://wikipedia.org/wiki/TRIZ (accessed 22 December 2017).
8. TRIZ Journal. https://triz-journal.com (accessed 22 December 2017).
9. Ilevbare, I.M., Probert, D., and Phaal, R. (2013). A review of TRIZ and its benefits and challenges in practice. *Technovation 33* (2-3): 30-37.
10. Rantanen, K. and Domb, E. (2008). *Simplified TRIZ,* 2e. New York: Auerbach Publications, Taylor and Francis Group.
11. TRIZ 40 Design Principles. https://www.triz40.com (accessed 22 December 2017).
12. Wikipedia. https://wikipedia.org/wiki/systematic_inventive_thinking (accessed 22 December 2017).
13. Goldenberg, J., Lehmann, D., and Mazursky, D. (2001). The idea itself and the circumstances of its emergence as predictors of new product success. *Management Science 47:* 69-84. doi: 10.1287/mnsc.47.1.69.10670.
14. Marshak, Y., Glenman, T., and Summers, R. (1967). *Strategy for R&D Studies in Microeconomics of Development.* New York: Springer-Verlag.
15. Connolly, T., Routhieaux, R.L., and Schneider, S.K. (1993). On the effectiveness of groups brainstorming: test of one underlying cognitive mechanism. *Small Group Research* 490-503.
16. Paulus, B.P. (1993). Perception of performance in group brainstorming: the illusion of group productivity. *Personality and Social Psychology Bulletin 19:* 78-89. doi: 10.1177/0146167293191009.

17. Goldenberg, J., Mazursky, D., and Solomon, S. Creative sparks. *Science 285:* 1495-1496. doi:10.1126/science.285.5433.1495.1999.

18. R. Horowitz, (1999). Creative Problem Solving In Engineering Design, PhD Dissertation, Tel Aviv University.

19. Goldenberg, J. (2002)."2-3". *Creativity-Product-Innovation*. Cambridge University Press.

20. Goldenberg, J., Levav, A., Mazursky, D., and Solomon, S. (2003). *Finding your Innovation Sweet Spot*. Harvard Business Review.

21. Levav, A. and Stern, Y. (2005). *The DNA of Ideas*. Bio-IT World Magazine.

22. Goldenberg, J., Mazursky, D., and Solomon,S.(1999).Toward identifying the inventive templates of new products:a channeled ideation approach. *Journal of Marketing Research 36:*200.

23. Unified Structured Inventive Thinking(USIT).https://wikipedia.org/wiki/Unified_structured_inventive_thinking(accessed 22 December 2017).

第 9 章
企业专利策略

9.1 引言

无论是大型企业还是小型企业，都需要借助专利策略来了解专利申请要达成的目标及需求。本章讨论企业专利策略[1-12]。

9.2 审核委员会体系

大型企业中的审核委员会可以决定哪些发明适合申请专利。在大型企业中，发明的目标要与企业目标同步[1-5]。

在半导体研发企业，有很多针对特定技术领域或专业的审核委员会。下面是其中一些委员会的例子：

- 半导体设备委员会
- 半导体互联委员会
- 电路设计委员会
- 制造委员会
- 系统委员会

每个委员会都由该领域的专业技术人员组成。委员会配有一名协调员和内部专利律师。发明交底书交给委员会，由其决定审核流程是否开始。

从发明到专利——科学家和工程师指南. 第一版. 史蒂文·H. 沃尔德曼 ©2018 John Wiley & Sons 公司，2018 年由 John Wiley & Sons 公司出版。

9.3 数据库专利追溯系统——世界专利追溯系统（WPTS）

在小型企业中，知识产权（IP）团队可能由单独的律师与协调员组成。技术交底手册用来记录发明。

在大型企业中，一般同时处理多项专利，拥有得力工具对于获取发明、简化提交过程、协调审核委员会及向专利律师递交结果十分重要。大型企业由专利律师、职员与软件的部门来推动这一流程。

在20世纪80年代，很多企业的发明交底书都记录在技术交底手册中。每页交底书完成后由发明人与见证人签名。早期的方法无法高效地处理发明事务。

在20世纪90年代，建立了员工、IP职员与专利律师可以访问的专利追溯系统数据库。所有发明均可以电子形式提交。专利追溯系统将发明人的个人信息存储到了数据库中（如图9-1所示）[5]。

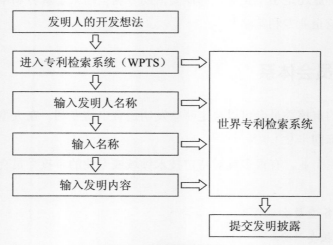

图9-1 专利检索系统

以下是流程总结，简要描述了如何将具有潜在价值的想法上报给企业，并基于该想法创建发明交底书，以提交给进行评估的IP或许可部门。

- 发明人提出别人从未提出的解决问题的想法。
- 发明人使用Lotus Notes或网络浏览器访问世界专利追溯系统（WPTS），以创建发明交底书。
- 发明人输入自己的想法的信息，如：

——该想法解决什么问题？

——该想法如何解决问题？

——谁是提出该想法的发明人？

- 发明人将发明交底书提交给 IP 处理部门。
- IP 部门将为发明交底书分配参考编号，称为交底书编号。
- IP 部门将为发明交底书分配一个发明研发团队（IDT）的评估人来决定其对公司的潜在价值。
- IDT 审核发明交底书并向 IP 部门进行初步推荐。
- 内部律师/专利专业人员审核评估结果，提交最终决定，并通知发明人。
- 如果最终决定是申请专利，那么将进行进一步检索，以保证对本发明交底书而言没有现有技术存在，且该想法尚未称为专利或在专利申请中。

发明提交过程中存在着不同的角色，包括发明人、专利律师、管理者与专利投资经理。每个角色在专利提交过程中都是唯一的。

共同发明人是指一同提出问题解决方案或面向市场研发新产品的员工或个人。发明人提出想法并交给其所对应的 IP 部门。现在有共同的界面供发明人把想法提交至发明流程。当他们试图获得专利时，便可以追溯自己的想法。

评估人是决定一个想法对于公司资产价值的技术人员。有些 IP 部门将评估人加入 IDT 部门内。这些部门的评估人可以组成一个团队。

评估人的范围可以从一线的发明人到管理人员、到律师或专利专业人员。他们的作用是评估一项发明对于公司的价值。他们将使用专利数据库文档控制设备以追溯评论与发明的评价。评估人将为发明提出结论。律师是指企业内部法律工作人员，其主要职责为配合本国专利主管部门开展答复审查意见等相关工作。

律师可负责向外部律师分配起诉责任。在这种情况下，律师或专利专业人员最终对专利事务所的行为结果负责。

在此过程中，管理人员负责律师或专利专业人员部门与特定国家专利事务所之间的文字工作。在发明分配给 IP 处理部门后，管理人员应能追溯一项发明的整个周期的任意时间点。管理人员确保工作流程不能有任何瓶颈。一旦发明提交后，管理人员应该在发明的整个周期中及时更新信息。管理人员也应为 IP 部门分配唯一的交底书编号。

专利组合经理（PPM）负责管理企业的专利组合。他们最有兴趣见证新发明发挥作用。PPM 仅在提交阶段访问并查看专利。

企业员工有责任追踪用于提供、出版和提交新发明的每个 IP 位置。他们将使用这些数据库用于延伸性报告的撰写。

企业员工虽然是政策制定者，但在保持数据完整性方面起到了积极的作用。企业员工在维护控制表上有重要责任，如处理位置和国家表。

遵守该过程的系统由 IBM IP 功能开发，即世界专利追踪系统（WPTS）。开发 WPTS 主要为了使发明能迅速提交给 IBM IP 功能部门。

它给予在专利过程中担当多种角色和职责的人们一种能够以组织和信息的方式去审查发明信息的能力。

9.4　记录发明想法与交底书

记录发明想法并以法律形式记录十分重要，以防发明想法被遗忘。最好以文字与图片的形式记录想法。可以参照以下步骤记录想法[2, 3]：

- 做出发明的图
- 在图页签字
- 为图片记录日期
- 提供至少一名见证人在图页签字
- 当撰写专利申请时，保存图片与文字并发送至公司资料库

一些企业为所有员工提供交底书的记录本，并加以保存。这些发明的记录本封面写有发明人的姓名、日期与材料编号。现在，有一些企业是以电子形式按照模板处理的。

9.5　发明交底书的提交

发明人应一同提交发明交底书。题目与日期应记录在系统中。系统允许在没有完成或正式提交交底书之前先创建发明。这种方式让发明人可以稍晚完成交底书。发明人也可以等到发明完全准备好之后再提交。为交底书或案卷建立"标志"，以控制其时限。一旦交底书正式提交，则形成了安排审核委员会的日期与时间。

技术交底书的记录应包括以下部分[2, 3]（如图 9-2 所示）：

- 案卷编号

- 发明标题
- 发明人姓名（法律上的全名）
- 首次交底书日期：
  —交底书的公开版
  —第三方版
  —含未公开条款的用户"NDA"
  —无 NDA 的用户或个人
  —包含交底书或发明产品的公开版
- 发明已知的现有技术
- 当前已知的实践
- 发明的领域
- 发明所解决的问题
- 发明的必要特征
- 相较于现有技术的优点
- 发明的新颖性特征
- 发明的规避——替代解决方案
- 发明的可检索性
- 付诸实践

```
案卷编号
发明标题
发明人姓名
首次交底书日期
发明已知的现有技术
当前已知的实践
发明领域
发明所解决的问题
发明的必要特征
新颖性特征
相较于现有技术的优点
发明的规避——替代解决方案
发明的可检索性
付诸实践
```

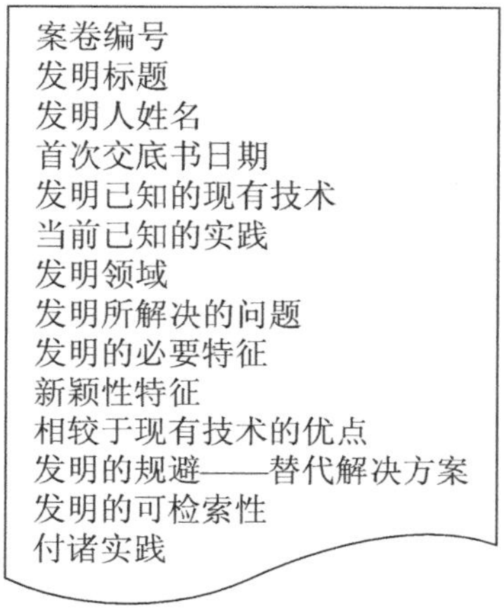

图 9-2　发明交底书

说明也可包含在交底书中。

9.6 发明审核与评估

当发明交底书提交后,交底书首先将被进行评估。评估人可以是审核委员会成员、管理部门或者其他企业成员。指派单人进行评估的方式可以在审核流程中节约时间。评估人可提出与交底书有关的关键问题并向委员会表达自己的观点。指派的人员可以是发明领域的专家。评估人可以在审核过程中为发明人辩护。

发明评估与发明交底书本身的形式类似,但解决不同的问题。发明审核与评估的形式[2-4]如图 9-3 所示。

9.6.1 标题

应包含交底书的标题。

9.6.2 交底书编号

案卷编号是必须的。

9.6.3 发明人姓名

所有发明人的姓名都应当列在评估表中。

> 标题
> 交底书编号
> 发明人姓名
> 审核人姓名
> 发明——清晰度
> 发明——范围
> 审核人已知的现有技术
> 发明的替代技术
> 现有技术的改进
> 可检索性
> 发明的必要技术特征
> 规避——替代的电路或方法

图 9-3 发明交底书审核记录

9.6.4 审核人姓名

审核人姓名与组织也很重要。

9.6.5 发明——清晰度

评估发明时,因为清晰度决定发明是什么,所以清晰度对于避免专利无效很重要。

9.6.6 发明——范围

发明范围对于权利要求撰写和说明书撰写很重要。

9.6.7 审核人已知的现有技术

记录审核人指出的、决定可专利性的、所有其他的现有技术十分重要。在交底书中,判定现有技术相对于新发明的不足也十分重要。

9.6.8　发明的替代技术

为撰写权利要求，了解替代技术十分重要。

9.6.9　现有技术的改进

为判断发明是否有优点，了解发明是不是现有技术的改进十分重要。

9.6.10　可检索性

根据发明是否能轻松判断出企业或个人是否专利侵权十分重要。因此，了解发明是否易于检索与如何判定侵权十分重要。

9.6.11　发明的必要技术特征

了解构成创新性或发明的必要技术特征很重要。撰写权利要求十分关键，因此在说明书或权利要求中不应遗漏必要的技术特征。

9.6.12　规避——替代的电路或方法

通过替代方法来规避发明是可行的。了解其他公司是否规避发明并提交替代的专利或发明很重要。

9.7　审核委员会

发明交底书提交后，交底书将会交至审核委员会[2, 3]。审核委员会由交底书涉及领域专家与一名专利律师组成。一般审核委员会会议会要求发明人准备短于20分钟的关于发明的陈述来阐述现有技术、发明与现有技术相比的不同之处以及其他事宜（如创新性、优点等）。有的企业审核委员会成员在会议之前与发明人会面，并作为发明人的推荐人向整个委员会解释发明。

发明人陈述之后，审核委员会就该发明交底书进行提问，并通过投票方式推进后续步骤。通常决议会有三种状态："归档""关闭"和"待定"。如果处于"归档"状态，该发明交底书会提交专利律师进行专利检索和撰写。"关闭"状态意味着委员会认为该发明交底书不需要进行后续程序。如果处于"待定"状态则意味着可以对此发明交底书进行修改或重新提交。

9.8 与专利律师合作

一旦发明交底书被评委归为"文件",发明人便开始就该发明与律师展开合作。

9.9 企业知识产权(IP)战略

对企业而言,拥有企业 IP 战略很重要。企业 IP 战略包括专利、商标、版权和商业秘密。IP 战略与企业产品路线相匹配的企业将能取得商业成功。企业战略[2,3]如图 9-4 所示。

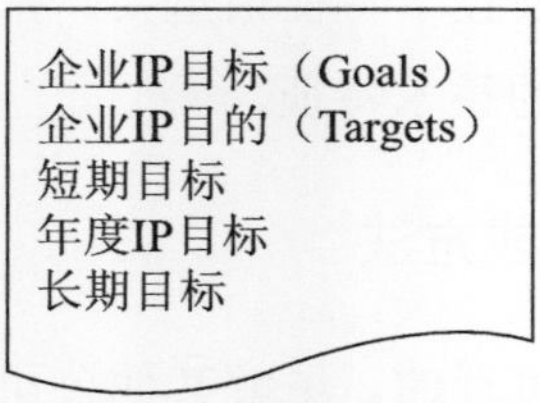

```
企业IP目标（Goals）
企业IP目的（Targets）
短期目标
年度IP目标
长期目标
```

图 9-4 企业 IP 战略

9.9.1 企业 IP 目标（Goals）

创新型企业构建企业 IP 目标很必要。企业应建立针对技术、创新与发展的 IP 技术路线图。随着企业 IP 目标的建立,用于企业 IP 法律团队与部门的计划与预算也将建立[2,3]。

9.9.1.1 组织目标

将 IP 整合到组织与管理的绩效计划中,是激励组织发展 IP 的一种方式。在有些企业中,管理者的考核包括每年提交一定数量的交底书或专利。由于这些考核通常是年度考核,所以可以整合到组织发展的团队目标中[2,3]。

9.9.1.2 个人目标

单个员工可以建立整合到组织目标的 IP 目标。有时个人的 IP 目标可整合至团队目标或企业目标。

9.9.1.3 将个人目标整合到绩效计划

激励员工发展 IP 的方式是将 IP 目标整合到员工的绩效计划中。有些企业，个人的考核包括每年的出版物或专利的要求。由于这通常是年度考核，所以可整合到对员工的要求中，也可以整合到组织目标中。

9.9.2 企业 IP 目的（Targets）

企业 IP 目的由部门、场所或发展团队构建，[2, 3] 如图 9-5 所示。

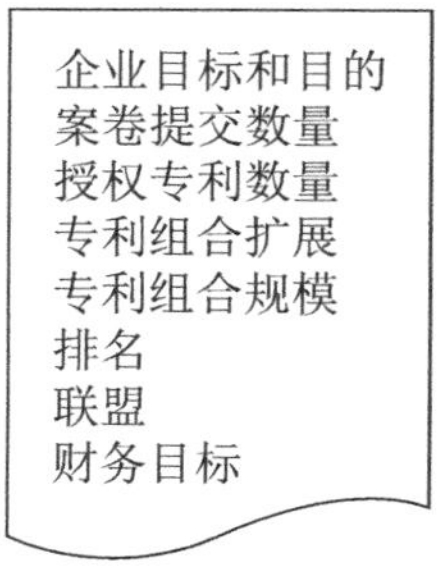

```
企业目标和目的
案卷提交数量
授权专利数量
专利组合扩展
专利组合规模
排名
联盟
财务目标
```

图 9-5　企业 IP 目的

案卷提交数量：企业目标之一是案卷提交数量。由此为将来的授权案卷建立起了专利队列。

授权专利数量：企业目标之一是授权专利的数量。由此建立起了专利组合的规模。

专利组合扩展：企业目标之一是扩展到新领域或新领域的 IP。这可与企业路线图匹配。

专利组合规模：企业目标之一是企业专利组合的规模。这能决定企业的 IP 资产。

排名：企业目标是以最多的专利数量位居专利局排名前列。曾经在美国拥有授权专利数量最多的前十企业中没有一家是美国企业。在 USPTO 中排名靠前，能够反映出企业是有创新性的。对于有些企业，专利数量排名靠前是一种品牌化的方式，这能够提升企业的产品价值。

联盟：企业目标可包括与其他企业的未来联盟。建立企业联盟的一个关键途径是通过交叉许可专利或共享专利组合来实现。

财务目标：企业目标可包括从 IP 专利组合产生多少收入或与企业价值相关的财务数字。

9.9.3 短期目标

建立短期目标用于保持知识产权储备的生产势头。短期目标可每月或每季度建立，能够为年度目标提供考核与平衡。

9.9.4 年度 IP 目标

年度目标可整合到企业的计划周期、财务周期及专利生产的目标。年度目标还能够整合到员工年度考核审核过程。

9.9.5 长期目标

长期 IP 目标可整合到企业发展、产品路线图与企业长期目标中。

例如，IBM 公司希望在美国的专利数量上排名第一。长期目标是保持这个排名多年。这将通过十年的时间达到该目标。

9.10 激励

为了有成功的发明过程，对于员工、管理者与组织而言，无论是财务还是升职的激励都是必不可少的。不提供高效的激励的企业在它们的 IP 发展过程中便很难取得成功。本节将展示成功的企业项目并可用到以下激励[2, 3]（如图 9-6 所示）。

```
发明交底书提交奖
发明专利授权奖
发明成就高原奖
额外20%专利奖
前5%专利奖
部门奖
企业奖
年度发明人晚宴邀请
顶级发明人企业技术认可奖邀请
大师级发明人奖
```

图 9-6 激励系统

9.10.1 发明交底书提交奖

提交的每个交底书文档获得批准后将获得该奖励，这是一种激励提交过程的方法。以前，奖励都是专利授权之后颁发，这不会起到激励发明人的作用，因为专利授权将耗费很长时间。当专利被律师提交到专利局处时，即可获得奖励，奖金为 700 美元。

9.10.2 发明专利授权奖

一些企业为授权专利提供奖励，每一个公开的专利被授权后都将获得奖励。当案卷接收到授权通知（NOA）时，奖励的条件成立。每个发明人将获得 500 美元的奖金。

9.10.3 发明成就高原奖

每四个交底书文档被批准后将获得额外的奖励。它也提供标注了发明人跨越申请数量的奖牌。发明人会得到 1 500 美元的额外奖金。

9.10.4 额外 20% 专利奖

每年，审查委员会就他们的企业价值对专利进行排名。如果能达到前 20%，将会奖励 2 000 美元的额外奖金。

9.10.5 前 5% 专利奖

排名前 5% 的专利将每年获得额外的奖励，这部分奖励比前 20% 奖励的奖金要高，约为 4 000 美元。

9.10.6 部门奖

奖励也会从部门级别获得。一般来说，在建立联盟或收入方面对部门有帮助的专利可以获得较大的奖励。

9.10.7 企业奖

对于企业成功有帮助的专利会获得企业奖励。如果企业通过该专利获得重要的联盟或收入，奖励级别将超过 10 万美元。

9.10.8 年度发明人晚餐邀请

发明人可以被邀请参加特殊的晚宴。很多年以前,我和发明人一起被邀请参加特殊的晚宴,使得发明人对管理层和企业感到满意。

9.10.9 顶级发明人企业技术认可奖邀请

另外,企业中前十名的发明人将被邀请度过一个特殊的周末,他们的配偶也同样被邀请。这个特殊事件是企业级别的奖励周末。

9.10.10 大师级发明人奖

大师级是指一个发明人完成了特定数量的专利和交底书。这一概念的成功在于设置了级别的功能。

9.11 总结

本章讨论了企业专利战略。建立企业现实目标的关键在于建立一个有效的、具有战略性的 IP 组合。本章回顾了企业流程,如审核流程、电子数据库、建档和提交流程。企业战略包括检索专利产品的方法、软件工具、目标和目的。企业目标和个人目标也在讨论之列。本章还讨论了企业对员工的激励方法。

第 10 章将会讨论专家证人。不同种类的专家证人、作证和非作证证人都在讨论之列。专家证人的角色和职责,咨询、教育和报告的专家证人的种类,以及专利诉讼的过程也在讨论之列。

问题

1. 如何记录你的发明?
2. 发明交底书如何使用?
3. 发明交底书包括哪些内容?
4. 发明交底书如何提交?
5. 什么是审核过程?
6. 你怎样准备以应对发明审核过程?

7. 你怎样处理发明的替代技术，解释为什么这很重要？
8. 说明本发明对于现有技术有何改进，这项内容为什么是很重要的部分？
9. 解释可检索性，为什么要了解可检索性？
10. 解释规避，为什么了解可替代的线路或方法很重要？
11. 你的个人发明目的是什么？
12. 你的短期目标是什么？
13. 你的长期目标是什么？
14. 你公司的 IP 战略是什么？
15. 企业的长期目标和短期目标是什么？
16. 如何定义企业年度目标？
17. 企业能提供什么样的规章制度以激励发明？
18. 什么样的奖励机制能激励员工进行发明？

案例研究

案例研究 A

一家企业希望建立专利组合。目前该企业没有专利。那么对于该企业而言，需要哪些步骤以开展它的发明进程？列举为了实现该目标，它需要进行的一系列步骤。

案例研究 B

一家企业希望建立专利组合。企业雇佣专利律师都需要具备哪些条件？期望的费用是多少？列举为了实现该目标，它需要进行的一系列步骤。

案例研究 C

一家企业想要开展发明。对员工而言，专利的撰写、归档与提交都需要做什么？

案例研究 D

一家企业想要以创新而闻名。为实现该目标，对该企业而言需要哪些步骤，请列出。

参考文献

1. Jackson Knight, H.(2013). *Patent Strategy for Researchers and Research Managers*, 3e. Chichester,England: Wiley.

2. Voldman, S.(2014). Short Course, *Innovating, Inventing, and Patenting*, Dr.Steven H.Voldman LLC, Ministry of Science Technology and Innovation(MOSTI), Putrajaya, Malaysia, May 2014.

3. Voldman, S.(2015). Short Course, *Innovating,Inventing, and Patenting*,Dr.Steven H.Voldman LLC,FITIS,Sri Lanka, February 2015.

4. Voldman,S.(2016). Short Course, *Writing and Generating Patents*, Dr.Steven H. Voldman LLC, FITIS, Sri Lanka, February 2016.

5. World Patent Tracking System(WPTS)(2005). *Invention Submission Procedure at IBM*. Thoughtcrafts,Sriks,6711.

6. U.S.Patent Office(USPTO). https://www.uspto.gov(accessed 21 December 2017).

7. European Patent Office(EPO). https://www.epo.org(accessed 21 December 2017).

8. U.S. Department of Commerce(2000). *Patents and How to Get One: A Practical Handbook*. U.S. Department of Commerce, Courier Corporation.

9. Adams, D.O.(2015). *Patents Demystified: An Insider's Guide to Protecting Ideas and Invention*. American Bar Association. ISBN:13 978-163425679.

10. Pressman, D. and Tuytschaevers,T.(2016). *Patent It Yourself: Your Step-by-Step Guide to filing at the U.S. Patent office*. Nolo.

11. Charmsson, H.J.A. and Buchaca, J. (2008). *Patents, Copyrights, and Trademarks for Dummies*. Wiley.

12. Amernick, B.A.(1991). *Patent Law for the Nonlawyer: A Guide for the Engineer, Technologist, and Manger*, 2e. Von Nostrand Reinhold. ISBN:13 978-0442001773.

第 10 章
专家证人

10.1 引言

本章主要聚焦于专利诉讼和科学家、工程师在专利诉讼过程中作为专家证人的职责[1-22]。就本人而言,我在超过 10 件诉讼中担任专家证人,发挥了不同的作用。每件案子都是独一无二的,需要不同的专利律师和律所。

10.2 专家证人

专家证人在专利诉讼中十分必要,主要作用是在与所述专利相关的技术和科学信息方面帮助专利律师。专家证人在不同专利诉讼中发挥不同的作用。专家证人可以看作站在法院角度作证的教育工作者[2-5]。

10.2.1 专家证人的定义

在美国,专家证人是在教育、培训、鉴定、技能、经验方面有特长的人,他们的意见会被法官采纳[2,3,5]。个人必须被法官接受为专家。法官会考虑专家证人在专业领域内就证据或事实而言的观点。专家证人还可以在其专业领域内传递"专业证据"。专业证据是在公开领域内的材料,以专利现有技术、专利申请、公司诉讼、技术出版物或技术诉讼的形式存在。他们的证词可能会被其他专家的证词或者其他证据或事实驳回。

从发明到专利——科学家和工程师指南. 第一版. 史蒂文·H. 沃尔德曼 ©2018 John Wiley & Sons 公司,2018 年由 John Wiley & Sons 公司出版.

10.2.2 专家证人的角色

一般，专家在受伤程度、神志清醒程度、损坏原因等方面发表意见。在知识产权案件中，可能会向专家展示书籍文本或电路板并确定他们的相似度[2,3,5,15]。

为了让法院对他们正在审理的事实或行为有全面的了解，法院本身或法官可以在一些系统中召集专家在技术上估计该特定的事实或行为。专家获得的数据具有法律价值。专家得出的结论将会和诉讼双方的专家的结论进行比较[10-14,20,21]。

10.2.3 专家证人的责任

在美国，根据联邦证据规则 702（FRE）的规定，专家证人需要有资格对主题作证。在决定专家证人是否有资格时，FRE 要求在本案的主题内专家有专门的教育、训练或时间经验[2,3,5,11-21]。专家的作证必须基于证据的事实，在描述结论时提供证据因果关系或相关性的观点。美国专家一般就他们的服务按小时收费，其中包括评估事实、准备报告，如果有必要，在审前质证或审判时作证。每小时的费用大约为 150～500 美元或更高，根据专家所处领域和专家素质及声誉的不同而不同。就上述讨论每小时的费用，总的固定费用因专家的领域、经验和声誉的不同而不同。专家的专业费用加上他/她的相关花费由雇佣的一方支付。在一般情况下，获胜方有权力要求败诉方赔偿。在一些案件中，给专家付的费用可能不视案件结果而定。

10.3 专家证人的类型

专家被雇佣为专家证人，可以担任不同的角色。专家证人可以是作证或不作证的证人。另外，他们还可以担任教育者的角色[3,5]（如图 10-1 所示）。

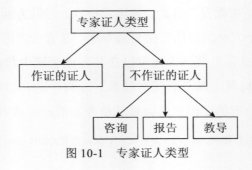

图 10-1　专家证人类型

10.3.1 非作证证人

非作证证人能在开庭或庭审中帮助律师对其他专家证人进行问询。与作证证人不同，一个非作证证人可以很容易地从案件中退出。非专家证人可以在专家公开日之前转为专家证人 [3,5,15]。

10.3.2 咨询证人

咨询证人是一种在诉讼案件中帮助专利律师的专家证人。咨询证人不在法庭上作证。咨询证人可以参与到案件的其他方面 [2,3,5,15]。

10.3.3 教育证人

教育证人是一种对重要科学理论及理论实现工具提供支持的专家证人。教育证人是一种提供他/她观点的专家证人，所述观点包括理论是有效的、实验是可重复和可重现的，也包括所涉及的工具是可靠的 [2,3,5]。作为专家证人必须是通过认证的，其中可能要求学历或特定的培训。

10.3.4 报告证人

证人的第二种形式叫报告证人 [2,3,5]。报告证人在教学证人作证之后站在证人席上。报告证人是能个人进行测量或试验的实验室技术员或工程师。报告证人可以描述一个测试、讨论测试过程，验证测试的可重复性和可重现性，验证测试设备处于正常工作状态。

10.3.5 专家证人

如果需要在法庭作证，专家证人不再拥有特权。专家证人的身份和几乎所有准备用于作证的材料将会公之于众。通常，有经验的律师将建议专家不要在材料上做笔记，因为所有材料将会向另一方公开。

专家在美国联邦法院作证时必须满足 FRE 702 的要求。一般来说，在条款 702 中，专家是具有"科学、技术或其他专业知识"的人，可以帮助事实的审判者，即陪审团 [1]。证人作为专家时首先必须通过他或她在相关领域内的资质检测 [1-3]。

为了证实证人的资格，法院允许对方律师对证人进行询问。如果得到法院的认证，证人可以以观点或其他形式作证,只要满足"(1)证词基于充分的事实或数据,(2)证词基于可靠的理论和方法而产生,(3)证人把理论和方法可靠地应用到案件的事实中。"

一些法院为专家提供指南。澳大利亚联邦法院为在澳大利亚法院出证的专家提供指南。所述指南包括专家书写证词的形式和他们在法院的行为。相似的过程应用在非法庭的论坛中,例如澳大利亚人权和平等机会委员会[6]。

英格兰为专家证人制定了法律。在苏格兰法中,Davie v Magistrates of Edinburgh (1953) 为在法院认证的领域内拥有特殊的知识或技能的证人提供权限,证人为帮助法院被要求出庭,在其所在的领域提供意见,证人可以对他/她在该领域内的观点提供证据[*]。

10.4 与专利律师在诉讼中共事

作为专家证人很重要的一部分是要和负责专利诉讼案件的专利律师和律所共事。专家证人可以在案件中担任不同的角色,这可能依赖于诉讼周期内的案件阶段[3-5,10,15]。

10.5 专家证人报告

专家可能会被要求完成专家证人报告,报告是一种正式的资料,可能会在法庭案件中使用[15,21,22]。专家证人报告可能会在法庭案件中向法庭提交。

10.6 策略

在专利诉讼案件中,可以采用不同的策略应对反对意见。对其他律所案件提反对意见的一种策略是展示现有技术从来没有出现过[22]。

10.6.1 专利无效

对其他律所案件提反对意见的一种策略是展示现有技术从来没有出现过。为了无效专利,现有技术的检索者要去找到在所述专利申请提交之前的公共领域内的专利或资料。如果能证明专利是现有技术,本不应该被申请,那么将会影响你侵权专利对方律

---

[*] 戴维斯诉爱丁堡地方法官案。1953 年,第 34 卷。

所的可信性。

10.6.2 无效主张文件

关于被诉权利要求和侵权主张的披露 30 天之内，反对基于非侵权、专利无效或专利不可执行的专利侵权起诉的每一方，就其非侵权、无效和不可执行主张向所有方披露，该披露应包括如下信息。

（b）无效主张必须包括以下信息并让主张无效方知晓。

（1）在设计或多种植物专利中，每一项能够预见被诉权利要求的内容或者证明权利要求的显而易见性的现有技术，包括从所有可行的角度和所有可行的实施例中得到的观点。每件现有技术专利可以通过号码、国别、公布日期识别；每件现有技术出版物可以通过题目、出版日期以及作者和出版商识别。在 35 U.S.C. 102（b）中规定的现有技术应被识别，通过详细说明许诺销售、公众使用或知晓的产品，许诺或使用发生的时间，或者信息被公众所知的时间，产品销售或公开使用的地点，使用或制造和接收许诺的人或实体的身份，让信息变成公知以及接受公知信息的人或实体。在 35 U.S.C. 102（f）中规定的现有技术应被识别，是通过提供人的名字、本发明或其中某部分来源于该现有技术的情况来实现的。在 35 U.S.C. 102（g）中的现有技术应被识别，是通过提供涉及的人或实体的身份、专利申请人面临的产生发明的情况来实现的。

（2）每项现有技术是否预见了被诉权利要求的内容或者证明了权利要求的显而易见性。如果涉及显而易见性，需要解释为什么现有技术证明了被诉权利要求的显而易见性，包括对显示显而易见性的现有技术组合的证据。本领域技术人员在发明提出时已经能够将参考文献组合的原因，对被告认定的主要参考文献进行识别。

（3）图表中可以识别，在现有技术中的哪部分能够找到每个被诉权利要求的内容或观点，以及对于发明专利，包括识别出诉讼方认定受 35 U.S.C.112（f）规定的，在每项现有技术中显示权利要求功能的设施、行为和材料。

（4）基于 35 U.S.C. 101 的任何无效理由，在 35 U.S.C.112（b）中规定的不确定性或 35 U.S.C. 112（a）下任何被诉权利要求的可行性或书面说明。

（c）基于后续发现事实的修改，不可执行的争辩将包括被控侵权人断言任何诉讼专利是不可执行的具体的原因。如果被控侵权人的权利要求的不可行性是基于不平等行为，被控侵权人应描述对于专利商标局（PTO）的每个遗漏或虚假陈述，应声明被控侵权人在庭审中辩述那些起诉专利的人想要欺骗PTO的所有理由，包括在专利起诉期内现有技术参考文献未向PTO公开的证据，任何表明了涉及诉讼专利的一个或多个人意识到该现有技术参考文献早于诉讼专利授权的事实，或任何与意图欺骗的元素相关的事实[22]（如图10-2所示）。

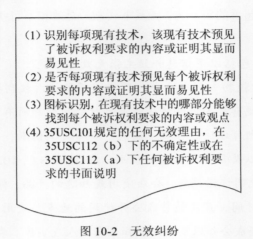

图10-2　无效纠纷

10.7　获取公知领域内的资料

在公知领域内的重要类别的资料，可以在专利诉讼中使用，用于展示专利无效和资料的存在早于专利授权（如图10-3所示）。

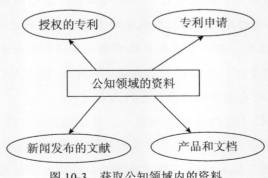

图10-3　获取公知领域内的资料

- 授权的专利
- 新闻发布的文献
- 专利申请
- 产品和文档

10.7.1 现有技术检索

现有技术检索可以通过建立搜索引擎或任何现存的专利办公室来启动。现有技术检索可以包括检索 USPTO、Google 专利或其他检索方式，目的在于检索到申请日之前公布的现有技术专利。

10.7.2 新闻发布的文献

从企业文献处发布的早于专利申请日的新闻，可以使专利的新颖性失效。新闻的发布可以是在公知领域内获得的硬拷贝或软拷贝。

10.7.3 客户信息

从企业文献处发布的早于专利申请日的客户信息注释（CIN），可以使专利的新颖性失效。在企业中，客户会发送可以用作现有技术的 CIN。

10.7.4 产品和文档

从企业文献处发布的早于专利申请日的产品说明文档，可以使专利的新颖性失效。通过使用产品说明文档或操作手册，可以直接获得信息。例如，管脚描述可以用来决定哪个电路板或功能模块包含在半导体芯片内。管脚描述还可以用来提供导电轨的数量、电压条件和芯片体系结构。

10.8 获取非公知领域的资料

在非公知领域内有重要种类的资料，可以在专利诉讼中使用，用于展示专利无效和资料的存在早于专利授权。还可以请求获取非公知领域内的资料（如图 10-4 所示）。这些包括：

- 发明技术交底书记录本
- 设计手册
- 设计物理布局（例如 GDSII）
- 工艺流程

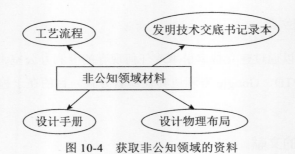

图 10-4　获取非公知领域的资料

10.8.1　发明技术交底书记录本

可以要求获得对方使用的发明技术交底书记录本，从而得到其中包含的日期和内容。法院可以公布交底书记录本以供查阅。在发明技术交底书记录本中，可以确定专利申请的发明人、讨论内容、新颖性、现有技术或其他相关信息。

10.8.2　设计手册

可以要求获得对方使用的技术设计手册和设计说明书，从而得到技术中包含的日期和内容。法院可以将其公布以供查阅。设计手册可以包含半导体流程步骤、元器件、流程条件、设计布局和设备特征。

10.8.3　设计物理布局

可以要求获得对方使用的设计物理布局，从而得到技术中包含的日期和内容。法庭会在硬盘上放置设计数据，并将硬盘交给对方查阅。电路的设计布局可以包含在专利中（如静态可读存取内存、SPAM 蜂窝布局）。

10.8.4　工艺流程

可以要求获得对方使用的技术流程文档和生产数据，从而得到技术中包含的应用和流程步骤。法院可以将其公布以供查阅。从流程上看，可以制作流程表以确认方法权利要求的步骤与技术关注点相一致。

10.9 科学证据

可接受的科学证据应是在学术圈内普遍接受的理论的结论。因此，专利诉讼过程的一部分就是评估科学证据[17]。

10.9.1 Frye 规则

弗赖伊规则，源于 Frye v. United States（1923）案，该案认为可接受的科学证据应是在学术圈内普遍接受的理论的结论。这个规则得出了与可接受性相关的统一结论。特别是 Frye 案中法官认为[17]：

只有当科学理论或发现跨越了实验和论证阶段的界限时才是难以定义的。在这个模糊地带，理论的证据效力需要被识别出来，法庭将花费精力去确认，从公知的科学原理或发现推导出的专家证词，上述推导出的结果必须充分建立在所属特定领域内，所获得的普遍接受的基础上。

这个规则由于误解了科学过程以及其是基于陪审团无法评估科学证词的假设之上而受到批评。本规则的目的是禁止采用通过过于可疑或有争议的科学理论而获得的证据，主要用于排除在最初案件中被告采用的测谎仪结果[17]。

10.9.2 Daubert 规则

Daubert 规则起因于美国高等法院的 Daubert v. Merrell Dow Pharmaceuticals, 509 U.S.579（1993）案。案件要求显示四件事[17]：

- 可测试：理论是可以测试的
- 同行评议：理论是被同行评议过的
- 可靠性和错误率：可靠性和错误率
- 可接受性：在科学圈内为公众接受的

在可测试性方面，问题是理论是否可以测试。在同行评议方面，问题是该理论是否被同行评议过，通过同行评议，可以减少理论中的错误概率。在可靠性和错误率方面，并非要求 100% 的可靠性和零错误率，但初审法官将会把概率作为考量因素。最后，在可接受方面，将会评估科学界内的普遍接受范围。

联邦证据规则采用了 Daubert 规则[1,17]。Frye 规则和 Daubert 规则是对提交给法院材料建立判断的方法。

10.10 总结

本章中，简短讨论了专家证人和专利诉讼的话题。回顾了不同种类的专家证人和他们在专利诉讼中承担的角色。专家证人可以是作证或非作证的，在不同的专利诉讼中起到不同的作用。下一章将讨论专家证人的角色和责任，咨询专家证人的类型：教育和报告专家证人。专利诉讼的过程中无效纠纷文档也将会被讨论。

问题

1. 专家证人的定义是什么？
2. 专家证人担任什么角色？
3. 专家证人的职责是什么？
4. 是不是有超过一种类型的专家证人？
5. 请说一下专家证人的类型。
6. 什么是咨询专家证人？
7. 什么是教育专家证人？他或她到底做什么？
8. 什么是专家证人作证？
9. 无效纠纷文档到底讨论的是什么？
10. 使一件专利无效时需要哪些材料？
11. 哪些在公知领域内的资料可以使用？
12. 哪些是法庭可以提供但却是非公知领域的资料？

案例研究

案例研究 A

在专利诉讼案件中，与芯片的数字和模拟部分之间的电子连接相关的专利存在争议。怎么从设计布局中知道这两部分的内在连接？

案例研究 B

在专利诉讼中，网络列表用于检索特定的电路类型。解释这是怎么做到的？怎样得到网络列表？可以把这个作为侵权证据吗？

案例研究 C

在专利诉讼中，专利律师的目的是证明在专利授权之前存在现有技术以及专利不应该被授权。该案件要获取能够证明现有技术已根据 35 USC 102 和 35 USC 103 得到授权的相关专利。解释为了满足这个要求我们需要做什么？上述两种论据哪种更有力，为什么？

参考文献

1. Federal Rules of Evidence-2011(2017).*Federal Evidence Review*. Michign Legal Publishing, Ltd.
2. Ganer,B.A.(2016).*Black's Law Dictionary*. Thomson Reuters.ISBN:10 0314613005.
3. Talve, M.(2013).*What Is an Expert Witness?* The Expert Institute.
4. G.H.Goldsholle(2015).*ExpertPages Expert Witness Fees& Practices Survey*. ExpertPages.
5. Christopher,E.(2013). *What are the Differences Between an Expert Witness and a Consultant Non-Testifying Expert*. Forensis Group.
6. Federal Court of Australia(2007). Guidelines for Expert Witnesses in Proceedings in the Federal Court of Australia, Practice Direction.
7. Worthington, T.(2005). *The Accidental Expert Witness*. Information Age(IDG).
8. R.L.Carlson,E.J.Imwinkelried, and E.J.Kionka.*Evidence in the Nineties: Cases, Materials, and Problems for an Age of Science and Statues. Michie Co*. 1991. ISBN:978-0-87473-740-0.
9. Berger, M.A(2011). The admissibility of expert testimony. In: *Federal Judicial Center; National Research Council, Reference Manual on Scientific Evidence,* 3e, 11-36. Washington, DC: National Academies Press, ISBN: 978-0-309-21421-6.
10. Bronstein, D.A.(1999). *Law for the Expert Witness*, 2e.CRC Press.
11. Dwyer, D.(2008). *The judicial Assessment of Expert Evidence*. Cambridge University Press.
12. (2011). Federal Judicial Center;National Research Council. Reference Manual on Scientific Evidence,3e. Washington, DC: National Academies Press. ISBN:978-0-309-21421-6.
13. Jasanoff,S.(1997).*Science at the Bar: Law, Science, and Technology in America*. Cambridge，MA: Harvard University Press.

14. Reynolds, M.P. and King, P.S.D.(1992). *The Expert Witness and his Evidence.* Blackwell.

15. Smith,D.(1993).*Being an Effective Expert Witness.* Thames Publishing.

16. Federal Judicial Center(2000). *Expert Testimony in Federal Civil Trials; A Preliminary Analysis.* Federal Judicial Center.

17. Michaels, D.(2006). Project on Scientific Knowledge and Public Policy: *Daubert-The Most Influential Supreme Court Ruling You've Never Heard Of.*

18. Yee,K.K.(2008). Dueling experts and imperfect verification. *International Review of Law and Economics* 28(4): 246-255.

19. Cole,S.A.(2007). *Where the Rubber Meets the Road:Thinking about Expert Evidence as Expert Testimony (Archive),* vol.52, Issue 4, Article 4, 803-840. Villanova Law Journal. Villanova University School of Law.

20. Amernick, B.A.(1991).*Patent Law for the Nonlawyer:A Guild for the Engineer, Technologist, and Manager,*2e. Van Nostrand Reinhold. ISBN:13 978-0442001773.

21. Durham,A.L.(2013). *Patent Law Essential: A Concise Guide,* 4e.Oxford: Praeger, ABC-CLIO,LLC.ISBN:13 978-1440828782.

22. Jackson Knight, H.(2013). *Patent Strategy for Researchers and Research Mangers,* 3e. Chichester:Wiley.

附　　录

附录 A　发明技术交底书

发明技术交底书表格包含公司提交发明构思所需的信息。

发明技术交底书案卷号：本发明技术交底书必须具有案卷号以供专利代理人跟踪提交。在提交过程中分配了案卷号。案卷号放在文档的头部。

发明名称：本发明的发明名称应该是简短的、具有描述性的。在发明技术交底书表格中，建议少于 15 个字。

业务部门：业务部门需要决定评估所需的适当组。用这种方式，选择正确的委员会来评估技术交底书。

发明人：发明技术交底书表格要求有发明人的姓名，以确定该发明的真正发明人。发明人姓名应在发明技术交底书表格中以大写字母书写。

批准专利／知识产权部门：如果一个共同发明人没有被授予专利的公司雇用，这是必须的。

附加页数：这是技术交底书所包含页码的文档。这对文档很重要。

审批部门主管：技术交底书要有部门主管姓名、签名和电子邮件地址。

发展状况：需要让公司了解发明的发展状况。确定它是处于概念状态、模拟状态还是原型状态是非常重要的。这可以包括概念验证、模拟结果或测量。

推荐申请国家：在提交的发明技术交底书中，发明人必须注明拟申请专利的国家。通常列出在本发明领域中有应用的国家。专利代理人建议给被推荐的申请国提供理由。

从发明到专利——科学家和工程师指南. 第一版. 史蒂文・H. 沃尔德曼 ©2018 John Wiley & Sons 公司，2018 年由 John Wiley & Sons 公司出版。

预期的战略价值：确定发明对专利组合的战略价值非常重要。以下是感兴趣的案例：
- 用于或即将用于产品
- 与未来产品相关：现有业务
- 与未来产品相关：新业务
- 保护现有专利的解决方案
- 抑制竞争对手的活动
- 其他兴趣

向公众或第三方（如客户）首次披露的日期：在发明过程中，公众或客户了解本发明。有时，本发明是与客户合作完成的。发布的信息可以采用以下形式：
- 示范
- 介绍
- 论文/文章

重要的信息可以包括披露的日期、披露的人或团体以及是否存在保密协议（NDA）。

首次销售或提供的日期：首次向公众销售的日期是发明披露文件的关键。

发明概念：了解本发明概念的手段是很重要的。可能的原因如下：
- 公务
- 工作经验
- 企业基础设施（如软件和测试设备）
- 公司联系人（如客户或代工厂）

现有技术：公开现有技术信息很重要。现有技术可包括以下内容：
- 发明人已知的专利文献
- 发明人已知的非专利文献

本部分可引用出版物、行业期刊、产品说明中的专利和文献清单。

发明的技术领域和现有技术：在该部分中，可以引用图和表格。

本发明要解决的技术问题：本节公开本发明解决的技术问题。

本发明的必要技术特征：在该部分中，公开使本发明起作用所需的本发明的特征是很重要的。这是确保在交底书中包含具有可操作性的发明特征的关键。

与现有技术相比，本发明的优点：这对于突出本发明的优点是重要的。

本发明的具体实施方式：本节讨论本发明的具体实施方式，用于扩展专利的说明书。

附录 B　发明技术交底书评审员表格

发明交底书评审员表格包含公司提交发明构思所需的信息。

发明技术交底书案卷号：本发明技术交底书必须具有案卷号，以供专利代理人跟踪提交。在提交过程中分配了案卷号，案卷号放在文档的头部。

发明名称：本发明的发明名称应该是简短的、具有描述性的。在发明技术交底书表格中，建议少于 15 个字。

评审员：评审员列在发明交底书评审员表格中。

第一阶段，技术审查：如果交底书良好，评估可以转移到交底书的第二阶段。第一阶段有以下类别：

- 清楚
- 范围
- 优点

现有技术：公开现有技术信息很重要。现有技术可包括以下内容：

- 发明人已知的专利文献
- 发明人已知的非专利文献

本部分可引用出版物、行业期刊、产品说明中的专利和文献清单。

第二阶段，技术评估：本节将在发明人接受评审员的反馈意见后完成。这些包括以下内容：

- 替代品数量
- 提升现有技术水平
- 可检测到的侵权

第一阶段：发明交底书评审员表格包括以下问题：

- 本发明的哪些特征至关重要，为什么？
- 哪些功能有利，为什么？
- 哪些功能可以推广？

附录 C　新颖性检索报告

在新颖性检索报告中,代理人向发明人和发明人所在的公司提供一份作为现有技术的材料报告。新颖性检索报告应包括以下内容:

- 律师事务所:做报告的律师事务所应列在信笺抬头。
- 律师事务所地址:律师事务所地址应在信笺抬头中列出。
- 日期:应列出新颖性检索报告的日期。
- 发明人姓名:发明人姓名应在新颖性检索报告中列出。
- 发明人公司:发明人的公司应该列在新颖性检索报告上。

正式的新颖性检索报告应列明以下事项:

专利代理人姓名:撰写报告的专利代理人应在文件上注明其姓名。

主题:主题行应说明"技术交底书案号的新颖性检索报告"和技术交底书的标题。

发明人姓名:发明人的姓名应当列在新颖性检索报告上。

接下来应该是专利检索的内容。标题应如下所示:

"以下是我们对主题发明交底书的新颖性检索的结果:"

接下来应该是检索结果。每个条目应包括以下内容:

- 专利类型(如专利或专利申请)
- 专利号
- 发明家
- 专利名称
- 相关内容的描述

应使用以下声明结束此部分:

"综上所述,上述引用的现有技术似乎存在可专利的差异。我们将开始准备专利申请。"

附录D 美国专利商标局审查意见通知书细节内容

附录是美国专利商标局（USPTO）审查意见通知书的一个例子。

在美国专利商标局审查意见通知书中，文件中包含以下内容：

- 申请号
- 申请日期
- 第一命名发明人
- 代理人案卷号
- 确认号码
- 审查员
- 技术单元
- 邮寄日期
- 递送模式

USPTO审查意见通知书的第二页包含以下内容：

- 状态
- 权利要求的处理
- 申请文件

UNITED STATES PATENT AND TRADEMARK OFFICE

UNITED STATES DEPARTMENT OF COMMERCE
United States Patent and Trademark Office
Address: COMMISSIONER FOR PATENTS
P.O. Box 1450
Alexandria, Virginia 22313-1450
www.uspto.gov

APPLICATION NO.	FILING DATE	FIRST NAMED INVENTOR	ATTORNEY DOCKET NO.	CONFIRMATION NO.

	EXAMINER

ART UNIT	PAPER NUMBER

MAIL DATE	DELIVERY MODE
	PAPER

Please find below and/or attached an Office communication concerning this application or proceeding.

The time period for reply, if any, is set in the attached communication.

Office Action Summary	Application No.	Applicant(s)
	Examiner	Art Unit	AIA (First Inventor to File) Status
			Yes

-- The MAILING DATE of this communication appears on the cover sheet with the correspondence address --

Period for Reply

A SHORTENED STATUTORY PERIOD FOR REPLY IS SET TO EXPIRE <u>3</u> MONTHS FROM THE MAILING DATE OF THIS COMMUNICATION.
- Extensions of time may be available under the provisions of 37 CFR 1.136(a). In no event, however, may a reply be timely filed after SIX (6) MONTHS from the mailing date of this communication.
- If NO period for reply is specified above, the maximum statutory period will apply and will expire SIX (6) MONTHS from the mailing date of this communication.
- Failure to reply within the set or extended period for reply will, by statute, cause the application to become ABANDONED (35 U.S.C. § 133).
- Any reply received by the Office later than three months after the mailing date of this communication, even if timely filed, may reduce any earned patent term adjustment. See 37 CFR 1.704(b).

Status
1) ☒ Responsive to communication(s) filed on _____.
   ☐ A declaration(s)/affidavit(s) under 37 CFR 1.130(b) was/were filed on _____.
2a) ☒ This action is FINAL.   2b) ☐ This action is non-final.
3) ☐ An election was made by the applicant in response to a restriction requirement set forth during the interview on _____; the restriction requirement and election have been incorporated into this action.
4) ☐ Since this application is in condition for allowance except for formal matters, prosecution as to the merits is closed in accordance with the practice under Ex parte Quayle, 1935 C.D. 11, 453 O.G. 213.

Disposition of Claims*
5) ☒ Claim(s) _____ is/are pending in the application.
   5a) Of the above claim(s) _____ is/are withdrawn from consideration.
6) ☐ Claim(s) _____ is/are allowed.
7) ☒ Claim(s) _____ is/are rejected.
8) ☐ Claim(s) _____ is/are objected to.
9) ☐ Claim(s) _____ are subject to restriction and/or election requirement.

* If any claims have been determined <u>allowable</u>, you may be eligible to benefit from the **Patent Prosecution Highway** program at a participating intellectual property office for the corresponding application. For more information, please see http://www.uspto.gov/patents/init_events/pph/index.jsp or send an inquiry to PPHfeedback@uspto.gov.

Application Papers
10) ☐ The specification is objected to by the Examiner.
11) ☒ The drawing(s) filed on <u>17 April 2015</u> is/are: a) ☒ accepted or b) ☐ objected to by the Examiner.
    Applicant may not request that any objection to the drawing(s) be held in abeyance. See 37 CFR 1.85(a).
    Replacement drawing sheet(s) including the correction is required if the drawing(s) is objected to. See 37 CFR 1.121(d).

Priority under 35 U.S.C. § 119
12) ☐ Acknowledgment is made of a claim for foreign priority under 35 U.S.C. § 119(a)-(d) or (f).
    Certified copies:
    a) ☐ All   b) ☐ Some**   c) ☐ None of the:
       1. ☐ Certified copies of the priority documents have been received.
       2. ☐ Certified copies of the priority documents have been received in Application No. _____.
       3. ☐ Copies of the certified copies of the priority documents have been received in this National Stage application from the International Bureau (PCT Rule 17.2(a)).

** See the attached detailed Office action for a list of the certified copies not received.

Attachment(s)
1) ☒ Notice of References Cited (PTO-892)
2) ☐ Information Disclosure Statement(s) (PTO/SB/08a and/or PTO/SB/08b) Paper No(s)/Mail Date _____.
3) ☐ Interview Summary (PTO-413) Paper No(s)/Mail Date _____.
4) ☐ Other: _____.

附录 E　美国专利商标局审查意见通知书部分

附录 E 是 USPTO 审查意见通知书的一个例子。

在 USPTO 审查意见通知书中，文件中包含下列内容：

- 说明书
- 权利要求
- 附图
- 文件编号

USPTO 审查意见通知书对文档中的项目进行编号。这些将在后续通信中提及。

从发明到专利——科学家和工程师指南. 第一版. 史蒂文·H. 沃尔德曼 ©2018 John Wiley & Sons 公司，2018 年由 John Wiley & Sons 公司出版。

附录 F　欧盟审查意见通知书

附录 F 是欧盟审查意见通知书的一个例子。

在欧盟审查意见通知书中，文件中包含下列内容：

- 说明书
- 权利要求
- 附图
- 文件编号

欧盟审查意见通知书对文档中的项目进行编号。这些将在通信中被提及。

附录 G　欧盟审查意见通知书答复

附录 G 是欧盟（EU）审查意见通知书答复的一个例子。

在欧盟（EU）审查意见通知书答复中，需要对欧盟（EU）审查意见通知书中列出的所有内容做出答复。欧盟（EU）审查意见通知书对文件中的条目进行编号。这些将在通信中被提及。

从发明到专利——科学家和工程师指南. 第一版. 史蒂文・H. 沃尔德曼 ©2018 John Wiley & Sons 公司，2018 年由 John Wiley & Sons 公司出版。

DATE:_____

TO: European Attorney

FROM: US Attorney

SUBJECT: Application No. XX
         Inventor
         Title:

Dear European Attorney,

In response to the office action issued for the above-referenced case, attached please find our suggested amendments and remarks.

Please feel free to contact us with any questions.

With best regards,

US Attorney

Amendments to the Claims are reflected in the listing of claims which begins on page X of this paper.

Remarks/Arguments begin on page Y of this paper.

Claims

What is claimed is:

1. (currently amended) A circuit, comprising

2. (original) The circuit of claim 1, wherein said XX

3. (original) The circuit of claim 1 wherein xxxx.

CLAIM LIST HERE !

REMARKS/ARGUMENTS

NOVELTY

The Examiner is thanked for pointing out various issues of novelty in the claims. These have been addressed as follows.

The examination proceedings are based on the original patent claims 1 to 21 of the application day, March 10, 2017, with US priority of February 29, 2017. Claim XX is cancelled.

Claim 1 is as amended:

While Document D1 (D1 Inventor Name et al US XXX) shows a method comprising of a XXXX.

All claims are now believed to be in condition for allowance, and allowance is so requested.

Claims 1-XXX remain in this application. Claim YYY was cancelled.

附录 H　美国对欧盟代理人信件——审查意见通知书答复

附录 H 是美国代理人与欧盟代理人就欧盟审查意见通知书答复进行沟通的一个例子。在通信中，信中应包含以下内容：

- 案卷号
- 日期
- 欧洲代理人姓名
- 美国代理人姓名
- 主题—申请号，发明人姓名和发明名称

该文件的内容应如下：

- 对权利要求的修改
- 评论/争辩

附录 I　申请提交彩色照片或附图

为了提交彩色照片或附图，需要填写申请表并提交给美国专利商标局（USPTO）。必须包含以下信息：

- 申请号
- 申请日期
- 申请人
- 申请标题
- 审查员
- 技术单元

表格包括以下内容：

- 专利专员的地址
- 彩色照片列表
- 需要彩色照片的原因

表格需要使用以下格式：

Commissioner for Patents

P.O.Box 1450

Alexandria, VA 22313-1450

　　Sir:

　　Applicant hereby respectfully petitions that the color photographs___ filed herewith _ already filed be accepted as formal drawings. The $130 petition fee is enclosed.

　　The color photographs or drawings are necessary because_____

　　The form must also include the names of the applicants and dates.

In the United States Patent and Trademark Office

App. No: _____
Filing Date: _____
Applicant: _____
App. Title: _____
Examiner: _____
Art Unit: _____

Petition for Submitting Color Photographs or Drawings

Commissioner for Patents
P.O. Box 1450
Alexandria, VA 22313-1450

Sir:

Applicant hereby respectfully petitions that the color photographs__ filed herewith _ already filed be accepted as formal drawings. The $130 petition fee is enclosed.

The color photographs or drawings are necessary because_____

_____

_____

_____

Very respectively,

Sole/First Applicant: _____

_____   _____
Sole/First Applicant Signature   Date

Joint/Second Applicant:_____

_____   _____
Joint/ Second Applicant Signature   Date

Joint/Third Applicant:_____

_____   _____
Joint/ Third Applicant Signature   Date

Joint/Fourth Applicant:_____

_____   _____
Joint/ Fourth Applicant Signature   Date

DATE: _____

TO: [European Attorney Name]

FROM: [US Attorney Name]

SUBJECT: Application No. 10 2015 225 095.5
Inventors: [Inventor Name]
Title: [Invention Name]

Dear [European Attorney] ,

In response to the office action issued for the above-referenced case, attached please find our suggested amendments and remarks.

Please feel free to contact us with any questions.

With best regards,

[US Attorney]

Amendments to the Claims are reflected in the listing of claims which begins on page X of this paper.

Remarks/Arguments begin on page Y of this paper.

Claims

1. (currently amended) A circuit, comprising:

   [elements of the claim]; and

   [<u>an amendment to the claim</u>]

2. (cancelled)

3. (original) The circuit of claim 1, wherein said [   ].

4. (original) The circuit of claim 3, wherein said [   ].

5. (original) The circuit of claim 1, wherein said [   ].

6. (original) The circuit of claim 1, wherein said circuit comprises of MOSFET transistors.

7. (original) The circuit of claim 1, wherein said supply voltage is greater than 2.5V.

8. (original) The circuit of claim 1, wherein said output voltage is greater than 1.2 V.

9. (currently amended) <u>The circuit of claim 1 wherein said</u> [   ], comprises:

   a first n-channel MOSFET current mirror <u>(110A, 110C, 110C)</u> configured to provide a current source;

   a first p-channel MOSFET current mirror <u>(160A,160B, 160C)</u> configured to provide a current source;

   a second p-channel MOSFET current mirror <u>(170A, 170B, 170C)</u> electrically coupled to said first p-channel MOSFET current mirror; and,

   a second n-channel MOSFET coupled to npn bipolar junction transistor (BJT) current mirror <u>(140A, 140B, 140C)</u>.

10. (currently amended) The circuit of claim 9 wherein said first p-channel MOSFET current mirror comprises high voltage transistors.

11. (currently amended) A method of circuit operation, comprising the steps of:

    (a) providing a circuit <u>(100)</u>;

    (b) Supplying current through a resistor <u>(105)</u>;

    (c) Setting a first current reference through said resistor <u>(105)</u>;

REMARKS/ARGUMENTS

UNITY

The Examiner is thanked for pointing out various issues of unity in the claims. These have been addressed as follows.

Point 4 has been addressed via amendments to the claims and the claims are now believed to be unified by modifying Claim 1 and Claim 9. The claims 1 to 8 are now unified with the claims 9 to 11.

Claim 1 is amended as follows:

1. (currently amended) A circuit, comprising:

  [elements of the claim]; and
  [an amendment to the claim]

Claim 9 is amended as follows:

9. (currently amended) The circuit of claim 1 wherein said [ ], comprises:
  a first n-channel MOSFET current mirror (110A, 110C, 110C) configured to provide a current source;
  a first p-channel MOSFET current mirror (160A,160B, 160C) configured to provide a current source;
  a second p-channel MOSFET current mirror (170A, 170B, 170C) electrically coupled to said first p-channel MOSFET current mirror; and,
  a second n-channel MOSFET coupled to npn bipolar junction transistor (BJT) current mirror (140A, 140B, 140C).

NOVELTY

The Examiner is thanked for pointing out various issues of novelty in the claims. These have been addressed as follows.

Points 5 including 5.1 have been addressed via amendments to Claim 1 and Claim 9, and the claims are now believed to be novel.

While Document D1 shows a [   ] it fails to disclose a circuit for initiating a [   ] circuit. Document D1 also fails to disclose integration of a circuit with the other functional blocks.

While Document D1 states a bipolar junction current mirror in FIG 2, and FIG. 3, the electrical connections are significantly distinct from the disclosure. Document D1 discusses in Page 1 Column 1 paragraph [0009] a first circuit. Document D1 shows a circuit with a resistor R_B electrically connected to the common base of the circuit, and electrically coupled to ground.

In the disclosure, FIG. 4 shows a current mirror (425A, 425B) with the collector connected to the base region, and no resistor element to ground, which is not the same as the initializing the voltage reference of Document D1. Additionally, a traditional current mirror also has the collector node of the current mirror transistors directly coupled to the base node of the BJT devices. D1 has a transistor separating the collector from the base, disconnecting the collector node from the base node.

Point 6 and 6.1 have been addressed via amendments to Claim 12.

Claim 12 is amended as follows:

12. (currently amended) A method for providing a bandgap voltage, comprising the steps of:

[steps of the method claim].

CLARITY

The Examiner is thanked for pointing out various issues of clarity in the claims. These have been addressed as follows.

Points 8 including 8.1 have been addressed via amendments to Claim 11.

All claims are now believed to be in condition for allowance, and allowance is so requested.

Claims 1-11 remain in this application. Claim 2 has been cancelled.

附录 J　专利合作条约

附录 J 是专利合作条约的一个例子。本文件包含以下内容：
- 申请人
- 申请人代理人档案号
- 国际申请号
- 国际检索
- 分类号
- 检索字段
- 相关文件

PATENT COOPERATION TREATY

From the INTERNATIONAL SEARCHING AUTHORITY

To:

PCT

NOTIFICATION OF TRANSMITTAL OF THE INTERNATIONAL SEARCH REPORT AND THE WRITTEN OPINION OF THE INTERNATIONAL SEARCHING AUTHORITY, OR THE DECLARATION

(PCT Rule 44.1)

Date of mailing (day/month/year)	14 August 2017 (14.08.2017)

Applicant's or agent's file reference

FOR FURTHER ACTION See paragraphs 1 and 4 below

International application No.

International filing date (day/month/year)	01 May 2017 (01.05.2017)

Applicant

1. ☒ The applicant is hereby notified that the international search report and the written opinion of the International Searching Authority have been established and are transmitted herewith.
   Filing of amendments and statement under Article 19:
   The applicant is entitled, if he so wishes, to amend the claims of the international application (see Rule 46):
   When? The time limit for filing such amendments is normally two months from the date of transmittal of the international search report.
   How? Directly to the International Bureau of WIPO preferably through ePCT or on paper to, 34 chemin des Colombettes 1211 Geneva 20, Switzerland, Facsimile No.: +41 22 338 82 70
   For more detailed instructions, see *PCT Applicant's Guide*, International Phase, paragraphs 9.004 . 9-011.

2. ☐ The applicant is hereby notified that no international search report will be established and that the declaration under Article 17(2)(a) to that effect and the written opinion of the International Searching Authority are transmitted herewith.

3. ☐ **With regard to any protest** against payment of (an) additional fee(s) under Rule 40.2, the applicant is notified that:
   ☐ the protest together with the decision thereon has been transmitted to the International Bureau together with any request to forward the texts of both the protest and the decision thereon to the designated Offices.
   ☐ no decision has been made yet on the protest; the applicant will be notified as soon as a decision is made.

4. **Reminders**

   The applicant may submit comments on an informal basis on the written opinion of the International Searching Authority to the International Bureau. These comments will be made available to the public after international publication. The International Bureau will send a copy of such comments to all designated Offices unless an international preliminary examination report has been or is to be established.

   Shortly after the expiration of **18 months** from the priority date, the international application will be published by the International Bureau. If the applicant wishes to avoid or postpone publication, a notice of withdrawal of the international application, or of the priority claim, must reach the International Bureau before the completion of the technical preparations for international publication (Rules 90*bis*.1 and 90*bis*.3).

   Within **19 months** from the priority date, but only in respect of some designated Offices, a demand for international preliminary examination must be filed if the applicant wishes to postpone the entry into the national phase **until 30 months** from the priority date (in some Offices even later); otherwise, the applicant must, **within 20 months** from the priority date, perform the prescribed acts for entry into the national phase before those designated Offices. In respect of other designated Offices, the time limit of **30 months** (or later) will apply even if no demand is filed within 19 months. For details about the applicable time limits, Office by Office, see www.wipo.int/pct/en/texts/time_limits.html and the *PCT Applicant's Guide*, National Chapters.

   Within **22 months** from the priority date, the applicant may request that a supplementary international search be carried out by a different International Searching Authority, that offers this service (Rule 45*bis*.1). The procedure for requesting supplementary international search is described in the *PCT Applicant's Guide*, International Phase, paragraphs 8.006-8.032.

Name and mailing address of the ISA/KR International Application Division Korean Intellectual Property Office 189 Cheongsa-ro, Seo-gu, Daejeon, 35208, Republic of Korea	Authorized officer COMMISSIONER
Facsimile No. 82-42-481-8578	Telephone No. 82-42-481-8751

* Attention

Copies of the documents cited in the international search report can be searched in the following Korean Intellectual Property Office English website for six months(expire date : **2018.02.14**   ) from the date of mailing of the international search report.

http://www.kipo.go.kr/cn/ => PCT Service => PCT Services

ID　: PCT international application number
PW :

Inquiries related to PCT International Search Report or Written Opinion prepared by KIPO as an International Searching Authority can be answered not only by KIPO but also through IPKC (Intellectual Property Korea Center), located in Vienna, VA, which functions as a PCT Help Desk for PCT applicants.

Homepage: http://www.ipkcenter.com

Email: ipkc@ipkcenter.com

PATENT COOPERATION TREATY

PCT

INTERNATIONAL SEARCH REPORT

(PCT Article 18 and Rules 43 and 44)

Applicant's or agent's file reference	FOR FURTHER ACTION	see Form PCT/ISA/220 as well as, where applicable, item 5 below.
International application No.	International filing date *(day/month/year)* **01 May 2017 (01.05.2017)**	(Earliest) Priority Date *(day/month/year)* 06 May 2016 (06.05.2016)
Applicant		

This International search report has been prepared by this International Searching Authority and is transmitted to the applicant according to Article 18. A copy is being transmitted to the International Bureau.

This international search report consists of a total of ___3___ sheets.

☐ It is also accompanied by a copy of each prior art document cited in this report.

1. **Basis of the report**
   a. With regard to the **language**, the international search was carried out on the basis of :

   ☒ the international application in the language in which it was filed

   ☐ a translation of the international application into _____, which is the language of a translation furnished for the purposes of international search (Rules 12.3(a) and 23.1(b))

   b. ☐ This international search report has been established taking into account the **rectification of an obvious mistake** authorized by or notified to this Authority under Rule 91 (Rule 43.6*bis*(a)).

   c. ☐ With regard to any **nucleotide and/or amino acid sequence** disclosed in the international application, see Box No. I.

2. ☐ **Certain claims were found unsearchable** (See Box No. II)

3. ☐ **Unity of invention is lacking** (See Box No. III)

4. With regard to the **title**,

   ☒ the text is approved as submitted by the applicant.

   ☐ the text has been established by this Authority to read as follows:

5. With regard to the **abstract**,

   ☒ the text is approved as submitted by the applicant.

   ☐ the text has been established, according to Rule 38.2, by this Authority as it appears in Box No. IV. The applicant may, within one month from the date of mailing of this international search report, submit comments to this Authority.

6. With regard to the drawings,

   a. the figure of the **drawings** to be published with the abstract is Figure No. ___9___

   ☐ as suggested by the applicant.

   ☒ as selected by this Authority, because the applicant failed to suggest a figure.

   ☐ as selected by this Authority, because this figure better characterizes the invention.

   b. ☐ none of the figures is to be published with the abstract.

INTERNATIONAL SEARCH REPORT	International application No.

A. CLASSIFICATION OF SUBJECT MATTER

According to International Patent Classification (IPC) or to both national classification and IPC

B. FIELDS SEARCHED

Minimum documentation searched (classification system followed by classification symbols)

Documentation searched other than minimum documentation to the extent that such documents are included in the fields searched
Korean utility models and applications for utility models
Japanese utility models and applications for utility models

Electronic data base consulted during the international search (name of data base and, where practicable, search terms used)
eKOMPASS(KIPO internal) & Keywords: transistor, source, drain, Schottky, contact

C. DOCUMENTS CONSIDERED TO BE RELEVANT

Category*	Citation of document, with indication, where appropriate, of the relevant passages	Relevant to claim No.

☐ Further documents are listed in the continuation of Box C.   ☐ See patent family annex.

* Special categories of cited documents:
"A" document defining the general state of the art which is not considered to be of particular relevance
"E" earlier application or patent but published on or after the international filing date
"L" document which may throw doubts on priority claim(s) or which is cited to establish the publication date of another citation or other special reason (as specified)
"O" document referring to an oral disclosure, use, exhibition or other means
"P" document published prior to the international filing date but later than the priority date claimed

"T" later document published after the international filing date or priority date and not in conflict with the application but cited to understand the principle or theory underlying the invention
"X" document of particular relevance; the claimed invention cannot be considered novel or cannot be considered to involve an inventive step when the document is taken alone
"Y" document of particular relevance; the claimed invention cannot be considered to involve an inventive step when the document is combined with one or more other such documents, such combination being obvious to a person skilled in the art
"&" document member of the same patent family

Date of the actual completion of the international search	Date of mailing of the international search report

Name and mailing address of the ISA/KR	Authorized officer
Facsimile No.	Telephone No.

PATENT COOPERATION TREATY

From the
INTERNATIONAL SEARCHING AUTHORITY

To:

PCT

WRITTEN OPINION OF THE INTERNATIONAL SEARCHING AUTHORITY

(PCT Rule 43bis.1)

| Date of mailing *(day/month/year)* | 14 August 2017 (14.08.2017) |

Applicant's or agent's file reference	FOR FURTHER ACTION
4261-001PCT	See paragraph 2 below

International application No.	International filing date *(day/month/year)*	Priority date *(day/month/year)*
	01 May 2017 (01.05.2017)	06 May 2016 (06.05.2016)

International Patent Classification (IPC) or both national classification and IPC

Applicant

1. This opinion contains indications relating to the following items:

   [X] Box No. I   Basis of the opinion
   [ ] Box No. II   Priority
   [ ] Box No. III Non-establishment of opinion with regard to novelty, inventive step and industrial applicability
   [ ] Box No. IV   Lack of unity of invention
   [X] Box No. V   Reasoned statement under Rule 43bis.1(a)(i) with regard to novelty, inventive step and industrial applicability; citations and explanations supporting such statement
   [ ] Box No. VI   Certain documents cited
   [ ] Box No. VII Certain defects in the international application
   [ ] Box No. VIII Certain observations on the international application

2. **FURTHER ACTION**

   If a demand for international preliminary examination is made, this opinion will be considered to be a written opinion of the International Preliminary Examining Authority ("IPEA") except that this does not apply where the applicant chooses an Authority other than this one to be the IPEA and the chosen IPEA has notified the International Bureau under Rule 66.1bis(b) that written opinions of this International Searching Authority will not be so considered.

   If this opinion is, as provided above, considered to be a written opinion of the IPEA, the applicant is invited to submit to the IPEA a written reply together, where appropriate, with amendments, before the expiration of 3 months from the date of mailing of Form PCT/ISA/220 or before the expiration of 22 months from the priority date, whichever expires later.
   For further options, see Form PCT/ISA/220.

Name and mailing address of the ISA/KR	Date of completion of this opinion	Authorized officer	
International Application Division Korean Intellectual Property Office 189 Cheongsa-ro, Seo-gu, Daejeon, 35208, Republic of Korea	14 August 2017 (14.08.2017)	KANG, Sung Chul	
Facsimile No. +82-42-481-8578		Telephone No. +82-42-481-8405	

**WRITTEN OPINION OF THE
INTERNATIONAL SEARCHING AUTHORITY**

International application No.

Box No. I Basis of this opinion

1. With regard to the **language**, this opinion has been established on the basis of:

    [X] the international application in the language in which it was filed

    [ ] a translation of the international application into _____ which is the language of a translation furnished for the purposes of international search (Rules 12.3(a) and 23.1(b))

2. [ ] This opinion has been established taking into account the **rectification of an obvious mistake** authorized by or notified to this Authority under Rule 91 (Rule 43bis.1(a))

3. [ ] With regard to any **nucleotide and/or amino acid sequence** disclosed in the international application, this opinion has been established on the basis of a sequence listing:

    a. [ ] forming part of the international application as filed:

       [ ] in the form of an Annex C/ST.25 text file.
       [ ] on paper or in the form of an image file.

    b. [ ] furnished together with the international application under PCT Rule 13ter.1(a) for the purposes of international search only in the form of an Annex C/ST.25 text file.

    c. [ ] furnished subsequent to the international filing date for the purposes of international search only:

       [ ] in the form of an Annex C/ST.25 text file (Rule 13ter.1(a)).
       [ ] on paper or in the form of an image file (Rule 13ter.1(b) and Administrative Instructions, Section 713).

4. [ ] In addition, in the case that more than one version or copy of a sequence listing has been filed or furnished, the required statements that the information in the subsequent or additional copies is identical to that forming part of the application as filed or does not go beyond the application as filed, as appropriate, were furnished.

5. Additional comments:

WRITTEN OPINION OF THE INTERNATIONAL SEARCHING AUTHORITY

International application No.

Box No. V Reasoned statement under Rule 43bis.1(a)(i) with regard to novelty, inventive step or industrial applicability; citations and explanations supporting such statement

1. Statement

   Novelty (N)   Claims 1-29   YES
                            Claims NONE   NO

   Inventive step (IS)   Claims 1-29   YES
                            Claims NONE   NO

   Industrial applicability (IA) Claims 1-29   YES
                                  Claims NONE   NO

2. Citations and explanations :

   Reference is made to the following documents:

   1. Novelty and Inventive Step
   1.1. Claims 1-15

   Claims 2-15 are directly or indirectly dependent on claim 1 and therefore meet the requirements of PCT Article 33(2) and (3).

   1.2. Claims 16-19

Continued on Supplemental Box

WRITTEN OPINION OF THE INTERNATIONAL SEARCHING AUTHORITY	International application No.

Supplemental Box

In case the space in any of the preceding boxes is not sufficient.
Continuation of : Box No. V

Claims 17-19 are dependent on claim 16 and therefore meet the requirements of PCT Article 33(2) and (3).

1.3. Claims 20-25

Claims 21-25 are dependent on claim 20 and therefore meet the requirements of PCT Article 33(2) and (3).

Claims 27-29 are dependent on claim 26 and therefore meet the requirements of PCT Article 33(2) and (3).

2. Industrial Applicability
Claims 1-29 are industrially applicable under PCT Article 33(4).

附录 K　补正书

附录 K 是补正书的一个例子。美国专利商标局（USPTO）补正书包含以下内容：
- 专利号
- 申请号
- 日期
- 发明人

美国专利商标局补正书声明

"经证明，上述专利中出现了错误，上述专利特此更正如下："

然后将 USPTO 的签名和印章放在文档上以供核实。

UNITED STATES PATENT AND TRADEMARK OFFICE
CERTIFICATE OF CORRECTION

PATENT NO. :
APPLICATION NO. :
DATED :
INVENTOR(S)

Page 1 of 1

It is certified that error appears in the above-identified patent and that said Letters Patent is hereby corrected as shown below:

In the Foreign Application Priority Data (30), add -- August 14, 2014 (EP) --.

Signed and Sealed this
Twenty-third Day of May, 2017

Michelle K. Lee

附录 L 补正通知书

附录 L 是美国专利商标局（USPTO）补正通知书的一个例子。本文件包括以下内容：
- 专利号
- 申请号
- 申请日期
- 第一命名发明人
- 案卷号
- 审查员
- 技术单元
- 递送模式

美国专利商标局补正通知书包括：
- 权利要求的申请日期
- 允许的权利要求列表
- 答复日期要求
- 专利审查员签名
- 主管专利审查员签名

UNITED STATES PATENT AND TRADEMARK OFFICE

AUG 0 3 2017

UNITED STATES DEPARTMENT OF COMMERCE
United States Patent and Trademark Office
Address: COMMISSIONER FOR PATENTS
P.O. Box 1450
Alexandria, Virginia 22313-1450
www.uspto.gov

APPLICATION NO.	FILING DATE	FIRST NAMED INVENTOR	ATTORNEY DOCKET NO.	CONFIRMATION NO.

EXAMINER

ART UNIT	PAPER NUMBER
2838	

MAIL DATE	DELIVERY MODE
	PAPER

Please find below and/or attached an Office communication concerning this application or proceeding.

The time period for reply, if any, is set in the attached communication.

Corrected Notice of Allowability

Application No.	Applicant(s)
Examiner	**Art Unit** 2838	**AIA (First Inventor to File) Status** Yes

-- The MAILING DATE of this communication appears on the cover sheet with the correspondence address --

All claims being allowable, PROSECUTION ON THE MERITS IS (OR REMAINS) CLOSED in this application. If not included herewith (or previously mailed), a Notice of Allowance (PTOL-85) or other appropriate communication will be mailed in due course. **THIS NOTICE OF ALLOWABILITY IS NOT A GRANT OF PATENT RIGHTS.** This application is subject to withdrawal from issue at the initiative of the Office or upon petition by the applicant. See 37 CFR 1.313 and MPEP 1308.

1. ☒ This communication is responsive to <u>claims filed on 3/10/2017</u>.
    ☐ A declaration(s)/affidavit(s) under 37 CFR 1.130(b) was/were filed on _____.

2. ☐ An election was made by the applicant in response to a restriction requirement set forth during the interview on _____; the restriction requirement and election have been incorporated into this action.

3. ☒ The allowed claim(s) is/are _____. As a result of the allowed claim(s), you may be eligible to benefit from the **Patent Prosecution Highway** program at a participating intellectual property office for the corresponding application. For more information, please see http://www.uspto.gov/patents/init_events/pph/index.jsp or send an inquiry to PPHfeedback@uspto.gov.

4. ☐ Acknowledgment is made of a claim for foreign priority under 35 U.S.C. § 119(a)-(d) or (f).
    Certified copies:
    a) ☐ All   b) ☐ Some   *c) ☐ None of the:
      1. ☐ Certified copies of the priority documents have been received.
      2. ☐ Certified copies of the priority documents have been received in Application No. _____.
      3. ☐ Copies of the certified copies of the priority documents have been received in this national stage application from the International Bureau (PCT Rule 17.2(a)).
    * Certified copies not received: _____.

Applicant has THREE MONTHS FROM THE "MAILING DATE" of this communication to file a reply complying with the requirements noted below. Failure to timely comply will result in ABANDONMENT of this application. **THIS THREE-MONTH PERIOD IS NOT EXTENDABLE.**

5. ☐ CORRECTED DRAWINGS (as "replacement sheets") must be submitted.
    ☐ including changes required by the attached Examiner's Amendment / Comment or in the Office action of Paper No./Mail Date _____.
    Identifying indicia such as the application number (see 37 CFR 1.84(c)) should be written on the drawings in the front (not the back) of each sheet. Replacement sheet(s) should be labeled as such in the header according to 37 CFR 1.121(d).

6. ☐ DEPOSIT OF and/or INFORMATION about the deposit of BIOLOGICAL MATERIAL must be submitted. Note the attached Examiner's comment regarding REQUIREMENT FOR THE DEPOSIT OF BIOLOGICAL MATERIAL.

Attachment(s)
1. ☐ Notice of References Cited (PTO-892)
2. ☐ Information Disclosure Statements (PTO/SB/08), Paper No./Mail Date _____
3. ☐ Examiner's Comment Regarding Requirement for Deposit of Biological Material
4. ☐ Interview Summary (PTO-413), Paper No./Mail Date _____.

5. ☐ Examiner's Amendment/Comment
6. ☐ Examiner's Statement of Reasons for Allowance
7. ☐ Other _____.

Examiner, Art Unit 2838	Supervisory Patent Examiner, Art Unit 2838

U.S. Patent and Trademark Office
PTOL-37 (Rev. 08-13)   Notice of Allowability   Part of Paper No./Mail Date

附录 M　授权通知书

附录 M 是美国专利商标局（USPTO）授权通知书的一个例子。本文件包括以下内容：

- 有关如何答复的说明
- 公共与预算管理办公室（OMB）许可
- 专利期限调整（PTA）
- 可授权权利要求的列表

USPTO 授权通知书包括：

- 权利要求的申请日期
- 可授权权利要求的列表
- 答复日期要求

从发明到专利——科学家和工程师指南．第一版．史蒂文·H.沃尔德曼 ©2018 John Wiley & Sons 公司，2018 年由 John Wiley & Sons 公司出版．

NOTICE OF ALLOWANCE AND FEE(S) DUE

UNITED STATES PATENT AND TRADEMARK OFFICE

JUN 19 2017

UNITED STATES DEPARTMENT OF COMMERCE
United States Patent and Trademark Office
Address: COMMISSIONER FOR PATENTS
P.O. Box 1450
Alexandria, Virginia 22313-1450
www.uspto.gov

28112   7590   06/14/2017

EXAMINER

ART UNIT	PAPER NUMBER

DATE MAILED: 06/14/2017

APPLICATION NO.	FILING DATE	FIRST NAMED INVENTOR	ATTORNEY DOCKET NO.	CONFIRMATION NO.

TITLE OF INVENTION:

APPLN. TYPE	ENTITY STATUS	ISSUE FEE DUE	PUBLICATION FEE DUE	PREV. PAID ISSUE FEE	TOTAL FEE(S) DUE	DATE DUE
nonprovisional	UNDISCOUNTED	$960	$0	$0	$960	09/14/2017

THE APPLICATION IDENTIFIED ABOVE HAS BEEN EXAMINED AND IS ALLOWED FOR ISSUANCE AS A PATENT. **PROSECUTION ON THE MERITS IS CLOSED.** THIS NOTICE OF ALLOWANCE IS NOT A GRANT OF PATENT RIGHTS. THIS APPLICATION IS SUBJECT TO WITHDRAWAL FROM ISSUE AT THE INITIATIVE OF THE OFFICE OR UPON PETITION BY THE APPLICANT. SEE 37 CFR 1.313 AND MPEP 1308.

THE ISSUE FEE AND PUBLICATION FEE (IF REQUIRED) MUST BE PAID WITHIN **THREE MONTHS** FROM THE MAILING DATE OF THIS NOTICE OR THIS APPLICATION SHALL BE REGARDED AS ABANDONED. **THIS STATUTORY PERIOD CANNOT BE EXTENDED.** SEE 35 U.S.C. 151. THE ISSUE FEE DUE INDICATED ABOVE DOES NOT REFLECT A CREDIT FOR ANY PREVIOUSLY PAID ISSUE FEE IN THIS APPLICATION. IF AN ISSUE FEE HAS PREVIOUSLY BEEN PAID IN THIS APPLICATION (AS SHOWN ABOVE), THE RETURN OF PART B OF THIS FORM WILL BE CONSIDERED A REQUEST TO REAPPLY THE PREVIOUSLY PAID ISSUE FEE TOWARD THE ISSUE FEE NOW DUE.

HOW TO REPLY TO THIS NOTICE:

I. Review the ENTITY STATUS shown above. If the ENTITY STATUS is shown as SMALL or MICRO, verify whether entitlement to that entity status still applies.

If the ENTITY STATUS is the same as shown above, pay the TOTAL FEE(S) DUE shown above.

If the ENTITY STATUS is changed from that shown above, on PART B - FEE(S) TRANSMITTAL, complete section number 5 titled "Change in Entity Status (from status indicated above)".

For purposes of this notice, small entity fees are 1/2 the amount of undiscounted fees, and micro entity fees are 1/2 the amount of small entity fees.

II. PART B - FEE(S) TRANSMITTAL, or its equivalent, must be completed and returned to the United States Patent and Trademark Office (USPTO) with your ISSUE FEE and PUBLICATION FEE (if required). If you are charging the fee(s) to your deposit account, section "4b" of Part B - Fee(s) Transmittal should be completed and an extra copy of the form should be submitted. If an equivalent of Part B is filed, a request to reapply a previously paid issue fee must be clearly made, and delays in processing may occur due to the difficulty in recognizing the paper as an equivalent of Part B.

III. All communications regarding this application must give the application number. Please direct all communications prior to issuance to Mail Stop ISSUE FEE unless advised to the contrary.

IMPORTANT REMINDER: Utility patents issuing on applications filed on or after Dec. 12, 1980 may require payment of maintenance fees. It is patentee's responsibility to ensure timely payment of maintenance fees when due.

Page 1 of 3

PTOL-85 (Rev. 02/11)

OMB Clearance and PRA Burden Statement for PTOL-85 Part B

The Paperwork Reduction Act (PRA) of 1995 requires Federal agencies to obtain Office of Management and Budget approval before requesting most types of information from the public. When OMB approves an agency request to collect information from the public, OMB (i) provides a valid OMB Control Number and expiration date for the agency to display on the instrument that will be used to collect the information and (ii) requires the agency to inform the public about the OMB Control Number's legal significance in accordance with 5 CFR 1320.5(b).

The information collected by PTOL-85 Part B is required by 37 CFR 1.311. The information is required to obtain or retain a benefit by the public which is to file (and by the USPTO to process) an application. Confidentiality is governed by 35 U.S.C. 122 and 37 CFR 1.14. This collection is estimated to take 12 minutes to complete, including gathering, preparing, and submitting the completed application form to the USPTO. Time will vary depending upon the individual case. Any comments on the amount of time you require to complete this form and/or suggestions for reducing this burden, should be sent to the Chief Information Officer, U.S. Patent and Trademark Office, U.S. Department of Commerce, P.O. Box 1450, Alexandria, Virginia 22313-1450. DO NOT SEND FEES OR COMPLETED FORMS TO THIS ADDRESS. SEND TO: Commissioner for Patents, P.O. Box 1450, Alexandria, Virginia 22313-1450. Under the Paperwork Reduction Act of 1995, no persons are required to respond to a collection of information unless it displays a valid OMB control number.

Privacy Act Statement

The Privacy Act of 1974 (P.L. 93-579) requires that you be given certain information in connection with your submission of the attached form related to a patent application or patent. Accordingly, pursuant to the requirements of the Act, please be advised that: (1) the general authority for the collection of this information is 35 U.S.C. 2(b)(2); (2) furnishing of the information solicited is voluntary; and (3) the principal purpose for which the information is used by the U.S. Patent and Trademark Office is to process and/or examine your submission related to a patent application or patent. If you do not furnish the requested information, the U.S. Patent and Trademark Office may not be able to process and/or examine your submission, which may result in termination of proceedings or abandonment of the application or expiration of the patent.

The information provided by you in this form will be subject to the following routine uses:
1. The information on this form will be treated confidentially to the extent allowed under the Freedom of Information Act (5 U.S.C. 552) and the Privacy Act (5 U.S.C 552a). Records from this system of records may be disclosed to the Department of Justice to determine whether disclosure of these records is required by the Freedom of Information Act.
2. A record from this system of records may be disclosed, as a routine use, in the course of presenting evidence to a court, magistrate, or administrative tribunal, including disclosures to opposing counsel in the course of settlement negotiations.
3. A record in this system of records may be disclosed, as a routine use, to a Member of Congress submitting a request involving an individual, to whom the record pertains, when the individual has requested assistance from the Member with respect to the subject matter of the record.
4. A record in this system of records may be disclosed, as a routine use, to a contractor of the Agency having need for the information in order to perform a contract. Recipients of information shall be required to comply with the requirements of the Privacy Act of 1974, as amended, pursuant to 5 U.S.C. 552a(m).
5. A record related to an International Application filed under the Patent Cooperation Treaty in this system of records may be disclosed, as a routine use, to the International Bureau of the World Intellectual Property Organization, pursuant to the Patent Cooperation Treaty.
6. A record in this system of records may be disclosed, as a routine use, to another federal agency for purposes of National Security review (35 U.S.C. 181) and for review pursuant to the Atomic Energy Act (42 U.S.C. 218(c)).
7. A record from this system of records may be disclosed, as a routine use, to the Administrator, General Services, or his/her designee, during an inspection of records conducted by GSA as part of that agency's responsibility to recommend improvements in records management practices and programs, under authority of 44 U.S.C. 2904 and 2906. Such disclosure shall be made in accordance with the GSA regulations governing inspection of records for this purpose, and any other relevant (i.e., GSA or Commerce) directive. Such disclosure shall not be used to make determinations about individuals.
8. A record from this system of records may be disclosed, as a routine use, to the public after either publication of the application pursuant to 35 U.S.C. 122(b) or issuance of a patent pursuant to 35 U.S.C. 151. Further, a record may be disclosed, subject to the limitations of 37 CFR 1.14, as a routine use, to the public if the record was filed in an application which became abandoned or in which the proceedings were terminated and which application is referenced by either a published application, an application open to public inspection or an issued patent.
9. A record from this system of records may be disclosed, as a routine use, to a Federal, State, or local law

UNITED STATES PATENT AND TRADEMARK OFFICE

UNITED STATES DEPARTMENT OF COMMERCE
United States Patent and Trademark Office
Address: COMMISSIONER FOR PATENTS
P.O. Box 1450
Alexandria, Virginia 22313-1450
www.uspto.gov

APPLICATION NO.	FILING DATE	FIRST NAMED INVENTOR	ATTORNEY DOCKET NO.	CONFIRMATION NO.

EXAMINER

ART UNIT	PAPER NUMBER
2838	

DATE MAILED: 06/14/2017

Determination of Patent Term Adjustment under 35 U.S.C. 154 (b)
(Applications filed on or after May 29, 2000)

The Office has discontinued providing a Patent Term Adjustment (PTA) calculation with the Notice of Allowance.

Section 1(h)(2) of the AIA Technical Corrections Act amended 35 U.S.C. 154(b)(3)(B)(i) to eliminate the requirement that the Office provide a patent term adjustment determination with the notice of allowance. See Revisions to Patent Term Adjustment, 78 Fed. Reg. 19416, 19417 (Apr. 1, 2013). Therefore, the Office is no longer providing an initial patent term adjustment determination with the notice of allowance. The Office will continue to provide a patent term adjustment determination with the Issue Notification Letter that is mailed to applicant approximately three weeks prior to the issue date of the patent, and will include the patent term adjustment on the patent. Any request for reconsideration of the patent term adjustment determination (or reinstatement of patent term adjustment) should follow the process outlined in 37 CFR 1.705.

Any questions regarding the Patent Term Extension or Adjustment determination should be directed to the Office of Patent Legal Administration at (571)-272-7702. Questions relating to issue and publication fee payments should be directed to the Customer Service Center of the Office of Patent Publication at 1-(888)-786-0101 or (571)-272-4200.

	Application No.	Applicant(s)		
Notice of Allowability	Examiner	Art Unit	**AIA (First Inventor to File) Status**	
				Yes

-- The MAILING DATE of this communication appears on the cover sheet with the correspondence address--

All claims being allowable, PROSECUTION ON THE MERITS IS (OR REMAINS) CLOSED in this application. If not included herewith (or previously mailed), a Notice of Allowance (PTOL-85) or other appropriate communication will be mailed in due course. **THIS NOTICE OF ALLOWABILITY IS NOT A GRANT OF PATENT RIGHTS.** This application is subject to withdrawal from issue at the initiative of the Office or upon petition by the applicant. See 37 CFR 1.313 and MPEP 1308.

1. ☒ This communication is responsive to *claim arguments filed on 3/10/217*.
   ☐ A declaration(s)/affidavit(s) under **37 CFR 1.130(b)** was/were filed on _____.

2. ☐ An election was made by the applicant in response to a restriction requirement set forth during the interview on _____; the restriction requirement and election have been incorporated into this action.

3. ☒ The allowed claim(s) is/are *1-3*_____. As a result of the allowed claim(s), you may be eligible to benefit from the **Patent Prosecution Highway** program at a participating intellectual property office for the corresponding application. For more information, please see http://www.uspto.gov/patents/init_events/pph/index.jsp or send an inquiry to PPHfeedback@uspto.gov.

4. ☐ Acknowledgment is made of a claim for foreign priority under 35 U.S.C. § 119(a)-(d) or (f).
   Certified copies:
   a) ☐ All   b) ☐ Some   *c) ☐ None of the:
      1. ☐ Certified copies of the priority documents have been received.
      2. ☐ Certified copies of the priority documents have been received in Application No. _____.
      3. ☐ Copies of the certified copies of the priority documents have been received in this national stage application from the International Bureau (PCT Rule 17.2(a)).
   * Certified copies not received: _____.

Applicant has THREE MONTHS FROM THE "MAILING DATE" of this communication to file a reply complying with the requirements noted below. Failure to timely comply will result in ABANDONMENT of this application.
THIS THREE-MONTH PERIOD IS NOT EXTENDABLE.

5. ☐ CORRECTED DRAWINGS (as "replacement sheets") must be submitted.
   ☐ including changes required by the attached Examiner's Amendment / Comment or in the Office action of Paper No./Mail Date _____.
   Identifying indicia such as the application number (see 37 CFR 1.84(c)) should be written on the drawings in the front (not the back) of each sheet. Replacement sheet(s) should be labeled as such in the header according to 37 CFR 1.121(d).

6. ☐ DEPOSIT OF and/or INFORMATION about the deposit of BIOLOGICAL MATERIAL must be submitted. Note the attached Examiner's comment regarding REQUIREMENT FOR THE DEPOSIT OF BIOLOGICAL MATERIAL.

Attachment(s)
1. ☐ Notice of References Cited (PTO-892)
2. ☐ Information Disclosure Statements (PTO/SB/08), Paper No./Mail Date _____
3. ☐ Examiner's Comment Regarding Requirement for Deposit of Biological Material
4. ☒ Interview Summary (PTO-413), Paper No./Mail Date *20170608*.
5. ☒ Examiner's Amendment/Comment
6. ☒ Examiner's Statement of Reasons for Allowance
7. ☐ Other _____.

Examiner, Art Unit 2838

U.S. Patent and Trademark Office
PTOL-37 (Rev. 08-13)   Notice of Allowability   Part of Paper No./Mail Date

附录 N 初次修改

附录 N 是美国专利商标局（USPTO）初次修改的一个例子。本文件包括以下内容：
- 专利案卷号
- 美国代理人、注册号、地址
- 发明人
- 发明名称
- 组技术单元
- 初次修改
- 邮寄证明
- 邮寄日期
- 对权利要求的修改
- 评论/争辩

December 28, 2015

To: Commissioner for Patents
P.O. Box 1450
Alexandria, VA 22313-1450

Fr: US Attorney, Reg. No. [Registration Number)
[Address – Street]
[Address - City, State, Zip code

Subject:

| Serial No. [Case]　　　　　　　[Date] |
| [Inventor] |
| "[Invention Title]" |
| |
| Grp. Art Unit: |

PRELIMINARY AMENDMENT

Dear Sir:

Please amend the above-identified application for patent as follows:

CERTIFICATE OF MAILING
I hereby certify that this correspondence is being deposited with the United States Postal Service as first class mail in an envelope addressed to: Commissioner for Patents, P.O. Box 1450, Alexandria, VA 22313-1450, on December 28, 2015.

US Attorney, [Registration No.]

Signature _____
Date   _____

Amendments to the Claims are reflected in the listing of claims, which begins on page **X** of this paper.

Remarks/Arguments begin on page **Y** of this paper.

Amendments to the Claims:

This listing of claims will replace all prior versions, and listings, of claims in the application:

Listing of Claims:

What is claimed is:

1. (original) A circuit providing regulation with an improved monitor, comprising:

    [elements of the circuit];

2. (original) The circuit of claim 1, wherein [   ].

3. (original) The circuit of claim 1, wherein [   ].

4. (original) The circuit of claim 2, wherein said circuit is configured to receive a signal from [   ].

5. (original) The circuit of claim 1, wherein said circuit is configured to receives a signal from [   ].

6. (cancelled)

REMARKS

Please enter the above amendment of the claims for the subject application. No new matter has been added.

Respectfully submitted,

US Patent Attorney, Reg. No. [Registration Number]

附录 O　更正附图提交

为了提交更正的附图,需要填写表格并提交给美国专利商标局(USPTO)。必须包含以下信息:

- 序列号
- 申请日
- 申请人
- 申请名称
- 审查员
- 技术单元

表格包括以下内容:

- 专利专员的地址
- 附图列表
- 邮寄证明

In the United States Patent and Trademark Office

Serial No: _____
Appn Filed: _____
Applicants: _____
App. Title: _____
Examiner: _____
Group Art Unit: _____

                                             Mailed:_____
                                             At:_____

Submission of Corrected Drawings

Commissioner for Patents
P.O. Box 1450
Alexandria, VA 22313-1450

Attn: Chief Draftsperson

Sir:

New drawing sheet (s) _____ for the above application is/are enclosed, corrected as necessary. Please substitute this/these for the corresponding sheet(s) on file.

The new drawing sheet (s) are all marked "Replacement Sheet, Ser. Nr.___/___,_____."

The new sheets contain the following changes; these changes do not introduce any new matter into the drawings:

Fig____:

[ ] Attached is/are copy (ies) of the original drawing sheets maked up to show the changes.

Very respectfully,

_____
Applicant

Certificate of Mailing

I hereby certify that this correspondence will be deposited with the United States Postal Service as First Class Mail, postage prepaid, in an envelope addressed "Commissioner for Patents P.O. Box 1450, Alexandria, VA 22313-1450" on the date below.

_____   _____
Person Depositing Paper   Date

术　语　表

组合：为了提供发明而把第一物品和第二物品合并在一起的过程。

申请：专利申请时为了获得专利而向专利局提交的文档。

背景：专利说明书的一部分，论述现有技术和专利申请的领域。

附图简要说明：专利说明书的一部分，提供专利申请中的附图列表。

权利要求：专利申请的一部分，论述发明要求保护的内容是什么。

部分延续案（CIP）：对现有专利申请的延续，在第二申请过程中添加另外的材料。

版权：对知识产权提供保护的一种标志。

附图详细说明：专利说明书的一部分，在说明书中对于列表中附图提供详细说明。

专家证人：介入到专利诉讼过程的领域内专家，提供专家证言。

欧盟（EU）：专利保护的欧洲组织。

发明领域：专利申请的背景部分，用于陈述发明领域。

最终驳回：专利申请过程中的最终审查意见，其中专利审查员描述为最终的审查意见驳回权利要求。

发明：组合、省略或重排的过程，以提供一个不在公知领域内或本领域技术人员熟知的新物品。

发明人：在宣誓声明中签字的个人，对最新权利要求有贡献。

共同发明人：对专利说明书中的至少一个权利要求有贡献的发明人。

非显而易见：一件发明对本领域内技术人员来说并不熟悉。

准许通知（NOA）：专利审查员或专利局发出的通知，说明专利申请已被接受成为授权专利。

从发明到专利——科学家和工程师指南．第一版．史蒂文·H.沃尔德曼 ©2018 John Wiley & Sons 公司，2018 年由 John Wiley & Sons 公司出版。

新颖性：可专利性的要求之一，要求专利权利要求是新颖的或新的。

宣誓声明：发明人签署的资料，证明他/她是提交专利申请的发明人。

审查意见：专利审查员返回的与提交的专利申请相关的意见。

专利：专利局批准并获得授权的资料。

专利申请：为了获得专利而向专利局提交的申请。

专利审查员：为专利局工作的审查专利申请的人员，说明专利申请什么时候初步适合成为专利。

专利审查中：产品上关于将在未来某日成为专利的说明。

现有技术：在公知领域内现存的早于专利申请提交的资料或材料。

移除：发明中为了简化而移除技术特征的过程。

重排：发明中对于第一技术特征向第二技术特征的方向调整的过程。

权利要求驳回：专利申请处理之后驳回该请求。

说明书：专利申请的主体。

总结：专利申请的一部分，总结发明领域。

商标：发明的标志，用于保护知识产权。

权利要求的撤回：删除说明书中的权利要求或重新提交不同的专利申请。

美国专利商标局（USPTO）：美国专利商标局。